P9-ARB-774

A HISTORY OF THE SECOND WORLD WAR IN 100 MAPS

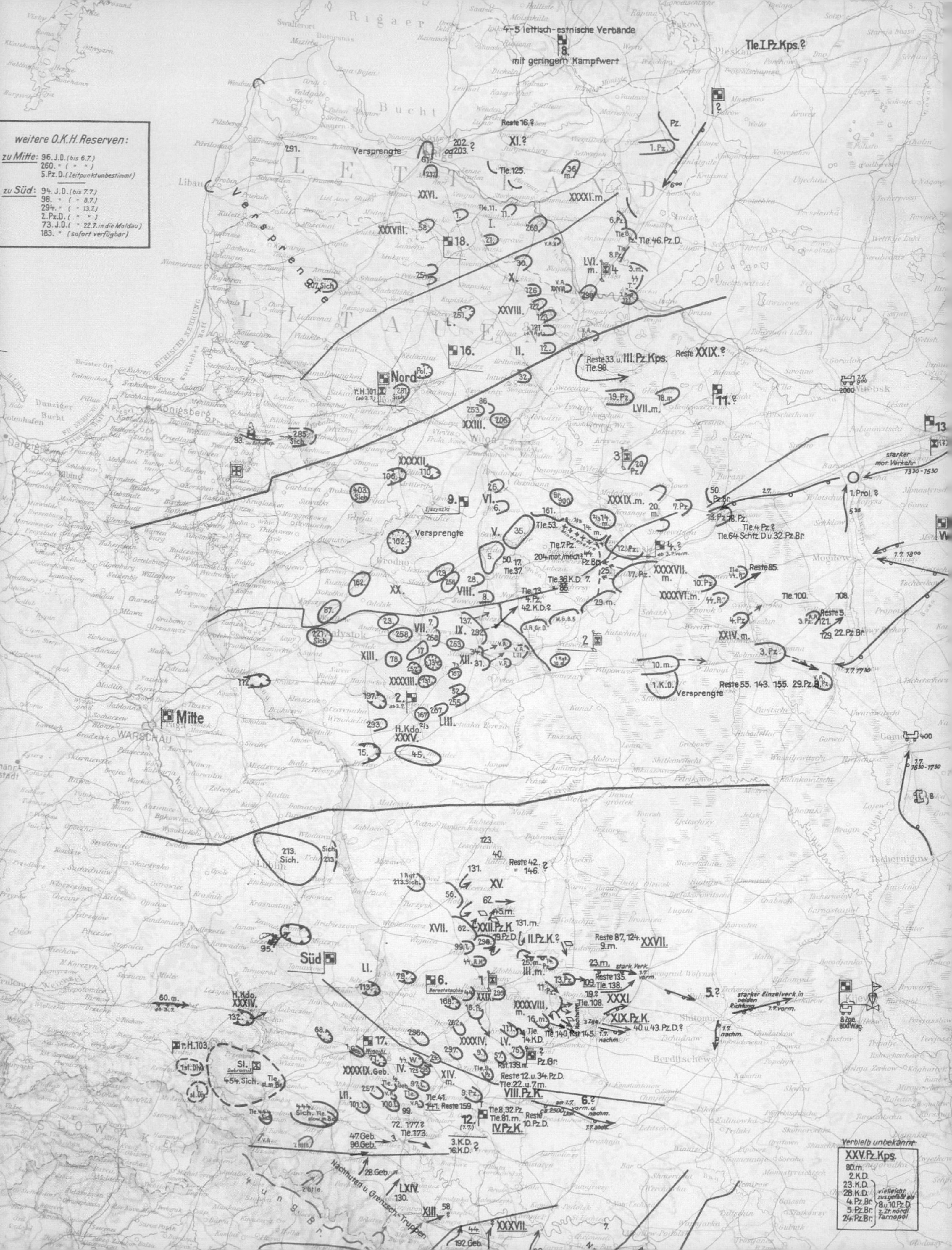

weitere O.K.H. Reserven:
zu Mitte: 96. J.D. (bis 6.7.)
260. " (" ")
5. Pz. D. (Zeitpunkt unbestimmt)
zu Süd: 94. J.D. (bis 7.7.)
98. " (" 8.7.)
294. " (" 13.7.)
2. Pz. D. (" ")
73. J.D. (" 22.7. in die Moldau)
183. " (sofort verfügbar)
Rigaer Bucht
4-5 lettisch-estnische Verbände
mit geringem Kampfwert
Tle.I.Pz.Kps.?
Reste 16.?
XI.?
202. od 203.?
Versprengte
LETTLAND
Tle.125.
XXXI.m.
XXVI.
XXXVIII.
18.
Tle.46.Pz.D.
LVI. m.
207. Sich.
Versprengte
LITAUEN
XXVIII.
16.
Reste 33.u. III.Pz.Kps. Tle.98.
Reste XXIX.?
Nord
19.Pz.
LVII.m.
11.?
XXIII.
13
Königsberg
Danziger Bucht
Danzig
Elbing
XXXII.
9.
VI.
XXXIX.m.
Versprengte
Tle.53.
Tle.7.Pz.
204 mot./mech.
Tle.64 Schtz.D.u.32.Pz.Br.
Tle.4.Pz.?
XXXXVII. m.
Reste 85.
Mogilew
Tle.36.K.D.
XX.
VIII.
Tle.13. 4.Pz. 42.K.D.?
XXXXVI.m.
Tle.100.
108.
Reste 3. 121.
22.Pz.Br.
VII.
IX.
XII.
XIII.
XXIV.m.
Reste 55. 143. 155. 29.Pz.B.
Versprengte
XXXXIII.
LIII.
H.Kdo. XXXV.
Mitte
WARSCHAU
213. Sich.
Reste 42. " 146.?
XV.
XVII.
XXII.Pz.K.
II.Pz.K.?
Reste 87, 124. 9.m.
XXVII.
Süd
LI.
6.
Reste 135. Tle.138.
Tle.108.
XXXI.
XXXXVIII. m.
XIX.Pz.K.
40.u.43.Pz.D.?
H.Kdo. XXXIV.
r.H.103.
17.
Sl. 454. Sich.
XXXXIX.Geb.
XIV. m.
Reste 12.u.34.Pz.D. Tle.22.u.7.m.
VIII.Pz.K.
12.
Tle.8.32.Pz. Tle.81.m.
IV.Pz.K.
47.Geb. 96.Geb.
3.K.D. 16.K.D.?
28.Geb.
Nachhuten u. Grenzsch.-Truppen
LXIV.
XIII.
XXXVII.
Kijew
Berditschew
Shitomir
Verbleib unbekannt
XXV.Pz.Kps.
80.m.
2.K.D.
23.K.D.
28.K.D.
4.Pz.Br.
5.Pz.Br.
24.Pz.Br.
vielleicht zusgefaßt als 8.u.10.Pz.D. 2. Pz. nordl. Tarnopol

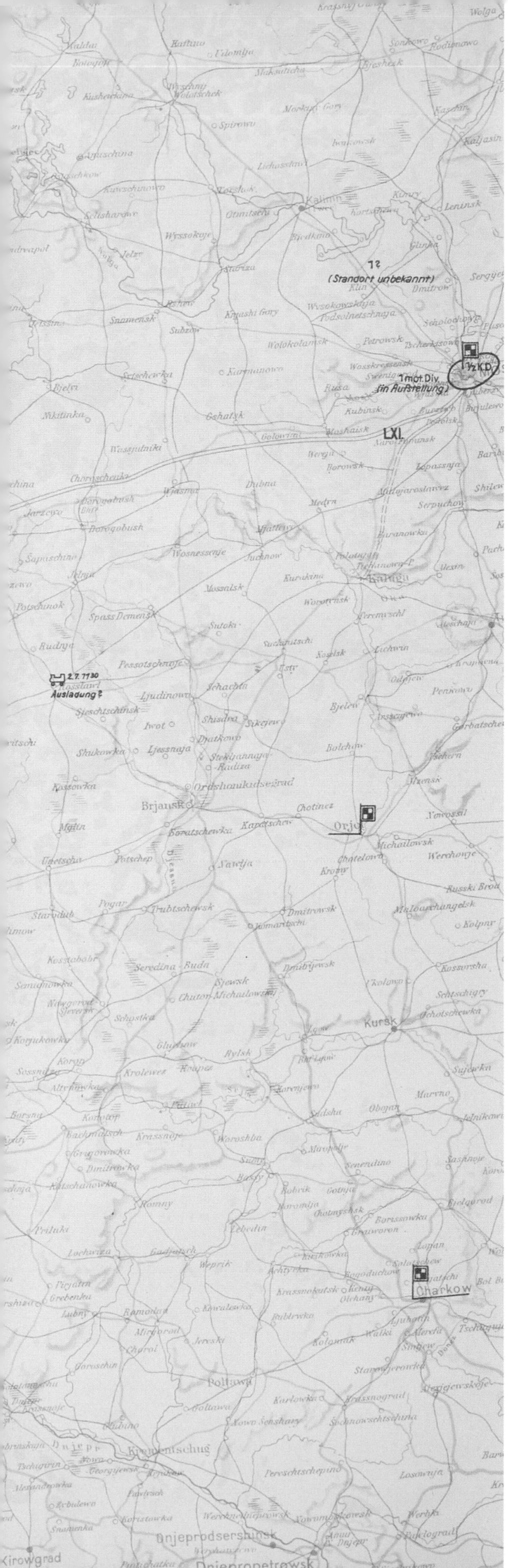

A HISTORY OF THE SECOND WORLD WAR IN 100 MAPS

JEREMY BLACK

The University of Chicago Press

For Kaushik Roy

The University of Chicago Press, Chicago 60637

Published 2020
Printed in Italy

29 28 27 26 25 24 23 22 21 20   1 2 3 4 5

ISBN-13: 978-0-226-75524-3 (cloth)
ISBN-13: 978-0-226-75765-0 (e-book)
DOI: https://doi.org/10.7208/chicago/9780226757650.001.0001

Published outside North and South America by the British Library, 2020.

Library of Congress Cataloging-in-Publication Data

Names: Black, Jeremy, 1955– author.
Title: A history of the Second World War in 100 maps / Jeremy Black.
Other titles: Second World War in 100 maps
Description: Chicago : The University of Chicago Press, 2020. | Includes bibliographical references and index.
Identifiers: LCCN 2020014429 | ISBN 9780226755243 (cloth) | ISBN 9780226757650 (ebook)
Subjects: LCSH: World War, 1939–1945—Maps. | Military maps—History—20th century. | Cartography—Political aspects—History—20th century. | Cartography—United States—History—20th century. | Cartography—Europe—History—20th century.
Classification: LCC G1038 .B55 2020 | DDC 940.53022/3—dc23
LC record available at https://lccn.loc.gov/2020014429

PAGES 2–3: Detail from a set of German military situation maps showing the dispositions of their forces on the Russian Front. The situation maps were produced on a day by day basis, and this map shows 2 July 1941. In the centre, the Germans had captured Minsk on 28 June. To the north, Riga fell on 30 June, while east of that German forces advanced toward Leningrad. In the south, German and Romanian forces had crossed the River Prut, while in west Ukraine Army Group South had taken heavy casualties before prevailing in frontier battles (see also pages 42–47 and 216–217).

Contents

Preface

The Second World War was a uniquely difficult war to map, and therefore a particularly interesting one to cover. Fought across an unprecedented range and in all the elements, the war saw fleets with aircraft carriers in combat, major airborne assaults, and large-scale insurgency and counter-insurgency campaigns, none of which had been the case with the previous world war, during which the fighting in East Asia, the Pacific and the Indian Ocean had been more limited, and the range and fighting capabilities of aircraft were far less. Moreover, the greater tempo of campaigning in the Second World War created practical problems for mapmakers seeking to provide tools for the military, as well as those informing publics at home. The range of combined operations meant that mapping was required in a more comprehensive and multi-purpose fashion than hitherto. Furthermore, the attempt to win and retain support, domestic and foreign, helped lead to particular pressures for favourable reportage and effective propaganda, both of which we show in this atlas.

This book, which is organised in terms of the key categories of mapping, provides guidance to the fascinating variety of maps produced during and about the war, illustrated by 100 significant examples of them. Countless more await your attention.

This book is dedicated to Kaushik Roy, India's leading military historian and a friend whose judgement I greatly value.

Introduction

'Like bombers and submarines, maps are indispensable instruments of war.'

John Kirtland Wright, Librarian at the American Geographical Society, *Geographical Review* (1942).

War inherently takes place in a spatial context and it is an activity that can only be conducted in that fashion. As a result, mapping is central to conflict, and at every level: from the most detailed (the tactical) to the most general (the strategic). Combatants order themselves in space, and have to maintain that sense whatever the strains of combat; doing so orientates them in terms of goal and means, and does the same for both allies and opponents. This sense of space, moreover, is important to point-of-contact engagement, however contested: whether hand-to-hand or by missiles and weapons at any range.

For most combat, we have no maps. Instead, mental mapping is the key, particularly in the shape of where is the enemy? Or where is cover? As, indeed, overwhelmingly remains the case today. That point is not one found in discussion of the subject, but any emphasis on mental mapping leads to the conclusion that the standard approach to war and cartography is teleological, in that it adopts a progressivist account, one predicated on the assumption that producing maps in physical copy is the desirable outcome and a necessary means. Instead, a needs-based assessment to mapping is appropriate, one that considers the idea of fitness for purpose in terms both of the maps and of their usage, and that is why this book is organised in terms of the key categories of mapping.

Such an approach, moreover, valuably complements that of considering the survival of artefacts, not least by asking what purpose is served? For maps in this context, there is the question not only of why they were retained, but, in particular, linked to that, their potential value on a recurrent and/or long-term basis, which is a crucial dimension of fitness for purpose. So also with the issue of accuracy, which is not a fixed, absolute value but a relative one to be gauged among other wartime pressures in terms of the opportunity costs. Accuracy requires both effort and time, but a map that lacks relevant accuracy is not fit for purpose.

Not all places were equal in coverage and significance. Indeed, that is a major difference between, on the one hand, the mapping of the world, in whole or part, and the use of such maps for war; and, in contrast, on the other hand, more specific and detailed mapping for particular military purposes. The latter process is integral to the point about places not being equal in mapping, both in terms of which places are mapped and to what level of detail.

Linked to that issue comes those of purpose and timing – in particular, maps produced prior to the period of immediate need, and those arising out of a national emergency with all the relevant pressures of need, production and timing. Indeed, as far as timing was concerned, there was a collapsing of the distinction between map and photograph, with the photograph from the world wars serving as a form of map, and one that could be used for immediate tactical purposes. This point underlines the difficulty of defining a map, but also, as a related point, the question of means versus ends in mapping. Thus, the trench maps of the First World War (1914–18) were substantially based on aerial photographs.

Returning to the point that not all places are equal, sites to be fortified attracted particular mapping attention, not least in order to plan how best to defend them. In this case, there was again an overlap with other forms of illustration, notably in the shape of diagrams and pictures. Although that contrast was very important it was not as clear as might be presumed because many maps included pictograms as devices. Moreover, some maps were simultaneously diagrammatic and pictorial, and each was regarded as enhancing the other.

The importance of maps for fortification was to be enhanced in the twentieth century as comprehensive front-wide systems developed during the First World War. In that conflict, trench warfare, to a degree, overwhelmed the strategic and operational dimensions of war in tactical problems – and notably so on the Western Front in France and Belgium – but that process also encouraged the mapping of the entire system. In the Second World War (1939–45, but 1937–45 in China) the fighting was less static in place and method but location, nevertheless, remained a key issue, not least due to the continual importance of artillery. Indeed, that was a powerful driver for mapping at the tactical level, for ballistics demanded a fixing of target location in order that the algorithms that determined the aiming of guns could apply.

Mapping therefore was clearly linked with capability, and the related requirement for maps for bombing was in effect another form of ballistics. The precise location of the target, and an understanding of the routes to get there, were both crucial. That meant that, in part, mapping for air warfare remained in effect two-dimensional, rather than focusing on the three dimensions that aerial conflict required. Instead, for a long time both bombing and aerial conflict relied on the visual identification of targets, although that changed greatly when radar became more significant, not least because it allowed effective night-fighting.

The First World War had made it clear that there would be a great need for maps in any future conflict. Not only in terms of the overall availability of maps, but also with reference to the number of types of fighting that had to be covered and with regard to the

THE SCALE OF PRODUCTION

In Britain, the Ordnance Survey produced about 193,775,000 maps for the Allied war effort, 35 private companies produced another 148,640,000 maps for the War Office, and the Geographical Section of the General Staff also had its own Survey Production Centre. The Survey of India, a major cartographic agency in the British Empire, produced about 22 million copies in total (of 2,483 different maps) in 1945 alone, compared with 750,000 (of 1,580) annually before the war. The United States' Army Map Service produced more than 500 million maps, and other branches of the United States military printed many more. After the Liberation from German occupation, in 1944 France's Institut Géographique National (IGN), or National Geographic Institute, produced 28 million maps in less than a year for Allied forces. This was valuable, not least because conflict continued within France, in Alsace, into 1945. In comparison, Germany printed about 1,300 million sheets.

Trends in production varied. By 1943, German production was declining, under the strain both of Allied bombing and of resource pressures, while that of the Allies was increasing.

WAR MAPS WHILE YOU WAIT

↑ Officers of the army War Plans Division study a world map as Brig. Gen. Leonard T. Gerow, division chief, points to a scene of operations. Map-making for the Army, Navy and Air Forces is a big industry in which American business machines and printing equipment have proved of outstanding value. Data for map-making are obtained by plane, patrol and reconnaissance, brought in by radio, telephone or courier and experts quickly compile and print the maps with duplicating machines, lithograph presses, mimeograph and multigraph. Some machines turn out multi-color maps. In the battle of Egypt quick changes in maps were made by using green carbon for allied positions, red for the enemy

↑ A sergeant of the U. S. Army engineers making a military map in the field prints a lithographic plate in an offset press. Fast duplication is necessary to supply all ground and air units with up-to-the-minute maps in quick time

← On the wall of field headquarters a master map of the battle zone is mounted, the lines changing as phone, radio or plane reports tell of advances or withdrawals. Strategic movements are plotted on the transparent overlay

46 POPULAR MECHANICS

'WAR MAPS WHILE YOU WAIT' FROM POPULAR MECHANICS, NOVEMBER 1943.

number of scales. In the inter-war years there were preparations for such a future conflict, which appeared increasingly likely from the mid-1930s both in Europe and in the Pacific. There were also technological advances that would affect mapping for war. In particular, there were further developments in aerial photography, including in the 1930s the use of colour and infrared film. Infrared images show colours which are invisible to the naked eye, and can also help the viewer to distinguish between camouflaged metal, which does not reflect strongly, and the surrounding vegetation, which does. Despite the obvious advantages of colour film, black-and-white film still had a place because it showed stronger contrasts than colour.

Aerial photography was not the sole sphere of development. All-weather instrument flying became possible. Moreover, in the United States of America (USA), the Engineer Reproduction Plant, a branch of the military, was instructed to develop maps that were legible under varied light conditions, a feature particularly necessary for night-flying and for long-distance missions. Meanwhile, hydrographic surveys employing sonar sounding were developed, first for deep water in the 1920s and then for shallow water in the 1930s.

From 1936 the British started work on a new map covering northeast France and Belgium because another German invasion of the region, attacked in 1914, appeared to be in prospect. The same year, the General Staff of the German Army established a section dedicated to military mapping and surveying, and the Heeresplankammer, the army map service, was re-established. In the event, the production and use of maps in the Second World War was on a far larger scale than in the First World War, large and unprecedented as that had been by its close; and maps played an important part in the impression of power and competence. Prominent in command headquarters and operations rooms, maps were depicted accordingly in photographs and films. The war introduced and diffused new technologies, both to the recording of information and to its presentation in map form. Moreover, speed in both the recording and presentation of maps was pushed to the fore (see box, opposite). One key element was the production and use of accurate and standardised base maps.

Map shortages

During the war the unprecedented range and scale of operations, and notably because they involved air power as well, were such that there were shortages of the necessary maps, particularly at the outset. For example, a dearth of maps in military collections when the United States entered the war in December 1941 necessitated extensive government borrowings, especially from the New York Public Library (NYPL). Walter Ristow from the NYPL's Map Division became head of the Geography and Map Section of the New York Office of Military Intelligence, while Lawrence Martin, the head of the Library of Congress Map Division, was assigned to the Map Division of the Office of Strategic Services (OSS), which built up a collection of over two million maps, and produced over 8,000 new ones. Advertisements were placed in popular magazines:

> *'The War Department, Army Map Service, is seeking maps, city plans, port plans, place name lexicons, gazetteers, guide books, geographic journals, and geologic bulletins covering all foreign areas outside the continental limits of the United States and Canada. Of particular interest are maps and guide books purchased within the last ten years, including maps issued by the U.S. government and the National Geographic Society.'*

Similarly, although they used much else, the British and Americans employed Michelin maps in preparation for the invasion of Normandy in June 1944. General maps were of great value for terrain and roads, while the 1939 *Guide Rouge* provided the Allies with detailed maps of French cities. The British also solicited, with success, postcards and photographs that provided very detailed pre-war views of the invasion beaches. The global extent of military movements made such information necessary and increased the pressure to create new maps as well as to accumulate and consolidate existing ones. Moreover, the large number of maps required led to a demand for innovations in production at every stage. The scribing of photographic negatives, a useful production technique, was generally done on glass, but scribing on plastic, which required less skill, was introduced experimentally by the US Coast and Geodetic Survey.

The pressure for mapping was felt by all powers, and the number of those involved helped ensure the scale of overall production. Thus, in response to being attacked by the Soviet Union (Union of Soviet Socialist Republics, USSR) in late 1939, the Finns finalised unfinished topographic maps and stepped up production by the topographic department of the General Staff, which was established on 18 October soon after Soviet pressure for significant frontier changes had begun. In the summer of 1940, after the Winter War was over, the Finns began producing multi-coloured maps in place of their earlier reprinting of black-and-white originals. These were useful when war resumed with the Soviet Union in 1941.

German–Soviet alignment

The Second World War was a global one from 1939, because it was then that Britain and France went to war with Germany, mobilising their global empires, the largest in the world, in doing so. Moreover, war between these powers and Germany added a far-flung naval dimension that was missing in the case of the war already underway between China and Japan, which had broken out in 1937. The Germans brought to an end the inter-war period, both militarily and politically, by rapidly conquering France in early 1940 and defeating British forces in France (and Norway) at the same time. The defeat of the key powers of Britain and France was also the end of the revived First World War that, in practice, had broken out in 1939, one in which Germany was resisted on the new Western Front.

Moreover, Germany and the Soviet Union had produced a de facto alignment in 1918 and for much of the 1920s, one that prefigured their alliance from August 1939 to June 1941. By the end of 1939 Germany was allied with Japan, Italy and the Soviet Union, and had co-operated with the latter in conquering Poland and determining specific spheres of influence in Eastern Europe. The independent states there, such as Bulgaria and Romania, were left with few options, other than to co-operate with Germany or the Soviet Union. The United States was neutral, as were the Latin American states. Britain and France, despite the support of their mighty empires, had been reduced to somewhat dubious hopes of long-term success, in particular through a blockade of Germany that was in practice not going to work due to the German–Soviet alignment.

By the summer of 1940 this British strategy scarcely seemed plausible. German successes in early 1940, first against Denmark and Norway, and subsequently against the Netherlands, Belgium, France and, to a lesser extent, British forces, were products of the existing geopolitical situation, because Germany was able to fight a one-front war and thus maximise its strength. In short, Stalin was the root cause of the German triumph in the West in 1940. In 1939 Stalin, by allying with Hitler, had followed in the footsteps of Lenin in 1918 by reaching an agreement with Germany that it was believed would advance the interests of the Russian/Soviet state and the cause of international communism. This was greatly facilitated by a shared hostility on the part of communist and Nazi leaders to Britain and its liberal democracy.

The need for superpower involvement

The past rarely repeats itself, as comparisons between the German offensives in 1870 and 1914, and between 1918 and 1940, indicate – or indeed between the Russo-German combination in 1939–41 and more recent relations between the two powers. German success in the field in 1940 owed much to the serious deficiencies of Allied strategy and planning, whereas in 1918 such deficiencies were more the case for Germany.

The defeat in 1940 of the armed forces of France and Britain on the mainland of Europe ensured that Germany would only be stopped as a superpower if the Soviet Union and the USA came into the war. A major American role would also be necessary to defeat Japan once it entered the war in December 1941: Britain and China would be able to deny Japan victory, which the British did by holding onto India, and the Chinese by continuing, despite heavy losses, the war begun in 1937. However, neither power would be able to defeat Japan.

The fall of France in June 1940 also marked the end of limited war because the new British government formed in May under Winston Churchill was not interested in a compromise peace dictated by a victorious Germany. Churchill's decision, and the inability of Germany to invade Britain or to bomb it into submission, meant that the conflict would continue until the Soviet Union and the USA could play a decisive role, but by the start of June 1941 it was by no means clear that Germany was bound to fail. Hitler was in a very strong position: dominant in Western and Central Europe, and allied to Italy (which was at war with Britain), Japan, the Soviet Union, Hungary, Romania, Slovakia, Bulgaria and Finland, and with the USA, Spain, Portugal, Switzerland, Ireland and Sweden neutral. Nevertheless, an unconquered Britain served as a base from which to launch air attacks on German-dominated Europe, diversionary operations in the Mediterranean and, eventually, in 1944, for the full-scale invasion of Western Europe. Britain's ability to act in this way was dependent on maintaining the security of sea routes, where Britain's particular strength, the Royal Navy, was up against Germany's weakest military arms: its navy and anti-shipping air power.

AN INFRARED PHOTOGRAPH OF LONDON FROM THE WEST, SHOWING THE RIVER THAMES TO THE NORTH SEA, 1934.

Infrared photography was a significant development of the 1930s. The visual recognition of ground features remained important to those mounting bombing raids, with rivers especially valuable because they could not be blacked out. German bombers attacking London in 1940 could orientate themselves by means of the Thames. So also with the River Exe and the German bombing of Exeter in 1942. The Thames also highlighted the docks, which were the key and repeated target for the bombers.

The war brought much more disruption and bombing to London than the First World War had done. There had been strong pre-war anxieties. 'The bomber will always get through,' Stanley Baldwin had announced in 1932 and there was much concern about the likely impact of bombing on civilian targets, a theme developed in books such as *The Menace of the Clouds* (1937) by Air Commodore L.E.O. Charlton. A savage air assault was anticipated at the beginning of the war, when there was widespread preparation for airborne gas attacks and large numbers of cardboard coffins were prepared. Under the apparent threat of such attacks, 690,000 Londoners were evacuated by rail from London in September 1939, while blackout regulations were implemented. In the event, the attacks did not come until the following autumn (by which time many of the evacuees had returned), and did not include gas attacks.

In the event, fortitude in the face of the 1940–41 Blitz became a key aspect of British national identity and the real and symbolic aspect of the assault on civil society accentuated the sense of the entire country being under attack, with London taking the central role. On the whole, morale remained reasonably high, with an emphasis on 'taking it'. Tea and cigarettes were palliatives for the fear and shock felt by many under bombardment and the dread of attack. By late September 1940, about 177,000 Londoners were sleeping in the Underground railway system.

The docks were particularly heavily bombed, and even in the early 1960s a river trip from Hungerford Bridge at Charing Cross down to beyond Tower Bridge went past serried ranks of burned-out warehouses. Much of the dockland landscape was destroyed, especially the interconnection of places of work and tightly packed terraced housing.

SINGAPORE

War Office, map of Singapore, 1944.

Singapore was developed as a major British naval base. Whereas Hong Kong was regarded as overly vulnerable to Japanese attack, Singapore was seen as sufficiently far from Japan to permit reinforcement without peril. Even so, it fell to the enemy in February 1942. The British hope that Singapore could hold out until the Royal Navy relieved it was wrong on both counts: it did not hold out, and there was no relief.

In March 1944, Winston Churchill instructed the Chiefs of Staff to delay sending naval help to the Americans in the Pacific because:

> *'It is in the interest of Britain to pursue what may be termed the "Bay of Bengal Strategy" at any rate for the next twelve months ... All preparations will be made for amphibious action across the Bay of Bengal against the Malay Peninsula and the various island outposts by which it is defended, the ultimate objective being the reconquest of Singapore. A powerful British fleet will be built up based on Ceylon, Adu Atoll and East India ports.'*

In the end, the naval priority was put onto helping the Americans in the Pacific, but planning nevertheless continued to be directed toward Singapore. In September 1944, Admiral Sir Geoffrey Layton, Commander-in-Chief Ceylon, wrote of:

> *'...the vital importance of our recapturing those parts of the Empire as far as possible ourselves. I would specially mention the recapture of Burma and its culmination in the recovery of Singapore by force of arms and not by waiting for it to be surrendered as part of any peace treaty.... The immense effect this will have on our prestige in the Far East in post-war years. This and only this in my opinion will restore us to our former level in the eyes of the native population in these parts.'*

Admiral Lord Louis Mountbatten, Supreme Allied Commander South East Asia Command, strongly agreed.

In the event, Operation Zipper, an amphibious invasion of western Malaya, was planned for 9 September 1945. It would probably have led to heavy casualties, both British and Japanese, but was made unnecessary by the Japanese surrender.

The surveying necessary to prepare for Zipper reflected extensive reconnaissance, both aerial and by means of agents landed by submarine. The onshore and offshore data are reflected in this map, which also notes the location of the camp at Changi, on a peninsula at the eastern end of the island, where Allied troops were kept as prisoners of war in harsh circumstances, until those who remained were moved to Changi Prison later in the war. The Japanese had captured Singapore itself in 1942 when they crossed the narrow Johore Strait.

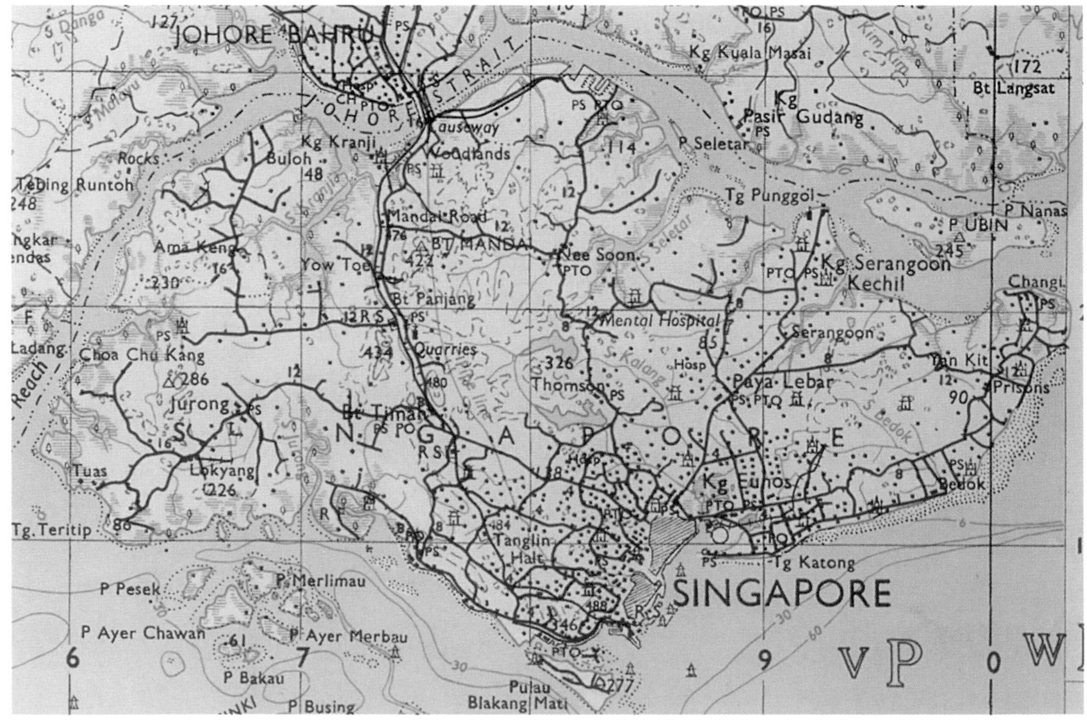

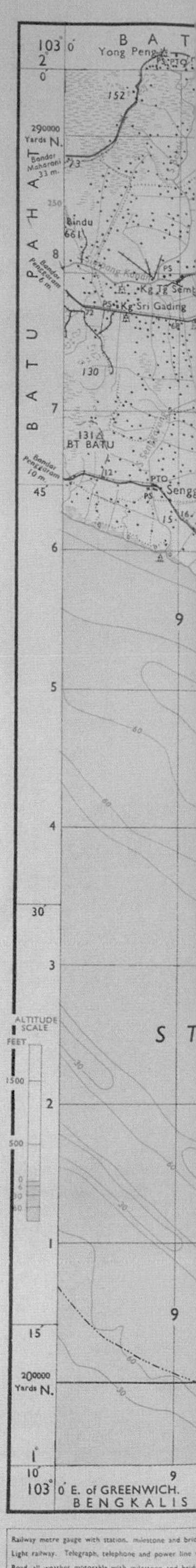

Refer to this map as G.S.G.S. 4218
Sheet A-48N
First Edition

3,440 SINGAPORE FIRST EDITION G.S.G.S. SHEET A-48N

Scale 1 Inch to 4 Miles **QUARTER INCH** or 1:253,440.

Miles 5 4 3 2 1 0 5 10 15 20 25 Miles
Yards 10000 0 10000 20000 30000 40000 Yards
Kilometres 5 0 5 10 15 20 25 30 35 40 Kilometres

HEIGHTS IN FEET

Contour Interval 250 feet
Form line interval on Singapore island 100 ft.
Bathymetric contours which are in feet have been taken from Admiralty Chart

Heights, triangulated station : point : approximate	△7186 723 .1295
Church. Mosque. Chinese temple. Hindu temple. Buildings.	
Buddhist temple. Fort. Lighthouse. Light beacon. Light buoy.	
Lopak. Mangrove swamp (tidal). Fresh water swamp. Submerged sand	
Contours with value. Form-lines. Mines	
Rest-house. Halting bungalow. Forest bungalow.	RH HB FB
Post office. Post and telegraph office. Police station.	PO PTO PS
Court house. Hospital.	CH Hosp

GLOSSARY

Bagan	Landing place, fishing village	Ladang (Ldg)	Cultivated clearing
Bendang	Rice swamp	Lopak	Impermanent swamp
Bukit (Bt)	Hill	Padang (Pdg)	Open space, moorland
Changkat (Ct)	Hillock	Parit (Pt)	Ditch, drain
Genting (Gtg)	Pass, col	Pasir	Sand
Gunong (G)	Mountain	Paya	Permanent swamp
Jalan	Road	Pulau (P)	Island
Jeram (J)	Rapids	Sungei (S)	River, stream
Kampong (Kg)	Village	**Tanjong (Tg)**	**Promontory**
Kuala (K)	River mouth	Telok (T)	Bay

Index to Sheets.

A-48G	A-48H	A-48I
	A-48N	A-48O

Index to Provinces.

JOHORE (MALAYA)
STRAITS OF MALACCA

1 SINGAPORE COLONY
2 REPUBLIC OF INDONESIA

Relative Reliability.

A Rigorous Survey
1-inch Malaya 1924-37 (HIND 1035) 1944
B Compiled from Sumatra 1:250,000 (HIND 1042)
Sheet No. 25 SINGAPORE 2nd Edition 1944
With additions & corrections from intelligence information

True North & Grid North
Magnetic North (1950)
Annual Change negligible

Reprinted from First Edition HIND/1076, 1945.
Reproduced by Ordnance Survey, 1950.

Grid references are given in thousands of yards East and North of the south-west corners of the lettered squares: thus the grid reference of BT TIMAH RS is VP 7714
for grid letters see body of map.

MOD

Projection:- Polyconic
JOHORE GRID (Yards)

Origin	2° 2′ 33″·30 N. & 103° 33′ 45″·93 E.
False Co-ordinates of Origin	450,000 Yards East
	300,000 Yards North

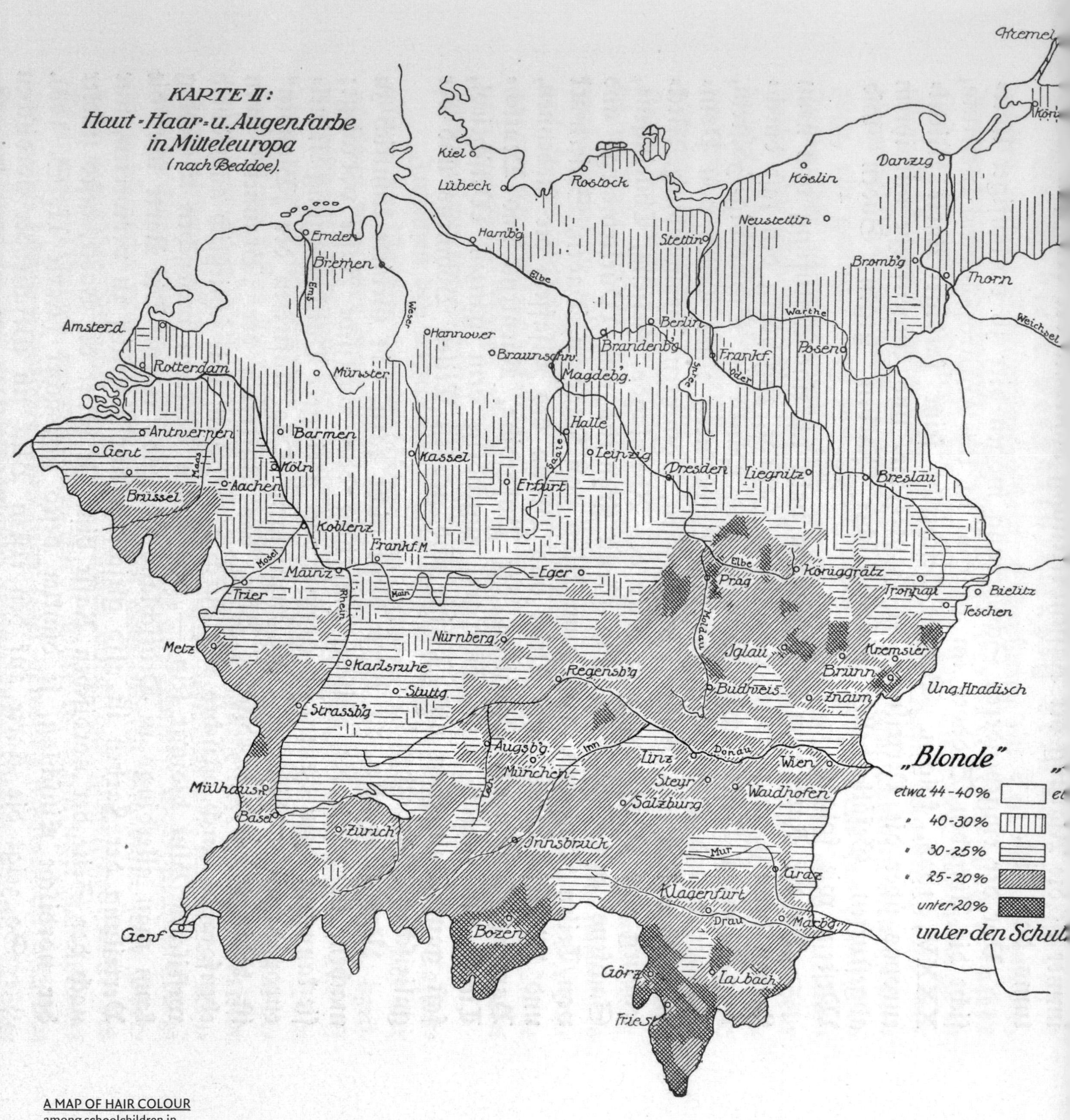

A MAP OF HAIR COLOUR among schoolchildren in 'MittelEuropa' from *Rassenkunde des deutschen Volkes* ('Racial Science of the German People') by Hans F.K. Günther, first published in 1922.

Geopolitics

1.

'Conquer the Heartland and you dominate the World Island.'

The Nazis Strike (1943).

The militarisation of information during the Second World War greatly involved geographers, just as it had earlier done in the First World War. Indeed, from 1942 to 1945 the US government employed two out of every five geographers who were members of the three national geographical associations. This specialist knowledge about understanding spatial relationships in the world was also put to use in the wartime economy: assessing how best to produce the necessary materiel, when it was likely to become available and helping to turn possibilities into realities. Mapmakers were part of a broader process of academics assisting the war effort. For example, economists employed by the US government to provide realistic production projections (notably Simon Kuznets, Robert Nathan and Stacy May) used statistics in an innovative fashion to understand and produce information on American national income. The resulting Gross National Product figures clarified the viability of planning a massive rise in production for the military without needing to cut the consumer economy. Cartographers played a role in planning new industrial capacity for military production by charting the best places for the new shipyards and for the inland manufacture of components, and helping with the relevant road and rail links.

Somewhat differently, maps were highly significant during the expansion of German rule in 1938–41, being important to the processes of land-grab by treaty that was employed to reward and influence – even control – German allies, notably Hungary, Romania, Bulgaria, Italy – including, it was hoped, the Soviet Union. In August 1939 the Nazi–Soviet Pact included a secret protocol in which Hitler and Stalin divided Eastern Europe into 'spheres of influence' in anticipation of 'political and territorial rearrangements'. This agreement included a map signed by Stalin that partitioned Poland, which was invaded the following month, first by Germany and then by the Soviet Union. The Allies subsequently found and published a microfilmed copy that had been seized from the archives of defeated Germany, but the Soviets claimed the document was a fake and denied the existence of the protocol until 1989. Three years later, Russia released the official protocol from the archives.

Ethnic and racial themes

The major theme of German geopolitics and mapping was ethnographical, pressing hard the alleged cause of the German people, many of whom, after the 1919 Versailles peace settlement, were under foreign rule – for example, the Sudeten Germans in Czechoslovakia (until the 1938 Munich settlement). This theme fuelled right-wing German geopolitics in the 1920s; it was claimed that the country's territorial losses under the peace settlement – for example placing Danzig (Gdansk) under League of Nations administration and the creation of the Polish Corridor – should be revised to reflect the fact that many Germans lived in these areas.

An ethnographical trend in geopolitics preceded the Nazi rise to power in 1933. Many Germans thought the Slavs of Eastern Europe inferior and held a racial theory of human development in which the Slavs were unsuited to lands even where they were in a clear majority. Indeed, objective geographic standards and values were abandoned in favour of tendentious presentation in a process that gathered fresh impetus and was given much more encouragement under Nazi rule.

The Nazi Party was dominated by ideas of race and conflict. A map displayed at the 1938 Nazi Party Convention showed the danger of a Soviet attack on Germany by indicating the routes that could be used. An earlier map had presented Czech air power as a threat to Germany. New research centres, notably the Northeast German Research Community at Berlin founded in 1933, produced research that reflected these Nazi racial preoccupations. As a result, maps presented these and related issues. For example, skin, hair and eye colour in Mitteleuropa (itself very much a German concept) were mapped (opposite) in Hans F.K. Günther's *Rassenkunde des deutschen Volkes* ('Racial Science of the German People', published in 1922, and which by 1945 had sold more than half a million copies). The blondeness of the Sudentenlanders in Czechoslovakia would have underlined its German character to the map's German readers. Mitteleuropa in this map was defined as encompassing Alsace and Lorraine, although these regions, seized in 1870–71, had been returned to France after the First World War. The map also incorporated the extensive territories lost to a restored Poland as part of the Versailles settlement.

Hitler's geopolitics and the Holocaust

With his construction of politics in terms of races rather than states, and of races supposedly engaged in an existential struggle, Hitler wanted a different geopolitics to Karl Haushofer (see box, page 16). Hitler would not countenance any limitation of German expansion on other than a short-term, tactical, basis that was motivated by opportunism in the shape of his reading of the

KARL HAUSHOFER

The USA, which had recruited geographers for its war effort, saw the Germans as determined to use geography to help conquer the world. The Americans presented that geography accordingly: *Plan for Destruction*, an American documentary film of 1943, showed Karl Haushofer (1869–1946, Germany's leading geopolitician, explaining geopolitics in front of a map centred on the North Pole. This was an exposition of the threat supposedly (and truly) posed by Germany to parts of the world and, implicitly, a suggestion that the American response should be global. In November 1939, *Life* magazine, a major American illustrated weekly, presented Haushofer as the 'philosopher of Nazism' and the 'inexhaustible Ideas Man for Hitler', while his son, Albrecht, was depicted looking at a globe. Readers were informed that in Haushofer's Institute of Geopolitics in Munich: '...the world is remade every day between breakfast and dinner.'

Haushofer argued in favour of the great powers expanding in their natural spheres of interest, presenting them as pan-regions: Pan-Europa, Germany's region, which would include expansion into Russia; Pan-Asien, which was designated for Germany's ally Japan, with which Haushofer had close personal links; and Pan-Amerika. He founded the *Zeitschrift für Geopolitik* (*Journal of Geopolitics*), and its many maps focused on 'expansion' and 'struggle'. Lebensraum ('living space') was a term that was much adopted, as was the idea of *wehr-geopolitik* ('geo-strategy'). Rudolf Hess, personal assistant and, from 1933, deputy to Hitler, was Haushofer's favourite pupil, and Hitler was influenced by these ideas of global racial expansionism. In 1934, Haushofer was appointed president of a council for those of German origin living outside Germany, while he also had links to the General Staff of the German Army. Within the Ministry of Propaganda, there were two study groups to ensure that teachers offered appropriate geopolitical education at school level.

During the Second World War, Germany's aggression was traced in American popular journals, such as *Reader's Digest*, in part to the impact of geopolitical thought. Directed by Frank Capra, the film *The Nazis Strike* (1943) depicted Haushofer's institute as seeking the 'military control of space'. Germany's intention was outlined by rephrasing Halford Mackinder's 1904 dictum that Europe is the 'world island': 'Conquer Eastern Europe and you dominate the Heartland. Conquer the Heartland and you dominate the World Island. Conquer the World Island and you dominate the World.'

In *Plan for Destruction*, Haushofer is depicted as advising Hitler and coordinating German intelligence-gathering around the world, information that is presented as helping the Germans achieve victory. Haushofer is shown as having a master plan to dominate the heartland, and then to join the Japanese in attacking the Americans. In reality, Haushofer strongly supported the August 1939 Nazi–Soviet (Ribbentrop–Molotov) Pact, believing that an alliance between Germany and the Soviet Union would be very potent, which was a theme adopted earlier in the century by Mackinder. To Haushofer, the Soviet Union offered Germany security from a two-front war, as well as access to vast natural resources, spatial mass and an alliance against Britain.

Both Karl and Albrecht Haushofer were discredited due to their link with Hess (see below). Albrecht went into hiding after his part in the Bomb Plot against Hitler in July 1944, but was captured and later shot. Karl was imprisoned in Dachau, but soon released. Although included on a list of war criminals by the International Military Tribunal in Nuremberg, Karl was not arraigned. After learning of Albrecht's fate, he and his wife committed suicide in 1946.

WEHR-GEOPOLITIK ('GEO-STRATEGY'), 1941.
Published in Berlin in 1941, this book was one of the last of a series of works by Karl Haushofer, the key figure in German geopolitics (see above). A retired major general and professor at the University of Munich, Haushofer pressed the case for geopolitics as a striving for survival and primacy between competing powers. To that end, he called for the revision of Germany's loss of territory and prestige in the 1919 Treaty of Versailles, a goal that was achieved with the defeat of France in 1940. In 1939 Haushofer received the Order of the Eagle of the German Reich from Hitler in recognition of his services to geopolitics.

However, Haushofer believed the invasion of the Soviet Union in 1941 was mistaken, while his son Albrecht, an adviser to Rudolf Hess, Hitler's deputy, wished to end the war with Britain and was closely linked to Hess's unsanctioned peace overture to Britain in 1941, a move that led to the disgracing of the Haushofers, although they were kept in play as possible intermediaries in any future negotiations with Britain.

On 10 May 1941, Hess flew to Britain on an unauthorised, uninvited and unsuccessful attempt to settle Anglo-German differences, that possibly reflected initiatives involving MI6, maybe some British politicians keen to avert a German invasion of Britain, and even the Polish government-in-exile. Hess thought there was a political faction in Britain that would negotiate peace with Germany; his mission was seen by Stalin as designed to isolate the Soviet Union, which may have been Hess's intention but it was not Churchill's response. Concerned that Italy and Japan would see the approach as an attempt by Germany to settle with Britain, Hitler disavowed Hess, who was imprisoned in Britain.

The Hess peace mission is evidence of the unstable nature of alliance links at this point, but Haushofer's geopolitics in 1941 would have required Germany to abstain from war, which would not have suited Hitler's expansionist purposes at all.

The map on the cover of *Wehr-Geopolitik* (right) reflected the interest of Haushofer in German dominance of what he identified as the region of Pan-Europa. The map also noted the importance of naval bases, a key theme in the bold expansionist plans of the Kriegsmarine.

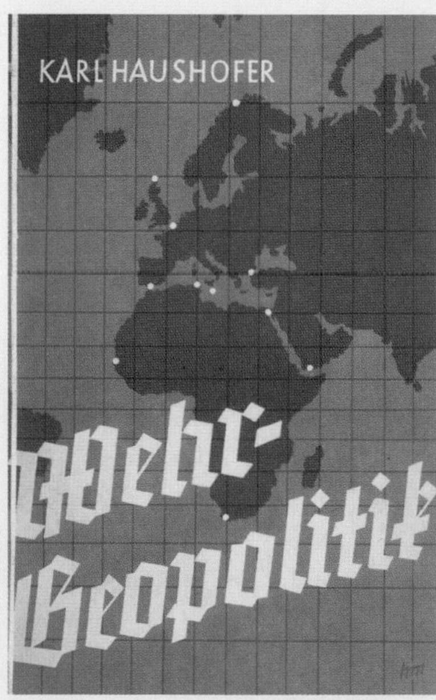

specific possibilities of the moment. Redressing the Treaty of Versailles was at best a tactic for Hitler, as was co-operation with Poland in 1935–38 or, even more, Russia in 1939–41. Hitler's focus was on struggle with Jews and Slavs, while Soviet expansionism, notably in 1940 at the expense of Romania, triggered Hitler's concerns about the situation in Eastern Europe.

Invading the Soviet Union in June 1941, in Operation Barbarossa, and declaring war on the United States that December, brought forward the millenarian strain in Nazism and encouraged Hitler to give deadly effect to his aspirations and fears. The removal of Jews from a German-dominated Europe moved to the fore in policy terms, and became a key aspect of German geopolitics, an extreme instance of seeing space in racial terms. The Holocaust was a geopolitical process, in terms of ensuring the 'biological eradication of the entire Jewry of Europe', which Alfred Rosenberg, the Minister of the Eastern Territories, promised in a press briefing in November 1941. Indeed, Germany was declared *judenfrei* ('free of Jews') in June 1943. The movement of Jews to Eastern Europe, where they were to be murdered, was crucial to that process and train movements on a pan-European scale were planned accordingly.

The Holocaust was a central part of a wider plan for a geographical 'New Order', with an enlarged Germany central to a new European system, and with the Germans at the top of a racial hierarchy. In September 1936, Hitler told a Nazi rally in Nuremburg:

> *'If I had the Ural Mountains with their incalculable store of treasures and raw materials, Siberia with its vast forests, and the Ukraine with its tremendous wheat fields, Germany and the National Socialist leadership would swim in plenty.'*

Centralised planning was a facet of German policy. This was the case with rearmament in the 1930s and with the control imposed on occupied territories, notably in Eastern Europe. The economy of Europe, both conquered and allied, was to be made subservient to German interests. The rest of Europe was to provide Germany with forced labour (both in situ and in Germany), raw materials and food, and, in turn, to receive German industrial products on German terms. Industrial plants in occupied areas were to be taken over. Plans for Eastern Europe had been developed by the Reichsarbeitsgemeinschaft für Raumforschung, an interdisciplinary scientific institution established in 1935, and directed by Konrad Meyer, a senior SS officer, that sought to coordinate spatial research. During the war, the Reichsamt für Landesaufnahme, the mapping service of the German government, mapped the occupied territories. Japan's plans for the 'Co-Prosperity Sphere' it intended to create in conquered areas were similar, but less developed.

Transport infrastructure

New routes and bridges, for rail and particularly road, were to provide the transport links in a German-dominated Europe, and were an expression in concrete and steel of the new geopolitics. Needs matched goals and the unusually harsh winter of 1941–42 revealed the serious inadequacies of the existing road system in the western Soviet Union, which the Germans had recently conquered and required to remain in good order to sustain their campaign. Many Soviet roads were not winter-ready tarmac routes, and the Soviet maps the Germans seized were apt to exaggerate the achievements of the Five-Year Plans of Soviet economic activity. These inadequacies created major problems for the German forces in sustaining their offensive and in resisting a large-scale Soviet counter-offensive mounted in December 1941, which pushed the Germans back from near Moscow. The Germans also totally failed to plan for the nature of the terrain and climate.

These deficiencies led the Germans in the spring of 1942 to decide to build a series of strategic roads to supply their forces and link their territories, the most crucial of which was the Durchgangstrasse IV ('Thoroughfare IV', see pages 32–33).

Allied geopolitics

Geopolitics was not simply a tool of the authoritarian powers. It was also significant for the Western powers. This was especially so for those with far-flung commitments, notably Britain. The topic had been developed and popularised by Halford Mackinder in *Britain and the British Seas* (1902) and *Democratic Ideals and Reality* (1919). In addition, C.B. Fawcett in 1933 produced maps centred on London, Canberra and Durban in order to convey the differing geostrategic vantages within the British Empire, the range of which made a good spatial understanding necessary. The maps illustrate the sheer power of distance in shaping strategic perspective. Fawcett used the technique of drawing an orthographic (one seen as if from outer space) globe, centred on say Canberra, and then 'peeling' the dark side of the planet around the edges from the back. The centring of maps as a cartographic device was seen in many maps – for example, Ernest Dudley Chase's 'Victory in the Pacific' (1944), an American work that carried the note 'Invest in Victory Bonds' (see pages 30–31).

The capacity for mobility and tactical advantage offered by air power ensured that, from the 1920s, it was increasingly seen as a major strategic asset and one that offered a new approach to geopolitics, a view that encouraged government support for the development of air services to link imperial possessions, notably in Britain. The Second World War was to bring these ideas to fruition in new views of a world at war. The resulting geopolitics was very much seen with the maps of Richard Edes Harrison, for example his 'One World, One War', which was published in *Fortune* in March 1943. Echoing the traditional British concern with India, and drawing on Mackinder's work, Harrison referred to India as where the Allies would 'have to fight one of the great decisive battles of history: The Battle for Asia', adding:

> *'If Japan and Germany are allowed to join hands in India, the Axis will have the advantages of "the inner line" – on a world scale. Uninterrupted Axis control of Eurasia's huge land masses, from Le Havre to Shanghai, would transform the New World into an island and the two surrounding oceans into highways of invasion.'*

INVASION GEOGRAPHY

Ewald Banse, *Raum und Volk im Weltkriege*, 1932.

On 29 June 1940 *The Illustrated London News* drew attention to the geopolitical background of the strategic challenge the country faced by reproducing as an inset a map from Ewald Banse's *Germany, Prepare for War!*, the 1934 translation of his *Raum und Volk im Weltkriege* (1932). An American edition, *Germany Prepares for War: A Nazi Theory of 'National Defense'*, also appeared in 1934.

A prolific geographer and professor of military science at the Technical College in Braunschweig (Brunswick), Banse (1883–1953) was a protégé of Karl Haushofer (see box, page 16) and pressed for the expansion of Germany to rule all Germans, which meant Germany extending beyond its 1914 frontiers. Banse regarded this goal as requiring war and praised warriors. In 1941 Banse published his last book, *Volk und Lebensraum*.

The larger map (below) captures the exaggerated use of terrain as a feature. Rail links not roads are shown. The inset map (opposite) provided a guide

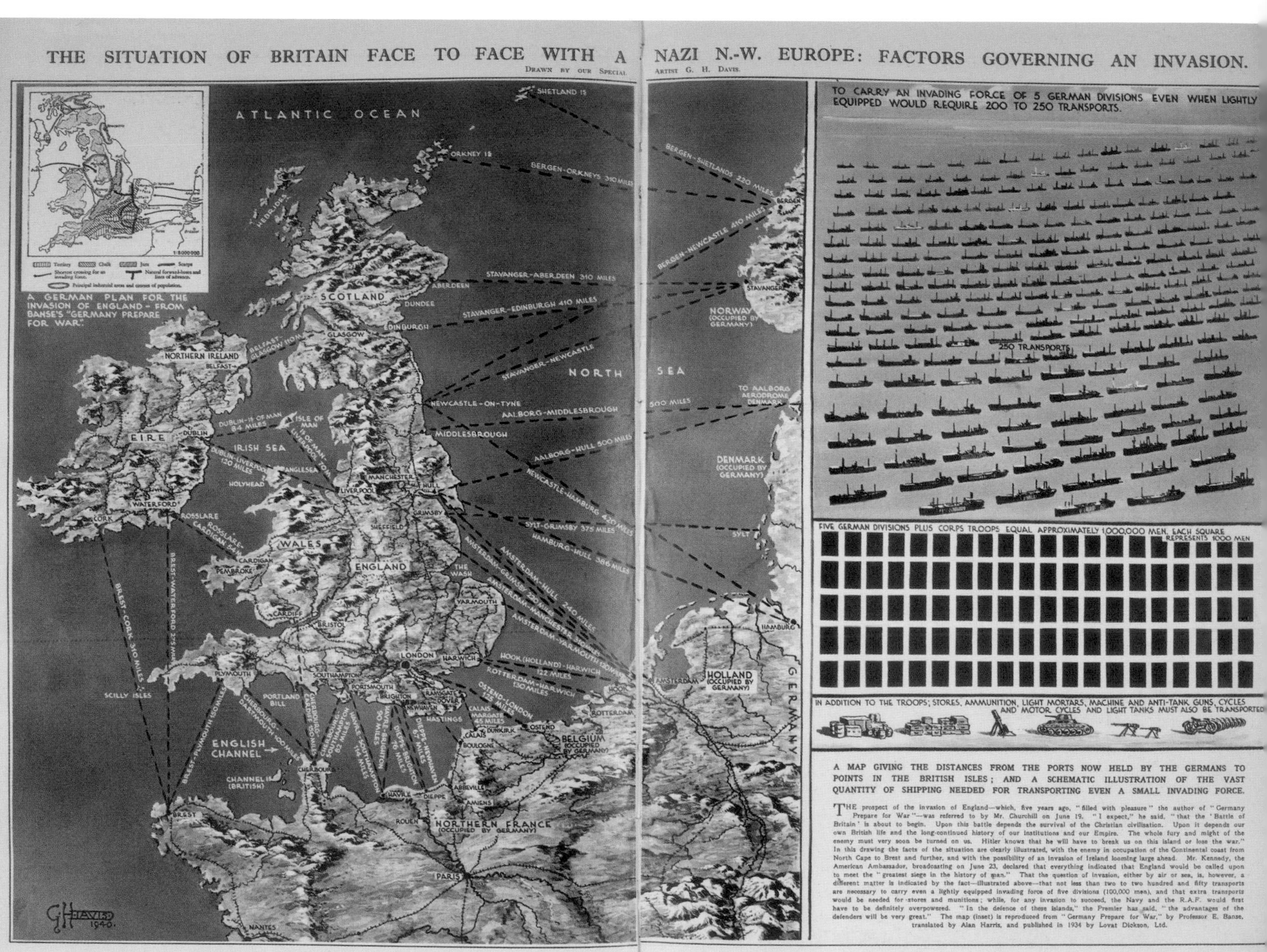

A MAP GIVING THE DISTANCES FROM THE PORTS NOW HELD BY THE GERMANS TO POINTS IN THE BRITISH ISLES; AND A SCHEMATIC ILLUSTRATION OF THE VAST QUANTITY OF SHIPPING NEEDED FOR TRANSPORTING EVEN A SMALL INVADING FORCE.

THE prospect of the invasion of England—which, five years ago, "filled with pleasure" the author of "Germany Prepare for War"—was referred to by Mr. Churchill on June 19. "I expect," he said, "that the 'Battle of Britain' is about to begin. Upon this battle depends the survival of the Christian civilisation. Upon it depends our own British life and the long-continued history of our institutions and our Empire. The whole fury and might of the enemy must very soon be turned on us. Hitler knows that he will have to break us on this island or lose the war." In this drawing the facts of the situation are clearly illustrated, with the enemy in occupation of the Continental coast from North Cape to Brest and further, and with the possibility of an invasion of Ireland looming large ahead. Mr. Kennedy, the American Ambassador, broadcasting on June 23, declared that everything indicated that England would be called upon to meet the "greatest siege in the history of man." That the question of invasion, either by air or sea, is, however, a different matter is indicated by the fact—illustrated above—that not less than two to two hundred and fifty transports are necessary to carry even a lightly equipped invading force of five divisions (100,000 men), and that extra transports would be needed for stores and munitions; while, for any invasion to succeed, the Navy and the R.A.F. would first have to be definitely overpowered. "In the defence of these islands," the Premier has said, "the advantages of the defenders will be very great." The map (inset) is reproduced from "Germany Prepare for War," by Professor E. Banse, translated by Alan Harris, and published in 1934 by Lovat Dickson, Ltd.

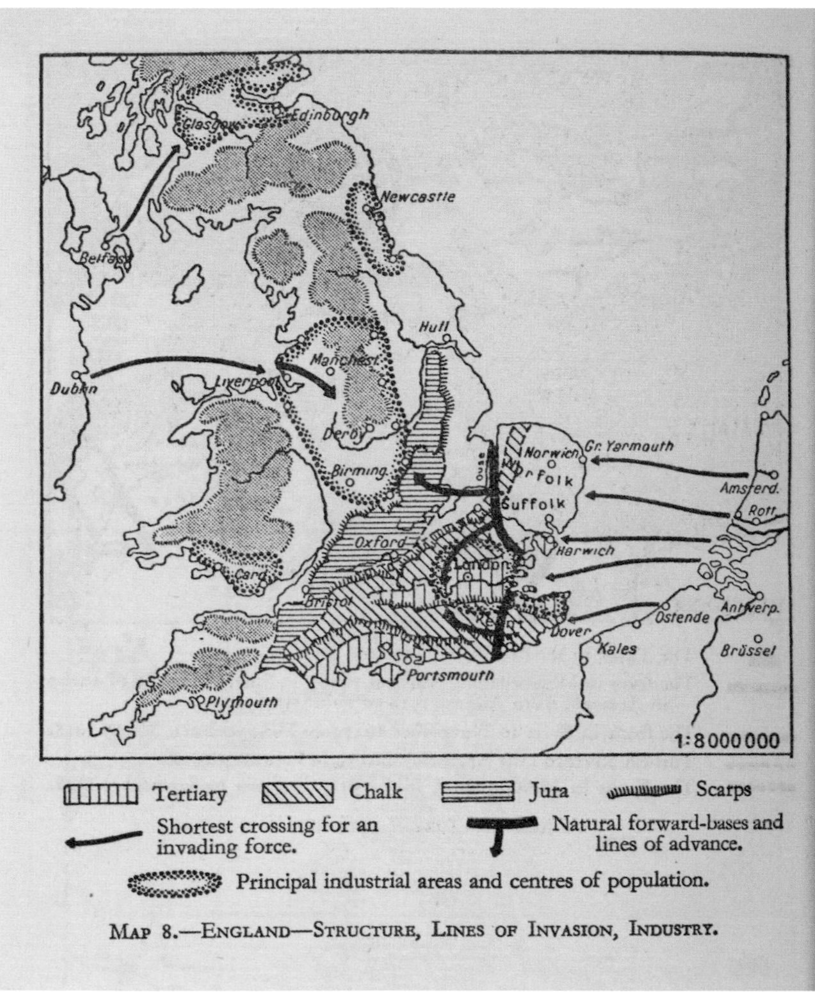

Map 8.—England—Structure, Lines of Invasion, Industry.

to the invasion geography of Britain. It suggested a landing from the Netherlands in East Anglia and from Belgium in Kent and then the establishment of positions from which London could be enveloped. There was no prospectus for a comparable invasion from France, which was overrun by the Germans in 1940. As a result, in Operation Sealion that year the Germans planned to land between Hythe and Rottingdean, with their forces despatched from Rotterdam in the Netherlands south to Le Havre in France. However, as was indicated by *The Illustrated London News*, the transport requirements for such an invasion were immense, 'while, for any invasion to succeed, the Navy and the RAF would first have to be definitely overpowered'. The Norway campaign, in which a number of German warships had been sunk or damaged, had indicated the serious vulnerability of amphibious assaults to British naval power, and on 11 July, the commander of the Kriegsmarine, Admiral Erich Raeder, expressed his scepticism to Hitler. Although Italy's entry into the war on the side of Germany on 10 June had greatly increased pressure on the Royal Navy in the Mediterranean, enough British warships remained in home waters to challenge any German naval attack.

However, rather than focusing on attacking British warships, Luftwaffe commanders were increasingly concerned to show that they could win by a bombing war on civilian targets, for which the prelude was to be an attack on the RAF and its supporting infrastructure.

The Germans lacked not only an adequate invasion strategy but also the necessary force structure, a situation that reflected and contributed to their ad hoc planning and make-do expedients. The lack of proper landing craft was a particular problem, but so also were relevant doctrine and training. Air superiority was necessary to hit the powerful and determined Royal Navy, but the Stuka dive bombers were highly vulnerable to Britain's modern fighters, and air superiority, even if achieved, would not have been effective against warships at night.

Timing was a key factor. It was deemed necessary to invade before mid-September, when the weather was likely to become hostile, but this timetable could not be met. Aside from the problems involved in assembling landing vessels, the German Army needed to be resupplied and moved to the embarkation zones. Victory in the previous campaign in France had involved losses. Unlike the well-prepared Allies in Normandy in 1944, a German invasion would have been improvised and resource-weak.

Unlike the prospectus offered by Banse, the Germans aimed to advance first across southeast England, to a line from Gravesend on the Thames estuary to Portsmouth, and then, having captured London, to a line from Maldon on the Essex coast to the Severn estuary. However, after delays, the operation was formally postponed on 17 October by when the focus was on attacking the Soviet Union. Conquering the latter was also seen as a way to force Britain into seeking peace by leaving it isolated.

WAR IN EAST AFRICA

S.J. Turner, 'Colour Relief Map of Abyssinia and War Zone', 1938.

This map, drawn by S.J. Turner for the *Daily Herald*, the leading British left-wing newspaper, is an aerial perspective map of the type more commonly associated with the American Richard Edes Harrison. The perspective both emphasises the proximity of the area of hostilities and provides guidance to the strategic significance of the Mediterranean and the Suez Canal. The competition between Britain and Italy emerges clearly, both in Africa and in the Mediterranean, and notably the threat posed by Italy to the Suez route to India. Massawa and Assab were the Italian naval bases in Eritrea.

The Italian conquest of Abyssinia (Ethiopia) in 1935–36 involved a major advance from Eritrea with a supporting one from Italian Somaliland. Addis Ababa, the capital, was captured on 5 May 1936. The Italians deployed nearly 600,000 men and used mustard gas bombs.

The Italians were greatly helped by the failure of other powers to act on the League of Nations' condemnation of the invasion. The British government considered oil sanctions against Italy, as well as closing the Suez Canal to Italian shipping. However, Stanley Baldwin, the Prime Minister, told the Cabinet on 6 April 1936 that doing so 'would involve war with Italy', which he was 'unwilling to envisage in the present state of Europe'. The British government wanted to ensure Italy stayed true to the Stresa Front against Germany, and did not want to provide Italy with immediate reasons to attack British shipping and bases in the Mediterranean. (The Stresa Front was an alliance formed by Britain, France and Italy in April 1935 to oppose Hitler's announced intention to rearm Germany and violate the Treaty of Versailles.) Moreover, both France and the United States were sympathetic to Italy's goals.

But there was planning for war. This included the need to protect British colonies from neighbouring Italian colonies. To provide the means for attacks on the latter, the RAF moved aircraft to Egypt, British Somaliland and Malta.

In the late 1930s, as Mussolini planned an invasion of Egypt in the event of war, the British anxiously considered how best to defend their position there, not least because they were trying to protect both their fleet base at Alexandria and the Suez Canal. The British were also concerned about the Italian naval presence in the Red Sea, Italian dealings with Yemen, which included the supply of arms, and the possibility that Italy might win over King Ibn Saud of Arabia.

After war broke out between Britain and Italy in June 1940, the Italians conquered British Somaliland but failed in their advance in Egypt. In turn, the British in 1941 regained British Somaliland and conquered Italian East Africa (Eritrea, Abyssinia and Italian Somaliland), a formidable achievement that provided additional strategic depth for the empire. The capture of the deep-water Red Sea quays at Massawa on 8 February 1941 and the Somali port of Kismayo on 14 February ended any prospect of naval co-operation between Italy and Japan.

Aerial views were to become increasingly common in the British press. Thus, Malta was presented as a target in a map by George Horace Davis published in *The Illustrated London News* on 8 February 1941, with the island depicted as viewed from the Sicilian bases of German and Italian bombers.

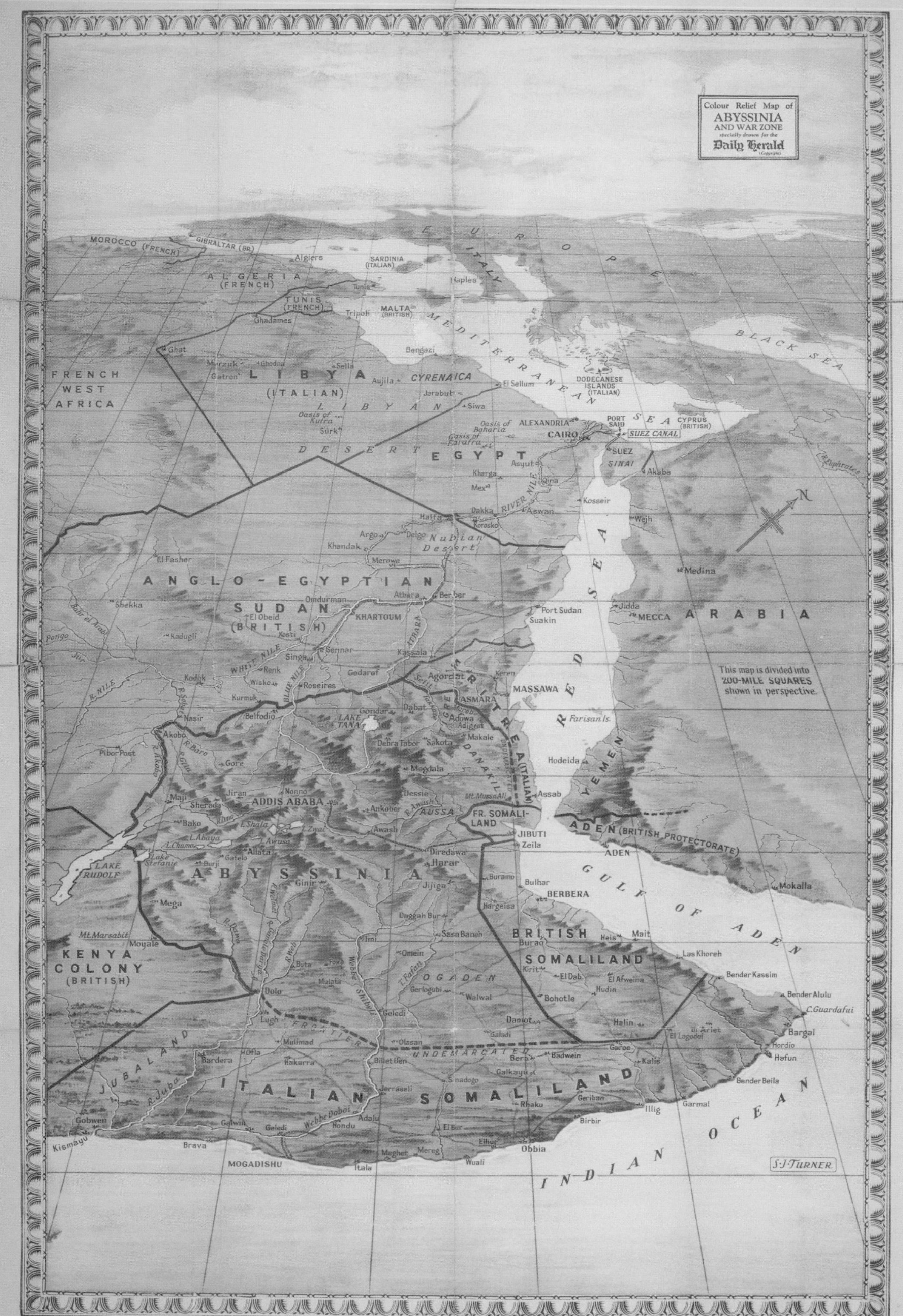

Colour Relief Map of
ABYSSINIA
AND WAR ZONE
specially drawn for the
Daily Herald
(Copyright)
E U R O P E
ITALY
Naples
SARDINIA (ITALIAN)
BLACK SEA
MOROCCO (FRENCH)
GIBRALTAR (BR)
Algiers
ALGERIA (FRENCH)
TUNIS (FRENCH)
Tunis
Tripoli
MALTA (BRITISH)
MEDITERRANEAN SEA
Ghadames
Ghat
FRENCH WEST AFRICA
Murzuk
Ghodna
Gatron
Sella
LIBYA (ITALIAN)
Aujila
CYRENAICA
Bengazi
El Sellum
Jarabub
Siwa
DODECANESE ISLANDS (ITALIAN)
CYPRUS (BRITISH)
LIBYAN DESERT
Oasis of Kufra
Sürk
Oasis of Baharia
Oasis of Farafra
ALEXANDRIA
CAIRO
PORT SAID
SUEZ CANAL
SUEZ
SINAI
Akaba
R. Euphrates
EGYPT
Asyut
Kharga
Mex
Qina
RIVER NILE
Kosseir
Aswan
Dakka
Korosko
Halfa
Wejh
Argo
Delgo
Nubian Desert
Khandak
Merowe
El Fasher
ANGLO-EGYPTIAN SUDAN (BRITISH)
Medina
Shekka
Omdurman
Atbara
Berber
El Obeid
KHARTOUM
ATBARA
Port Sudan
Suakin
Jidda
MECCA
ARABIA
Bahr el Arab
Kadugli
Kosti
Sennar
Kassala
Singa
WHITE NILE
Renk
BLUE NILE
Gedaref
Agordat
ERITREA
Keren
This map is divided into 200-MILE SQUARES shown in perspective.
Kodok
Wisko
Roseires
R. NILE
Setit
MASSAWA
ASMARA
Kurmuk
Gondar
Dabat
Adowa
Adigrat
RED SEA
Farisan Is.
Belfodio
LAKE TANA
Nasir
Akobo
R. Baro
Debra Tabor
Sokota
Makale
DANAKIL
Pibor Post
Gore
Magdala
Hodeida
YEMEN
Jiran
Nonno
Dessie
Mt. Mussa Ali
Assab
Maji
Sheroda
ADDIS ABABA
Ankober
AUSSA
FR. SOMALILAND
Bako
L. Shala
Awash
JIBUTI
ADEN (BRITISH PROTECTORATE)
L. Abaya
Zeila
ADEN
L. Chamo
Allata
Diredawa
Gatelo
Harar
LAKE RUDOLF
Lake Stefanie
ABYSSINIA
Buramo
Bulhar
GULF OF ADEN
Ginir
Jijiga
BERBERA
Mokalla
Mega
Hargeisa
Daggah Bur
Sasa Baneh
BRITISH SOMALILAND
Burao
Heis
Mait
Mt. Marsabit
Moyale
Imi
Omein
Las Khoreh
KENYA COLONY (BRITISH)
Buta
Foke
OGADEN
Kirit
El Dab
El Afweina
Bender Kassim
Mulata
Gerlogubi
Walwal
Hudin
Dolo
Bohotle
Bender Alulu
C. Guardafui
Lugh
Geledi
Danot
Galadi
Halin
El Lagodei
Bargal
FRONTIER
Olasan
Muimad
UNDEMARCATED
Badwein
Garoe
Hordio
Hafun
JUBALAND
Ofla
Bardera
Hakarra
Billet Uen
Beri
Galkayu
Kalis
R. Juba
ITALIAN SOMALILAND
Jerraseli
Sinadogo
Bender Beila
Rhaku
Geriban
Illig
Garmal
INDIAN OCEAN
Gobwen
Galwin
Geledi
Webbe Doboi
Nondu
Adalu
El Bur
Birbir
Kismayu
Brava
Elhur
Obbia
MOGADISHU
Meghet
Mereg
Wuali
Itala
S. J. TURNER

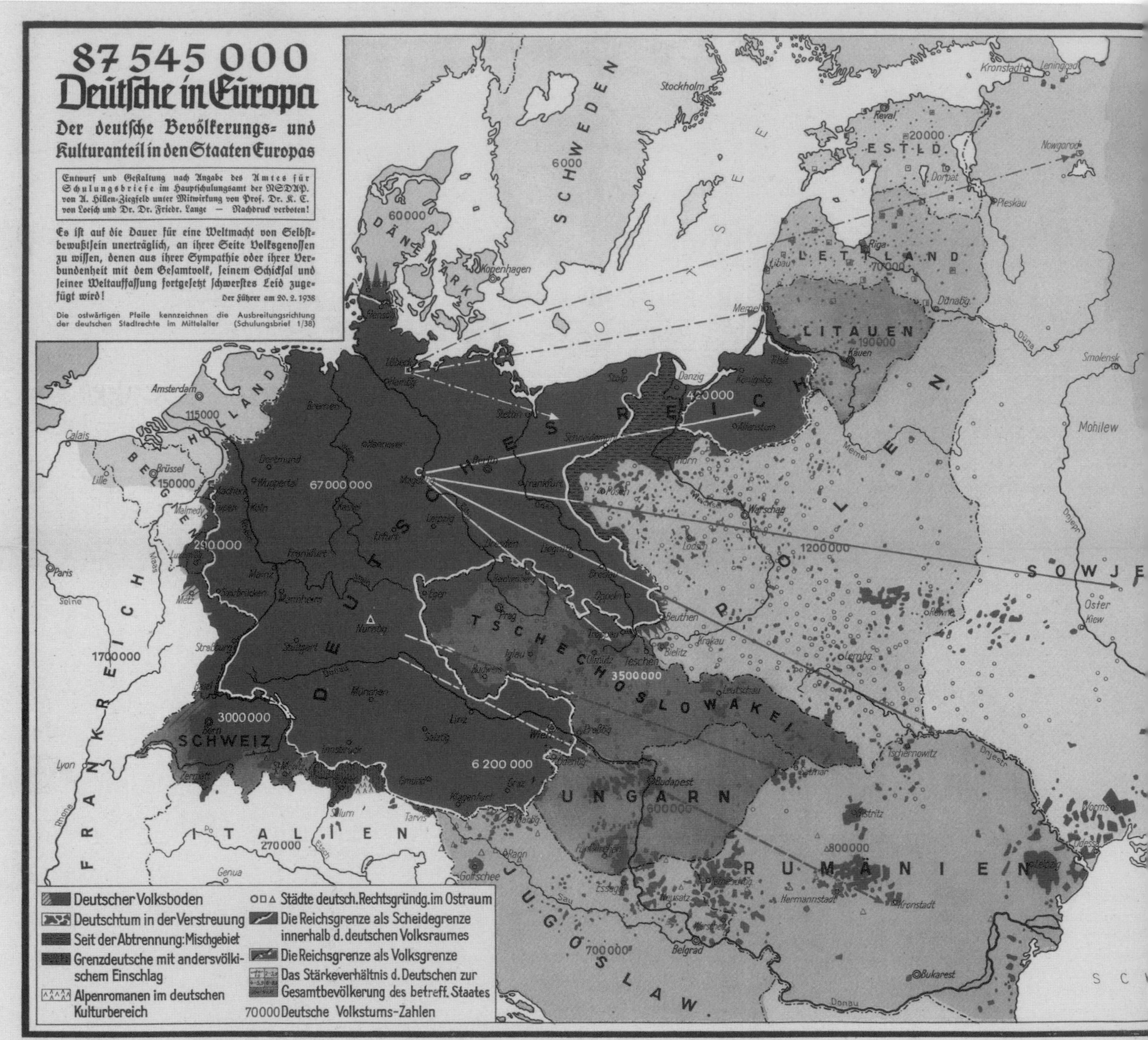
87 545 000
Deutsche in Europa
Der deutsche Bevölkerungs= und Kulturanteil in den Staaten Europas
Entwurf und Gestaltung nach Angabe des Amtes für Schulungsbriefe im Hauptschulungsamt der NSDAP. von A. Hillen-Ziegfeld unter Mitwirkung von Prof. Dr. K. C. von Loesch und Dr. Dr. Friedr. Lange — Nachdruck verboten!
Es ist auf die Dauer für eine Weltmacht von Selbstbewußtsein unerträglich, an ihrer Seite Volksgenossen zu wissen, denen aus ihrer Sympathie oder ihrer Verbundenheit mit dem Gesamtvolk, seinem Schicksal und seiner Weltauffassung fortgesetzt schwerstes Leid zugefügt wird!
Der Führer am 20. 2. 1938
Die ostwärtigen Pfeile kennzeichnen die Ausbreitungsrichtung der deutschen Stadtrechte im Mittelalter (Schulungsbrief 1/38)
Deutscher Volksboden
Deutschtum in der Verstreuung
Seit der Abtrennung: Mischgebiet
Grenzdeutsche mit andersvölkischem Einschlag
Alpenromanen im deutschen Kulturbereich
Städte deutsch. Rechtsgründg. im Ostraum
Die Reichsgrenze als Scheidegrenze innerhalb d. deutschen Volksraumes
Die Reichsgrenze als Volksgrenze
Das Stärkeverhältnis d. Deutschen zur Gesamtbevölkerung des betreff. Staates
70000 Deutsche Volkstums-Zahlen
SCHWEDEN
6000
Stockholm
OSTSEE
ESTLD.
20000
Reval
Dorpat
Leningrad
Kronstadt
Nowgorod
Pleskau
DÄNEMARK
60000
Kopenhagen
LETTLAND
70000
Riga
Libau
Dünaburg
Memel
LITAUEN
190000
Kauen
Tilsit
Smolensk
Mohilew
HOLLAND
115000
Amsterdam
Calais
BELGIEN
150000
Brüssel
Lille
Malmedy
Eupen
DEUTSCHES REICH
67000000
420000
Danzig
Königsberg
Allenstein
Stettin
Lübeck
Bremen
Dortmund
Wuppertal
Köln
Erfurt
Leipzig
Frankfurt
Dresden
Liegnitz
Breslau
Mainz
Saarbrücken
Metz
Nürnberg
Stuttgart
Straßburg
München
Linz
Wien
Graz
Innsbruck
Donau
POLEN
1200000
Warschau
Posen
Thorn
Lodz
Memel
Krakau
Bielitz
Teschen
Lemberg
Beuthen
Oppeln
SOWJE
Oster
Kiew
Dnjepr
Dnjestr
Düna
TSCHECHOSLOWAKEI
3500000
Prag
Eger
Iglau
Budweis
Troppau
Preßburg
Leutschau
Paris
Seine
FRANKREICH
1700000
Lyon
Rhone
SCHWEIZ
3000000
Bern
Zermatt
ITALIEN
270000
Po
Etsch
Genua
Tarvis
Klagenfurt
6200000
UNGARN
600000
Budapest
RUMÄNIEN
800000
Czernowitz
Bistritz
Hermannstadt
Kronstadt
Bukarest
Odessa
JUGOSLAW.
700000
Gottschee
Belgrad
Neusatz
Essegg
Save
Donau

A EUROPE OF VOLK

Arnold Hillen Ziegfeld, map of the German diaspora, 1938.

This map was commissioned by a Nazi Party periodical in 1938 and produced by Arnold Hillen Ziegfeld. Europe is classified in terms of its German population, which stretches to the Caucasus (covered in one of the inset maps). The quote from a Hitler speech of 20 February 1938 described how 'unbearable' it is for a 'world power' to witness the suffering inflicted on 'national comrades' because of 'their sympathy or loyalty to the people as a whole'. Such maps were a feature of German conservative geopolitics, which had strong support after the losses of territory under the 1919 Versailles peace settlement. The sentiments it reflected came to serve expansionist Nazi interests.

Ziegfeld was a major practitioner and theorist of propaganda mapping. In 'Kartengestaltung, ein Sport oder eine Waffe' published in Haushofer's *Zeitschrift für Geopolitik* in 1935, Ziegfeld suggested the term *kartographik* for the newly dynamic character of German mapping, with its emphasis on arrows as a means to demonstrate relationships between particular areas – often relationships of challenge and threat. This term referred to the new development of *gebrauchsgraphik* (commercial art), and drew on artistic and intellectual developments such as Futurism, an understanding of visualisation, and a cult of modernity that challenged conventional themes and means.

Hitler sought to control Eastern Europe, where Lebensraum ('living space') for Germans was to be sought, a theme he advanced in his book *Mein Kampf*. Although there was a tension between Nazi views and conservative geopolitics, Hitler's arguments drew on a longstanding nationalist belief that Germany's destiny included domination of Eastern Europe. This belief drew, in part, on Prussian attitudes, notably toward Poland, but also entailed a transference of Austrian assumptions. The racial inflection of these beliefs focused on a supposed struggle between Germans and non-Germans, one in which there was no uncertainty where virtue, progress and destiny lay. Lebensraum involved both geopolitical and racist objectives. Although Hitler was willing in the short term to co-operate with Poland, notably against Czechoslovakia in 1938, and potentially against the Soviet Union, he hated Poles and had long planned the destruction of Poland.

Atlases presented Germany as under threat from communists and Jews. The opening page of maps in the 1931 edition of F.W. *Putzgers Historischer Schul-Atlas*, the standard school historical atlas, included one of Germany as the bulwark of European culture against the Asiatic hordes, the latter depicted in terms of Huns, Avars, Arabs, Mayars, Turks, Mongols, Jews, Tsarist Russians and communists. Thus, the past was recruited to the service of the present, and with racial groups and ideologies melded together.

By August 1940, German maps reflected success in the May–June 1940 campaign and asserted the 'cultural' claims of Germany. Germany annexed Alsace-Lorraine, which was treated as part of the Third Reich, with conscription accordingly. Beyond that, Germany stipulated a *zone réservée* in France where there was to be ultimate German colonisation. Near Aachen, the Belgian frontier cantons of Eupen, Malmédy and St. Vith, all transferred from Germany in 1919, were reincorporated into the Third Reich. Luxemburgers were declared German nationals and in 1942 conscription was introduced.

Ziegfeld's assistant on this map was Friedrich Lange, who in 1940 produced a 'Map of German Culture in the West' to show western border areas that it was claimed had been beneficiaries from German expansion in the past, with a double-eagle symbol of the 'Reichsdeutsche Städte' placed alongside a number of cities in France and Belgium together with dates when they were German, including Liege (Lüttich, 1801), Verdun (Wirten, 843), Nancy (Nanzig, 1766), Épinal (Epieneln, 1766) and Besançon (Bisanz, 1032, 1410). Yellow lines within French territory are labelled 'Natural Defence Stages of the Paris Basin'.

GERMANY UNCONSTRAINED

Ernst Adler, from *Der Krieg 1939/40 in Karten*, 1940.

These maps about the geopolitics of Germany, 1940, are from *Der Krieg 1939/40 in Karten*, which was translated into English by the German Library of Information in New York as *The War in Maps* (1941), and they contrast 1914, when the Central Powers faced a two-front war, with 1939–40 when 'Germany has broken free'. The situation is given as at the end of April 1940, with the Soviet Union an ally; Poland, Denmark and Norway all conquered; Italy's alliance assured; and Spain supportive. The map underlined British fears of the Nazi–Soviet Pact because it revealed the extent to which Germany believed itself at liberty to act. In the English-language edition the two map captions are translated as '1914: Central Powers Encircled by Entente' (below) and '1940: No Manacles this Time' (opposite).

The introduction to *The War in Maps* explains the indoctrinatory value of a map in helping a mass audience make sense of information that would otherwise be lost to them, or 'put into clear focus the hazy impressions gained by laymen from communiques and newspaper dispatches', revealing 'formerly incomprehensible phases of the war ... as simple problems in geopolitics'. The two maps used in the introduction are intended to invite comparison: in this case, a concerted effort by the Nazis to convince readers that this war was not a repeat of the previous one, which had ended in defeat. By showing the encirclement of Germany in the First World War, a situation (in 1940) in which the Nazis have avoided a war on two fronts makes for a much more favourable impression.

In a 1941 article called 'Magic Geography', published in the journal *Social Research*, Hans Speier pointed out how certain visual illusions are achieved in the map by the use of colour. For example,

1914: Einkreisung der Mittelmächte durch die Entente

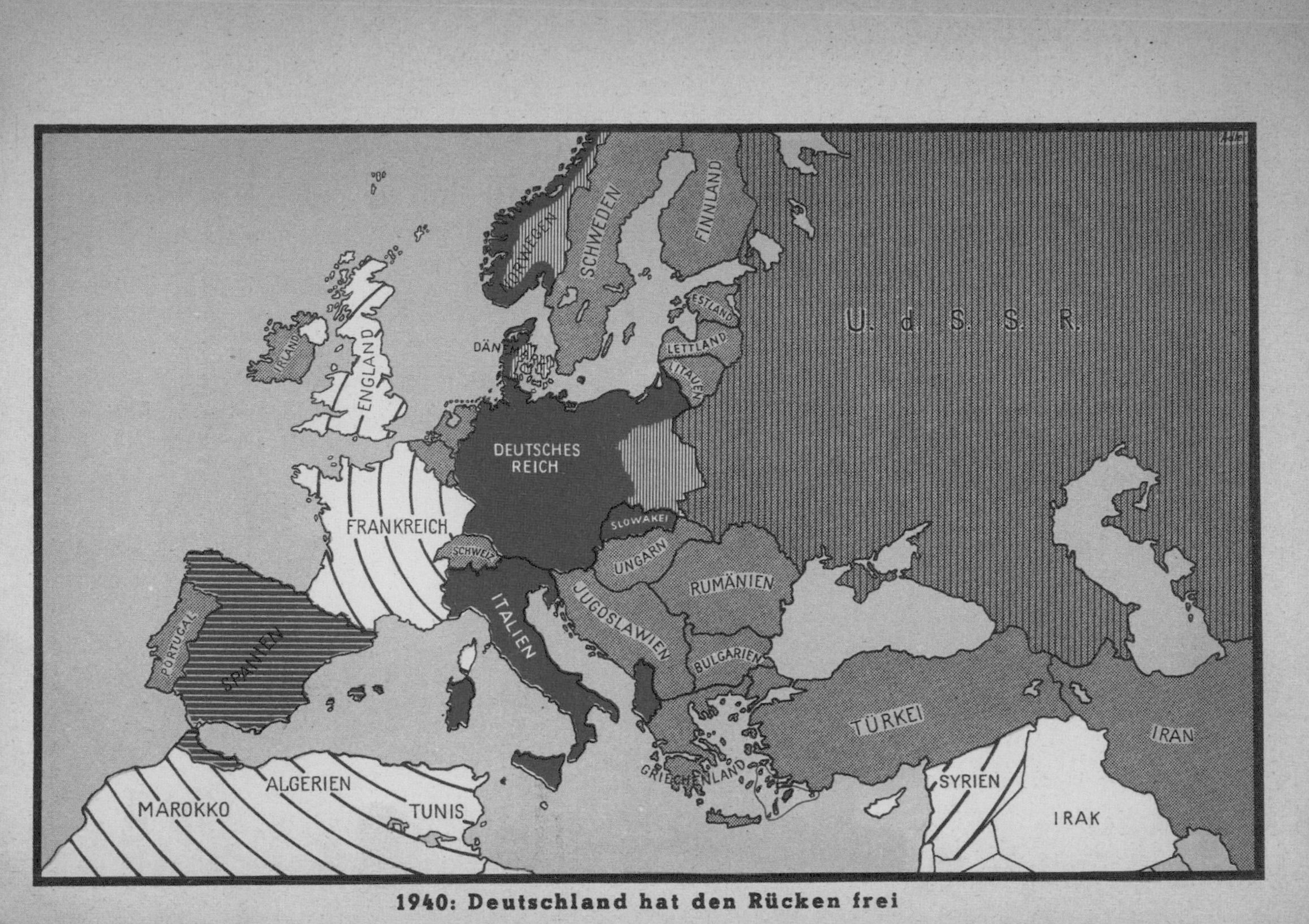

1940: Deutschland hat den Rücken frei

Germany appears in red in each map and its enemy countries in yellow, with neutrals in grey. A greater contrast is achieved in the first map by depicting Germany's allies in pink, which makes Germany look more vulnerable by reducing the relative impact of the Central Powers. In the second map, Italy, Slovakia and Albania are also solid red, which enlarges the German bloc, while the use of grey with red striped lines for the Soviet Union disproportionately reduces the relative size of the former enemy country (likewise 'friendly' Spain). The use in both maps of the curved lines on the yellow (enemy) countries re-emphasizes an encirclement focused on Germany, but the visual impression this gives on the second map is to greatly reduce the sense of encirclement and to leave the viewer with the belief that, as the caption says, Germany is no longer constrained.

The War in Maps refers to itself as an *Atlas of Victory* 'because it portrays the practical expulsion of Great Britain from the European Continent, where she is an unwanted intruder'. The 'birth of the new Europe', as the book puts it, arises out of the fact that Germany has escaped its historic encirclement, which it argues was a 'tool of the West in its efforts to keep Central Europe in a state of political impotence' – in other words, preventing Germany from fulfilling its destiny of Continental leadership and domination.

THREE APPROACHES TO THE U.S.

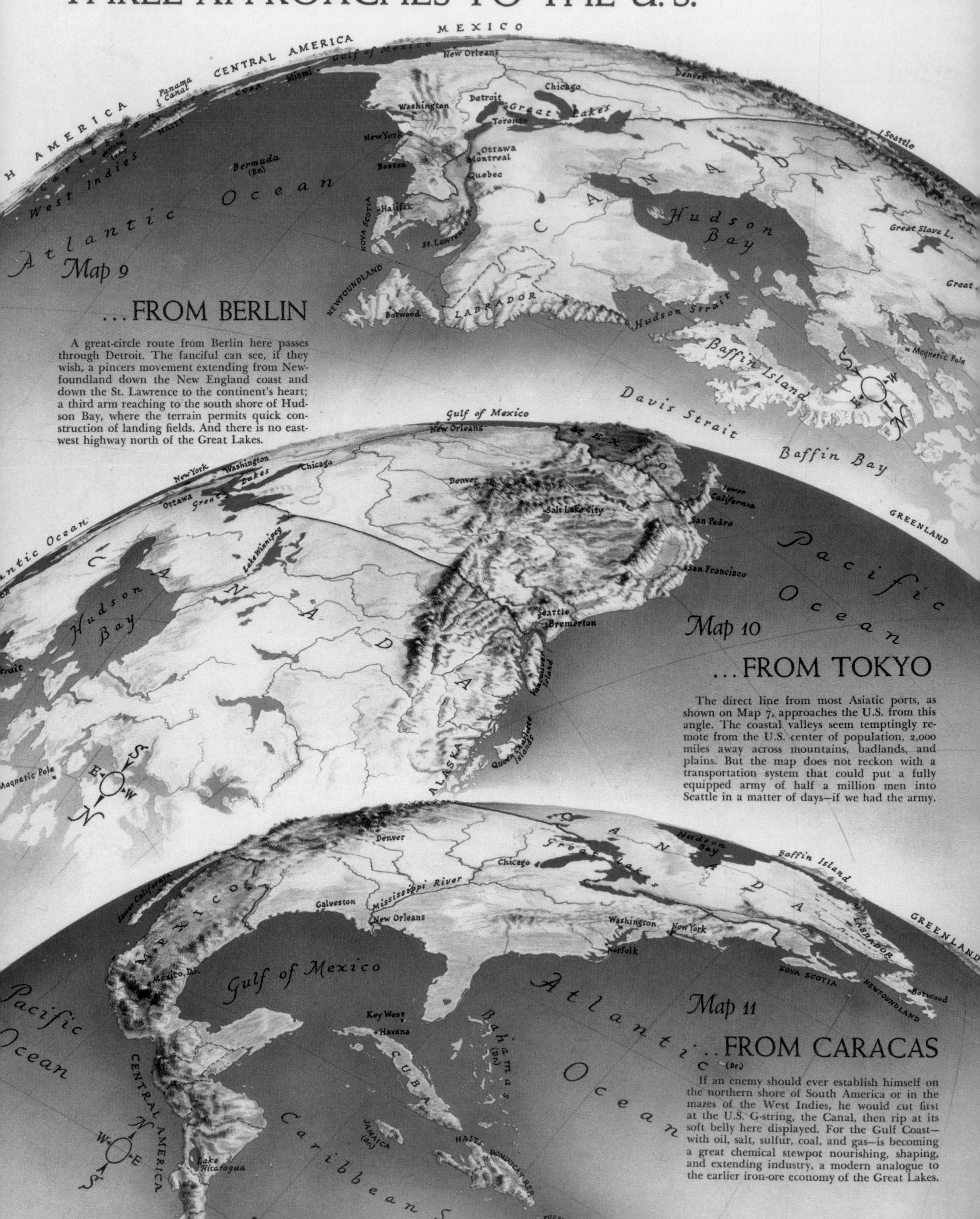

Map 9

...FROM BERLIN

A great-circle route from Berlin here passes through Detroit. The fanciful can see, if they wish, a pincers movement extending from Newfoundland down the New England coast and down the St. Lawrence to the continent's heart; a third arm reaching to the south shore of Hudson Bay, where the terrain permits quick construction of landing fields. And there is no east-west highway north of the Great Lakes.

Map 10

...FROM TOKYO

The direct line from most Asiatic ports, as shown on Map 7, approaches the U.S. from this angle. The coastal valleys seem temptingly remote from the U.S. center of population, 2,000 miles away across mountains, badlands, and plains. But the map does not reckon with a transportation system that could put a fully equipped army of half a million men into Seattle in a matter of days—if we had the army.

Map 11

...FROM CARACAS

If an enemy should ever establish himself on the northern shore of South America or in the mazes of the West Indies, he would cut first at the U.S. G-string, the Canal, then rip at its soft belly here displayed. For the Gulf Coast—with oil, salt, sulfur, coal, and gas—is becoming a great chemical stewpot nourishing, shaping, and extending industry, a modern analogue to the earlier iron-ore economy of the Great Lakes.

GLOBAL AMBITIONS

Richard Edes Harrison, 'Three Approaches to the U.S.', September 1940.

Responding to the German conquests in the West and the growing threat from Japan, Henry Luce, the publisher of *Time*, *Life* and *Fortune* magazines, sought to warn about the United States' exposed position. These maps, by the innovative cartographer Richard Edes Harrison, illustrated an article in *Fortune* magazine about the readiness of the American military asking: 'Our line of minimum physical security stretches from Alaska to the Galapagos, from Greenland to the Amazon Valley. Have we the "with what" to hold it?' The depiction of an attack via Canada was somewhat fanciful. The plans of the German Kriegsmarine focused, instead, on an attack in warmer waters using Spanish and Portuguese island territories – the Canaries and the Azores – as bases. Chesapeake Bay, near Washington, DC, appeared particularly vulnerable to the Americans. In contrast, Japan lacked the 'lift' and logistical systems to transport a large army to the Pacific Coast. Concern about the prospect of an attack from the Caribbean was exaggerated.

The sense of relative menace contributed strongly to the 'Germany First' strategy, which was in line with the United States Army War College exercises from 1935. This had already been outlined by Admiral Harold Rainsford Stark, the Chief of Naval Operations from 1939 to 1942, in a memorandum drawn up in November 1940 known as Plan Dog, as well as in pre-war plans by the American and British military staffs, in the Rainbow 5 war plan, which, once revised by November 1941, became the actual war plan, and in the Anglo-American-Canadian ABC-1 Plan talks of 29 January–27 March 1941. These had envisaged a defensive strategy in the Pacific in the event of war with the three Axis powers. In a clear instance of prioritisation, and of its centrality to strategic planning, Roosevelt had supported this approach because of concern that Britain might collapse.

Hitler's declaration of war on the United States on 11 December 1941 undercut any alternative idea of a 'Japan First' strategy, although it did not automatically end it. Instead, the German declaration, combined with Roosevelt's policies, helped lead to a 'Germany First' strategy on the part of the United States, the priority already settled – in very different and much more urgent circumstances – by Britain. Under this strategy, the bulk of American land and air assets were allocated to preparing for an invasion of Europe, and that commitment underlined the significance of securing the safety of Atlantic shipping lanes.

However, this preference was to be controversial to some Americans at the time, notably those involved in the Pacific War, especially General Douglas MacArthur and some others in naval circles, and has remained so, as every conversation with Americans about wartime strategy appears to bear out. This is linked to a tendency, as a result of the focus on Pearl Harbor, to forget that Germany declared war, or to treat that declaration as not amounting to a real threat. In reality, in early 1942 more German submarines were in American coastal waters mounting attacks than Japanese ones.

The logic was clear: Germany, the stronger adversary, and the power with the more globally ambitious ideology, had a greater potential than Japan to overthrow its opponents, as well as to intervene in South America. In contrast, Japan was faced by the far-greater distances of the Pacific, and not able to draw on an economy comparable to that of the German-run sphere. The Japanese navy and amphibious forces could only achieve so much. This was very much geopolitics as a guide to grand strategy. Although colonies might be conquered, as they were in 1941–42, Britain was far less vulnerable to Japanese power. In practice, however, the pressures of the war in the Pacific were to encourage the allocation of more resources there than might have been anticipated under 'Germany First'.

Blatt 7
Ausgabe Wien, 1941

35° 48° PRESZBURG

35° 49° LUNDENBURG

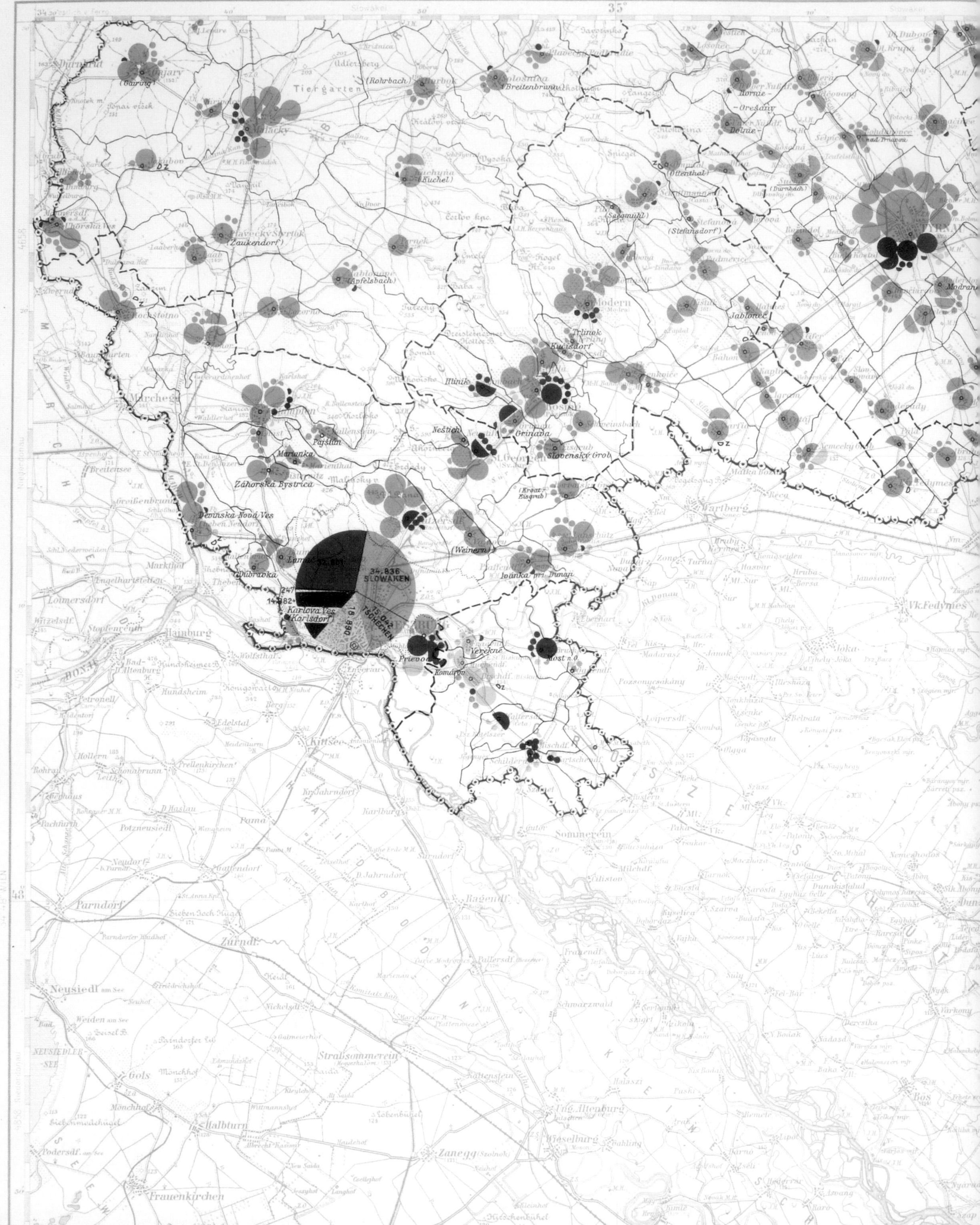

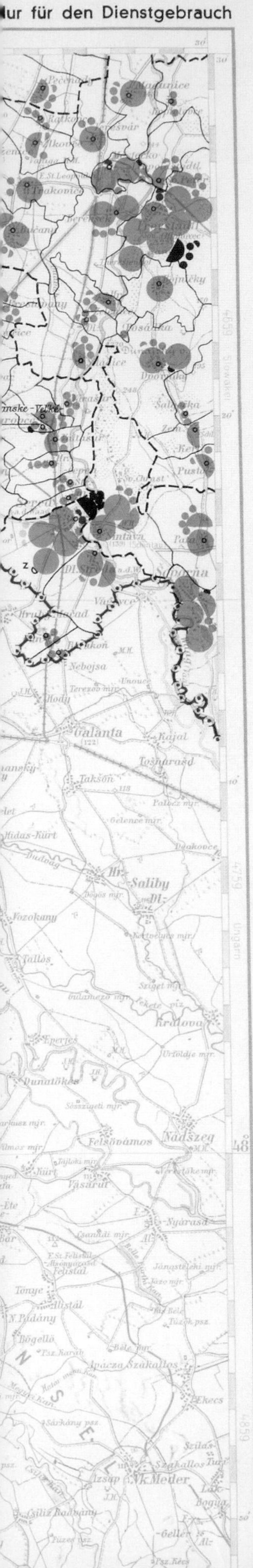

ETHNICITY IN SLOVAKIA

Wilfried Krallert, ethnic map of Slovakia, 1941.

The map is from *Volkstumkarte der Slowakei*, published in Vienna, 1941. It draws upon official Czechoslovak statistics and uses proportionately sized symbols and colours to depict the numbers of residents from different groups: red for Germans, blue for Slovaks and Czechs, yellow for Hungarians and black for Jews, explained in a legend (below). There are also separate classifications for Roma (*zigeuner*, or 'gypsies'), Poles and Ukrainians. The Slovaks had a nationalist movement opposed to Czechoslovakia and it took advantage of German sponsorship and the Munich Agreement to gain autonomy in 1938. Independence followed in March 1939 when the rest of the Czech lands were seized by Germany, but this 'independence' was very much a matter of subordination in policy and economic terms to Germany. Hungary had gained territory from Czechoslovakia in 1938, creating the border shown on the map, while the German minority in Slovakia was granted special rights. Josef Tiso (1887–1947), a Catholic priest, ran the country and actively supported German policies. Anti-Semitic legislation and forced labour was followed by participation in the Final Solution. From 1942 the regime deported its 100,000 Jewish population, including baptised Christians of Jewish origin, to the extermination camps, rather than slaughtering them in Slovakia: 10,000 Jews were sent to Auschwitz between 28 August and 27 October 1944 alone. The Germans did not need to put much pressure on Tiso to support the deportations, and there were few exemptions. Tiso was convinced that he had a duty to free Slovakia from Jews in order to ensure a regenerated Christian nation, one in which the Catholic Church would be a leading force. Tiso promoted corporatist social politics to try to create a Christian Slovak middle class that was able to provide self-respect and not be subordinate to Hungary or to supposedly foreign interests, the last a policy that was interpreted in an anti-Semitic fashion.

Krallert, the Austrian editor of the maps, was a high-ranking Nazi as well as an academic (geographer and historian). After an early career in right-wing paramilitary activity, he became a member in 1934 of the Schutzstaffel (SS) and the Security Service of the Reichsführer SS (SD), conducting intelligence work in the Balkans. Having studied in Vienna he gained a doctorate in 1935 and became Secretary of the Southeast German Research Society (SGRS), then a SS Untersturmführer in 1938 and a member of the SS foreign intelligence branch the following year, when he also attained the rank of SS Obersturmführer. While at the SGRS Krallert used a card file from the Vienna Hofkammer Archives about 80,000 German emigrants into southeastern Europe to prepare for the Nazi ethnic policy in the Balkans. Producing ethnographic maps and overseeing 'research' during the war, he was arrested in 1945, and after his release in 1948 he worked for the French and German intelligence agencies.

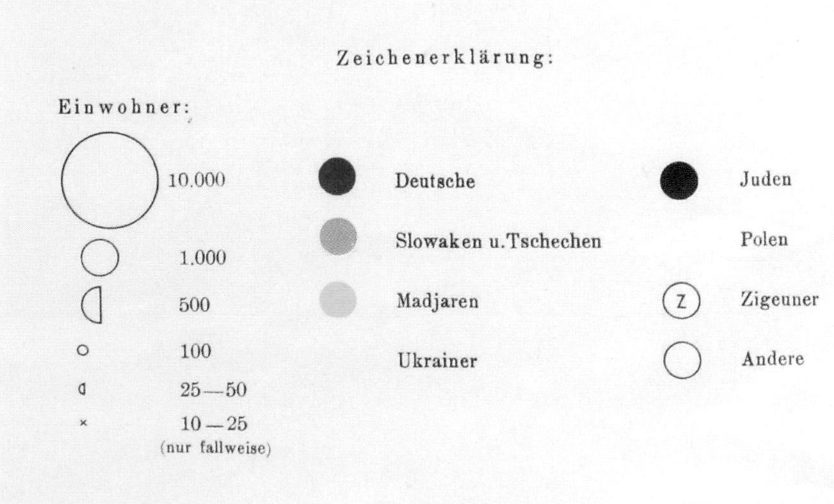

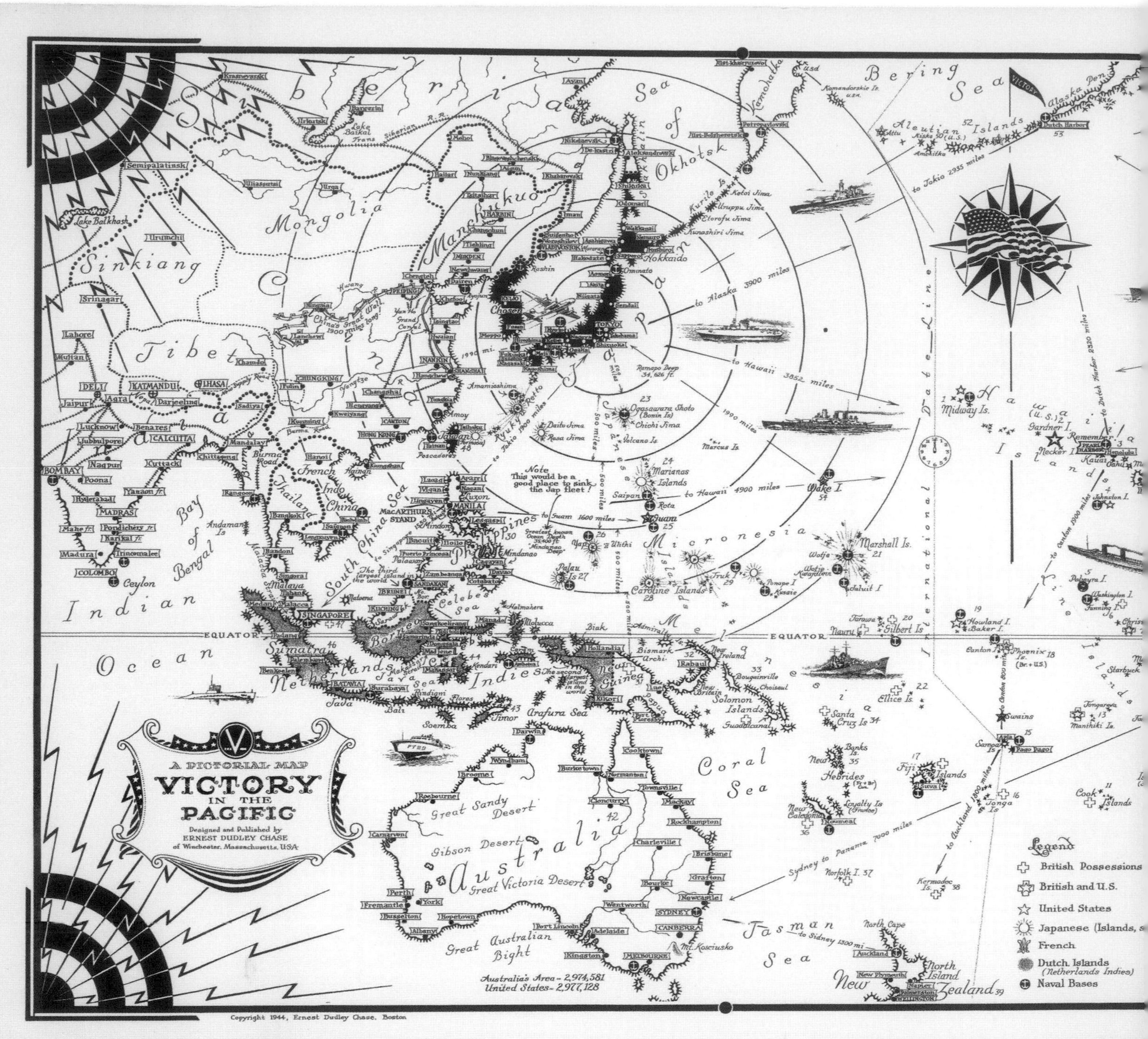
A PICTORIAL MAP
VICTORY
IN THE
PACIFIC
Designed and Published by
ERNEST DUDLEY CHASE
of Winchester, Massachusetts, U.S.A.
Legend
British Possessions
British and U.S.
United States
French
Dutch Islands (Netherlands Indies)
Naval Bases
Note
This would be a good place to sink the Jap fleet!
MacARTHUR'S STAND
Australia's Area - 2,974,581
United States - 2,977,128
Copyright 1944, Ernest Dudley Chase, Boston

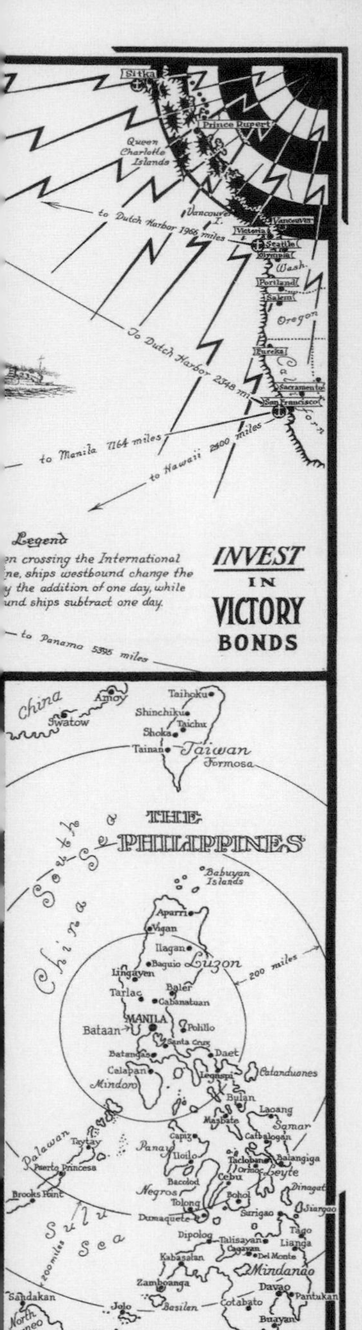

THE PACIFIC WAR

Ernest Dudley Chase, 'Victory in the Pacific', 1944.

Ernest Dudley Chase (1878–1966), a graphic artist who lived in Winchester, Massachusetts, was an expert illustrator of greetings cards, who later wrote the first history of the industry, *The Romance of Greeting Cards*. He contributed to the developing pre-war international trend of pictorial or illustrated maps and rose to the challenge of producing interesting ones about the Second World War. After the wit of greetings cards, he had moved on to create pictorial maps of his travels, which were based on ink drawings he made under magnifying glasses to render the embellishments and detailed, informative vignettes of interesting landmarks he incorporated. The perfectionist wartime map layout and design were intended to produce a sense of optimism, and he provided a patriotic account of American power as a benign process and product. The linework is superb. Chase produced several maps on the war, including the rare, two-sided 'Total War Battle Map… Certain Victory Will be Ours!' (1942), which has a European theatre 'The Victory War Map' on its verso. His 'Victory in the Pacific' map focused on Tokyo as a target, with the inset doing the same for Manila, and encouraged its American purchasers to buy Victory Bonds, which were presented as a form of investment. The Mindanao Deep, a long underwater trench to the east of the Philippines, was described as 'a good place to sink the Jap fleet'. The depiction of converging shipping encouraged a sense of Japan as being under attack.

The loss of Japanese offensive capability at Midway (left, west of Hawaii) on 4–5 June 1942 – as a consequence of the sinking of four of their carriers, the destruction of many aircraft (much of their fleet strike force), and the loss of aircrew and support staff – made thoughts of further Japanese advances in the Indian or Pacific oceans implausible, let alone ideas of joint action with the Germans. Thus, the events of a few minutes at Midway reset strategic parameters, both then and subsequently, and not only for the war in the Pacific, as is usually thought, but also in the Indian Ocean and, indeed, further afield. A novel feature of the war was the significant role that Japan played in it. As a consequence, Allied planners were obliged to confront challenges on a far greater scale than in the First World War. The trade-off with regard to Japan was not simply between the conflict in the Pacific and that in the Indian Ocean. China was also important, but it was not a factor at sea, not even as a diversion to Japanese naval strength.

Midway was made more significant by a more general emerging pattern of increasing American naval effectiveness and the attrition of the Japanese navy. Thanks to Midway and, more generally, the American-Japanese war in the Pacific, on which the Japanese navy overwhelmingly concentrated until the end of the war, the British needed only to deploy limited naval strength against Japan in the Indian Ocean after April 1942. This situation did not change until the closing year of the war, when, after covering D-Day on 6 June 1944, much of the British fleet was transferred so as to be able to play a major role in the war with Japan. Doing so provided Britain with a strategic capability in the Pacific that was lacking in the case of land forces and, in particular, a means to regain imperial prestige, to impress allies, notably the United States and Australia, and to be an influence in the re-creation of European colonial empires (French, Dutch, British), particularly the British Empire.

Prior to that, the British had had two carriers in May 1942 to cover the successful attack on Diego Suarez, the main port in Vichy-held Madagascar, but from January 1943 there were no British carriers in the Indian Ocean until October, when an escort carrier arrived. No British warship was lost in the Indian Ocean in 1943, but no major attempt was made by Britain to launch a 'second front' against Japan on the land perimeter of its newly conquered empire south of Burma, and thus no effective British military pressure was brought to bear to assist the United States and Australia.

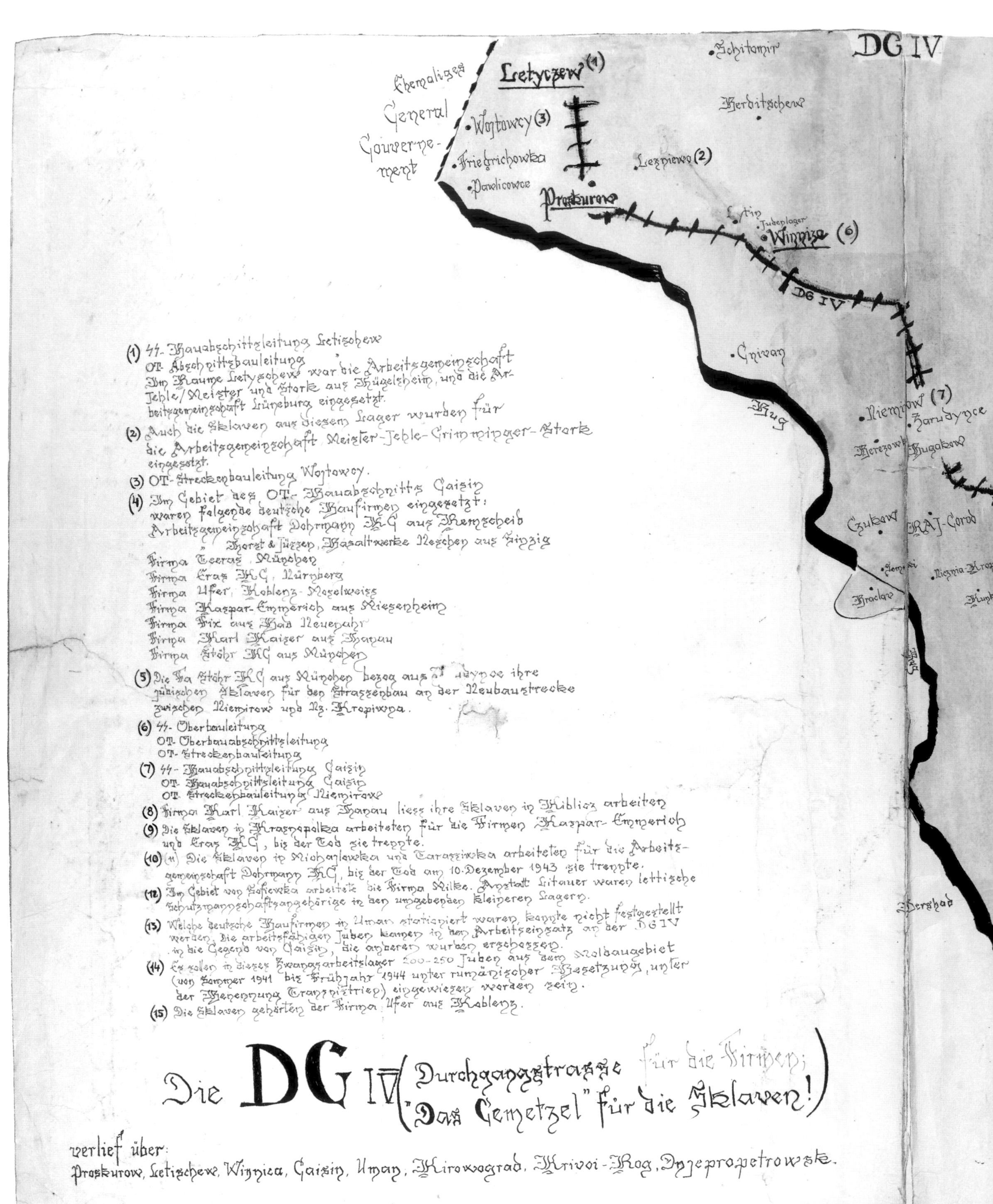
DG IV
Schitomir
Ehemaliges General Gouvernement
Letyczew (1)
Wojtowcy (3)
Friedrichowka
Pawlicowce
Berditschew
Lezniewo (2)
Proskurow
Lytin
Judenlager
Winniza (6)
DG IV
Gniwan
Bug
Niemirow (7)
Zarudynce
Berezowka
Bugakow
Czukow
RAJ-Gorod
Hraclaw
Nemirowski
Niznia-Kropiwna
Hershad
(1) SS-Bauabschnittsleitung Letischew
OT-Abschnittsbauleitung
Im Raume Letyschew war die Arbeitsgemeinschaft
Jehle/Meister und Stork aus Hügelsheim, und die Arbeitsgemeinschaft Lüneburg eingesetzt.
(2) Auch die Sklaven aus diesem Lager wurden für
die Arbeitsgemeinschaft Meister-Jehle-Grimminger-Stork
eingesetzt.
(3) OT-Streckenbauleitung Wojtowcy.
(4) Im Gebiet des OT-Bauabschnitts Gaisin
waren folgende deutsche Baufirmen eingesetzt:
Arbeitsgemeinschaft Dohrmann KG aus Remscheid
" Horst & Jüssen, Basaltwerke Neschen aus Sinzig
Firma Teeras München
Firma Eras KG, Nürnberg
Firma Ufer, Koblenz-Moselweiss
Firma Kaspar-Emmerich aus Riesenheim
Firma Fix aus Bad Neuenahr
Firma Karl Kaiser aus Hanau
Firma Stöhr KG aus München
(5) Die Fa Stöhr KG aus München bezog aus [illegible]udynce ihre
jüdischen Sklaven für den Strassenbau an der Neubaustrecke
zwischen Niemirow und Nz. Kropiwna.
(6) SS-Oberbauleitung
OT-Oberbauabschnittsleitung
OT-Streckenbauleitung
(7) SS-Bauabschnittsleitung Gaisin
OT-Bauabschnittsleitung Gaisin
OT-Streckenbauleitung Niemirow
(8) Firma Karl Kaiser aus Hanau liess ihre Sklaven in Kiblicz arbeiten
(9) Die Sklaven in Krasnopolka arbeiteten für die Firmen Kaspar-Emmerich
und Eras KG, bis der Tod sie trennte.
(10) (11) Die Sklaven in Michajlowka und Tarassiwka arbeiteten für die Arbeitsgemeinschaft Dohrmann KG, bis der Tod am 10. Dezember 1943 sie trennte.
(12) Im Gebiet von Sofiewka arbeitete die Firma Milke. Anstatt Litauer waren lettische
Schutzmannschaftsangehörige in den umgebenden kleineren Lagern.
(13) Welche deutsche Baufirmen in Uman stationiert waren, konnte nicht festgestellt
werden. Die arbeitsfähigen Juden kamen in den Arbeitseinsatz an der DG IV
in die Gegend von Gaisin, die anderen wurden erschossen.
(14) Es sollen in dieses Zwangsarbeitslager 200–250 Juden aus dem Moldaugebiet
(von Sommer 1941 bis Frühjahr 1944 unter rumänischer Besetzung, unter
der Benennung Transnistrien) eingewiesen worden sein.
(15) Die Sklaven gehörten der Firma Ufer aus Koblenz.
Die DG IV (Durchgangsstrasse für die Firmen;
"Das Gemetzel" für die Sklaven!)
verlief über:
Proskurow, Letischew, Winnica, Gaisin, Uman, Kirowograd, Krivoi-Rog, Dnjepropetrowsk.
172

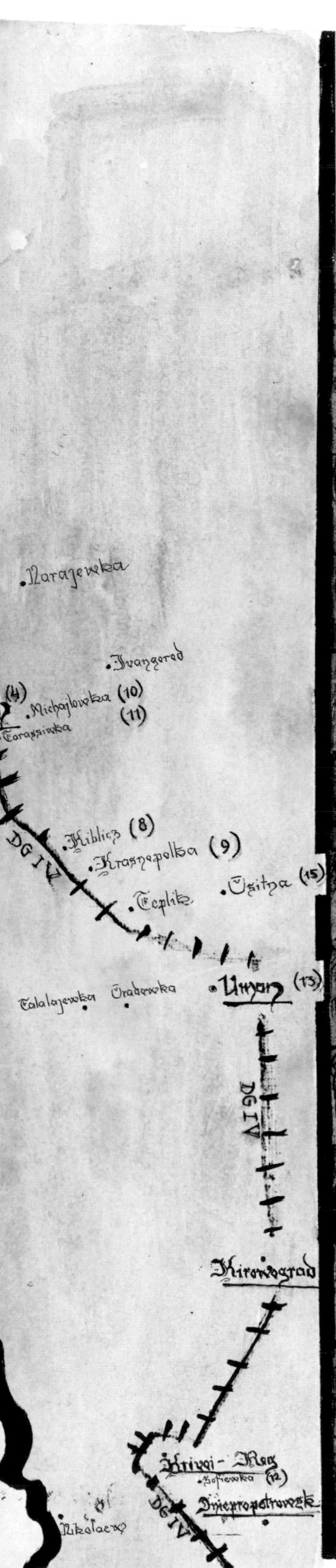

HIGHWAY OF THE SS

Arnold Daghani, manuscript map of Durchgangstrasse IV, 1947.

Arnold Daghani (1909–85) was born into a German-speaking Jewish family and survived the camps at Mikhailowka and Gaisin. He kept a diary and made drawings that were first published in Romania in 1947. His map of Durchgangstrasse ('Thoroughfare') IV, DG IV, was created after the war and shows its labour camps and the organisation of its construction zones.

DG IV redeveloped an earlier route linking Lvov in Poland and Stalino, a major centre of coal and steel production in the Soviet Union, and was the most important highway that the Germans decided to build in 1942 to help deal with their logistical crisis supporting German forces operating further east in southern Russia, which was the principal area of intended operations later that year. Heinrich Himmler's personal involvement, and his interest in a road from Romania to the Sea of Azov, led to the road being called the 'Highway of the SS'. The cruel use of forced labour to build such roads was also a way to exterminate Jews, notably Ukrainian Jews, and the route was marked by mass graves ('the slaughter of the slaves' as Arnold called it in the map).

A spur of DG IV was to cross the Crimean Peninsula and also bridge the Kerch Strait to its east, a span of 4.5 kilometres. From there, the new road would continue into the Caucasus to serve German strategic interests in the area, notably gaining control of oil production in the region, but also providing the possibility to advance against the British in the Middle East and Iran. Hitler took a personal interest in the Kerch bridge, which was seen as a symbol of German control and engineering prowess.

Envisaging the establishment of a colony called Gotengau, the Germans planned to settle Crimea with South Tyroleans, displaced from Italy to satisfy Mussolini's desire to consolidate control over the Alto Adige. Although Ukraine was to contain settler colonies of Germans, including Hegewald, much of Ukraine was to be devoted to SS *latifundia* (estates) supported by subjugated peasants. DG IV was also seen as a setting for new model towns. The Pripet Marshes were to be drained (an impossible task), the inhabitants removed and the area settled by German colonists. Beyond German control, Siberia was to be maintained as a rump state to which those deemed undesirables, appropriate neither for Germanisation nor for extermination, were to be forcibly transferred.

However, bold plans for major new communication links did not match the scale of the problem. Even land

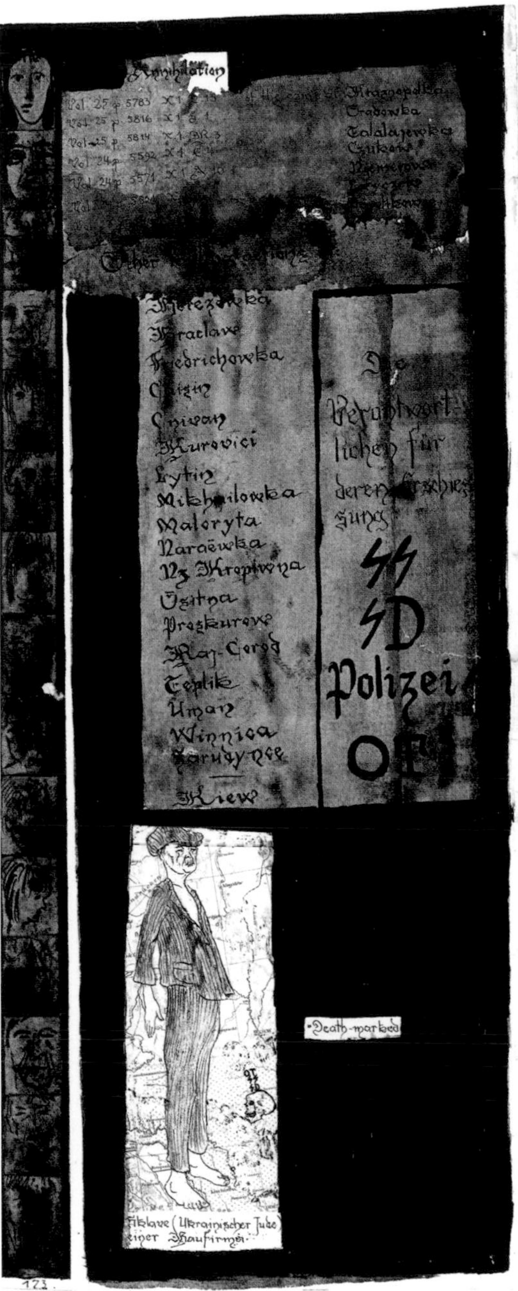

for settlement posed issues, not least the provision of a controllable workforce. In the event, DG IV and the bridge were never completed because the Red Army advanced back into Ukraine in 1944. Instead, the bridge sections stockpiled at Kerch enabled the Soviets to complete it as a railway bridge, which reflected their particular logistical needs and capabilities. From October 1944 to February 1945, the bridge supported the Soviet advance by providing a transport route into Crimea, until it broke up under the pressure of ice. A road bridge across the strait was finally opened in 2018.

For use by
War and Navy Department Agencies only
Not for sale or distribution

CYRENAICA 1:100,000

This sheet is gridded with PURPLE GRID based on long. 27° E., and numbered thus .. 395

TOBRUCH

Cutting lines of the "LIBYAN GRID" are numbered thus......1560

A. M. S. 2

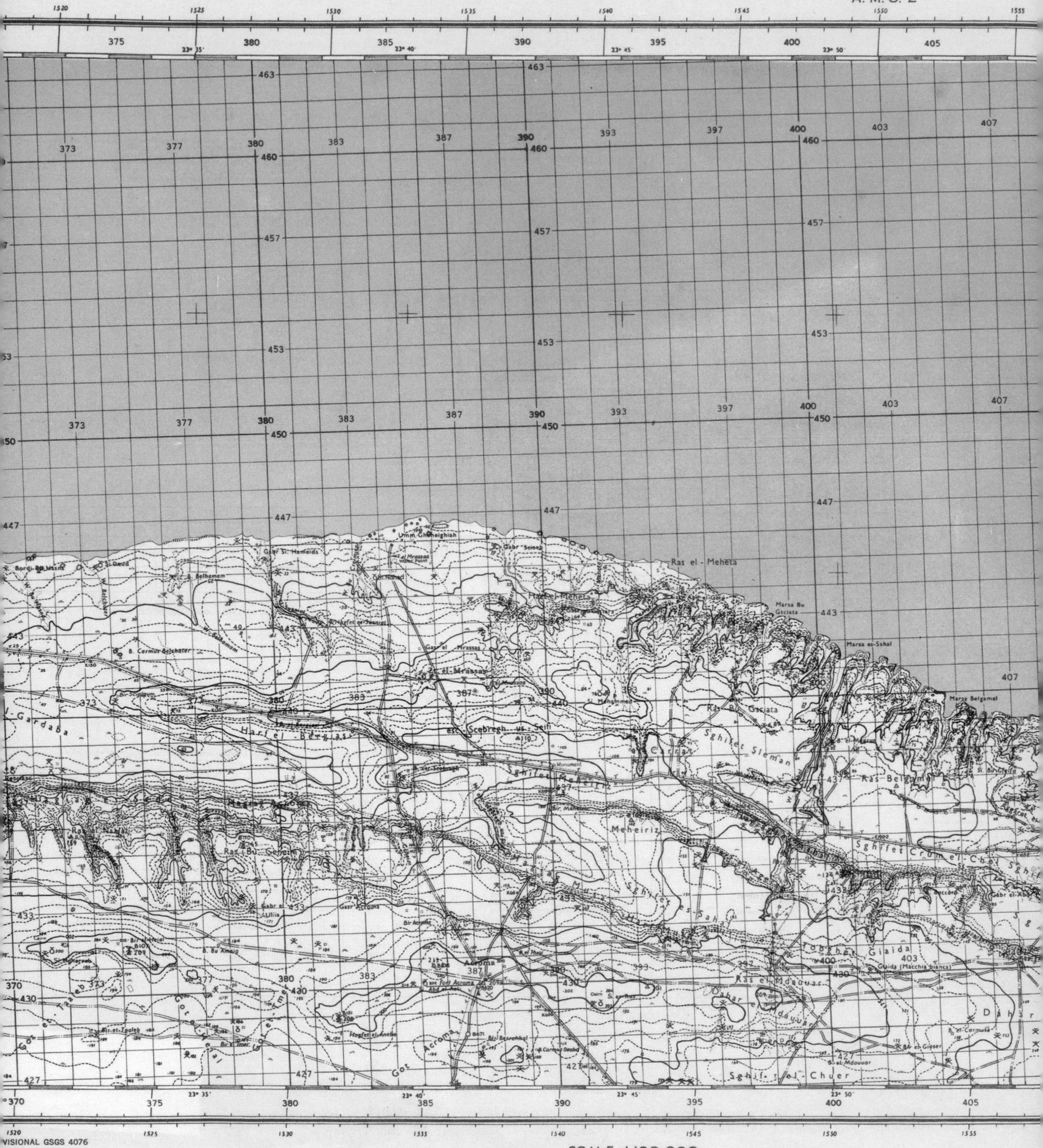

VISIONAL GSGS 4076

000/1515

FOR REFERENCE SEE OVER

SCALE 1:100,000

METRES 1000 500 0 1 2 3 4 5 6 7 8 9 10 KILOMETRES

(1 MILLIMETRE = 100 METRES)

MILES 1 ¾ ½ ¼ 0 1 2 3 4 5 6 MILES

YARDS 1000 500 0 1000 2000 3000 4000 5000 6000 7000 8000 9000 10000 YARDS

Vertical Interval: 25 metres

Dotted form lines approx.

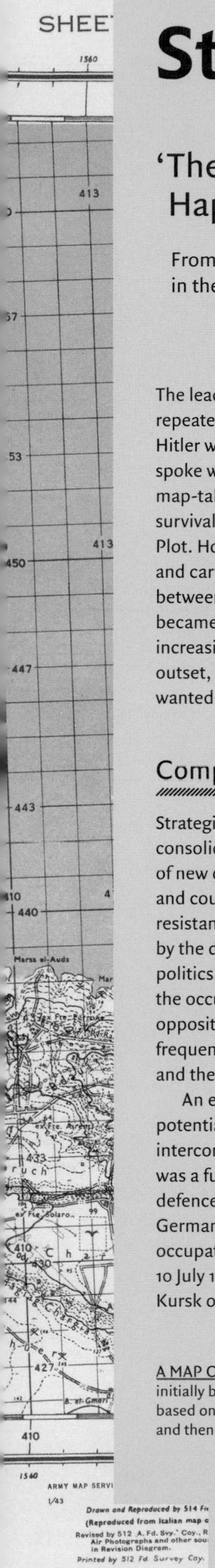

Strategic

2.

'The maps are of time, not place, so far as the army Happens to be concerned—'

From '*Judging Distances*' (1943) by Henry Reed, a poem on his wartime basic training in the British Army.

The leaders, both civilian and military, of the combatants were repeatedly depicted with maps (see photograph, page 37), while Hitler was frequently shown giving maps close attention as he spoke with his military advisers and notably when bent over a map-table in his headquarters. Indeed, this was important to his survival in the Wolf's Lair headquarters during the July 1944 Bomb Plot. However, Hitler lacked any serious education in geography and cartography, and his understanding of the relationship between symbols on the map and relative capability in the field became far weaker with time. That was an aspect of Hitler's increasingly delusional approach to the war; although, from the outset, he had always tended to look at maps and see what he wanted to see, rather than reading maps.

Complexity and interconnectedness

Strategic and operational planning required the accumulation, consolidation and use of existing maps; as well as the production of new ones. It was also important to understand what maps could, and could not, provide. For example, rather than being uniform, resistance to the Germans (as to the Japanese) was greatly affected by the detailed configurations of local geography, ethnicity, politics, religion and society, which were related to the nature of the occupation and of the complex dynamic of collaboration and opposition. These situations could be made more complex by frequently ambivalent relations between sections of the resistance and the occupying power.

An engagement with priorities and an understanding of potential consequences were crucial to strategic planning; and interconnectedness, which was best understood through mapping, was a fundamental aspect of this process. For example, the defence of British India gained added depth – against a possible German advance on India via the Caucasus – with the Anglo-Soviet occupation of Iran in August 1941. The Allied invasion of Sicily on 10 July 1943 helped lead Hitler to cancel his already unsuccessful Kursk offensive on the Eastern Front on 13 July.

A MAP OF TOBRUK, created initially by Britain's War Office, based on an Italian map of 1936, and then shared with the USA.

Japan's entry into the war on 7 December 1941 had greatly increased the challenges faced by the Western Allies, and also the complexities of the war. As a consequence, there was the need to think in broad geographical terms and to understand their political implications. Aside from the difficulties of doing so successfully, however, there were particular deficiencies to address, not least in accurate maps.

The United States put a major emphasis on producing intelligence reports and maps. The Office of Strategic Services (OSS) developed a formidable collection of original (about half a million) and copied maps, which was greatly expanded as the war continued. The Americans produced strategic planning maps in a systematised fashion, and the consistency of the latter process ensured that production could be stepped up. The systematisation in part represented a practice of more scientific mapping than hitherto, and it was to be continued by the Central Intelligence Agency (CIA), the successor to the OSS.

The National Geographic Society (NGS) was highly active during the war, both in creating maps – not least using new projections, such as the azimuthal equidistant projection – and in distributing them (see box, page 36). In addition, the War Department ordered over one million maps from the NGS, while its Pacific map was widely used by military and civilian agencies, and its Germany map played a role in military planning.

As a response to the outbreak of war in Europe in 1939, the United States' Army Map Service was expanded prior to the country's entry into the war. This service was responsible for the United States Army Air Corps as well as for land operations, and efforts were made to extend its geographical coverage, a process accelerated by a survey of the existing situation which suggested that solely one-tenth of the world had been adequately covered, and that some of this coverage was out of date. In 1941, again prior to the war, work began on aeronautical charts for the Americas, long-range charts for the world, and then planning charts for the world at 1:5,000,000, and charts at the same scale for presenting meteorological data.

To help address the need for information, existing maps were acquired, but the emphasis was on new ones. For example, Armin Lobeck, a Professor of Geology at Columbia University, New York, who had worked in the Geography Section of the Versailles delegation in 1919, was employed by the Military Intelligence Service of the General Staff and by the Army Map Service. He produced a set of topographic maps for Europe, as well as more

PRESIDENT ROOSEVELT AND THE MYSTERIOUS MAP

President Roosevelt saw mapping both as a threatening device of America's enemies and as a beneficial resource for the Allies. In his speech at the Mayflower Hotel in Washington, DC, on 27 October 1941, given while the USA remained ostensibly neutral, he referred to the Germans mapping a new order in Latin America. He claimed: 'I have in my possession a secret map, made in Germany by Hitler's government – by the planners of the new world order. It is a map of South America and a part of Central America as Hitler proposes to reorganize it.' The map, he explained, divided South America, including the Panama Canal, into 'five vassal states' under German domination. The map did exist but it had been created as part of a British Intelligence plot in North America to encourage American support for Britain in the war.

Preparing for that war, Roosevelt obtained his maps from the National Geographic Society (NGS) – both Roosevelt and Churchill had special NGS-supplied wall-mounted map cabinets hidden by enlarged photographs – and drew on the geopolitical skill of Isaiah Bowman, who had played a major role at the Versailles peace conference in 1919.

In January 1942 the President created a Map Room in the White House, the precursor to the modern-day Situation Room. The large globe (see photograph, below) that was gifted to Roosevelt for Christmas in 1942 enabled him to gauge distances over the oceans to allocate personnel and material in support of the war effort worldwide against the Axis Powers. Matching globes were created for both Prime Minister Churchill and General George C. Marshall, which they used as a common reference point when communicating about the war. It is thought that the total number of these globes made was a dozen or so. According to the American geographer and cartographer Arthur H. Robinson, who was the director of the map division of the Office of Strategic Services (OSS) during the war, the OSS mapmakers who worked on the project left their 'signatures' by inserting their hometowns among the 17,000 place-names on the globe.

A HUGE GLOBE, 50 inches round and weighing 750 pounds, manufactured by the Weber Costello Company, under the supervision of the map division of the Office of Strategic Services and the War Department, was presented in 1942 to President Roosevelt by the Chief of Staff, General George C. Marshall. This photographed occasion was designed to show Roosevelt's interest in maps. The area of the map shown demonstrated the linkage from the eastern Mediterranean to Southeast Asia. The photograph depicts the part of the world of which the Americans knew least. Roosevelt's attention is drawn toward East Asia but the photograph shows that the British position in India and the Middle East is part of the equation and that the whole of Eurasia is a linked unit.

WINSTON CHURCHILL was well travelled and used maps constantly; indeed, as in the First World War, he was obsessed with them. The crucially important figure in sustaining this was Captain Richard Pim (pictured) of the Royal Naval Volunteer Reserve (RNVR), the supervisor of the War Room at the Admiralty when Churchill was its First Lord (1939–40), before heading the Map Room in No. 10 Downing Street and the Cabinet War Rooms for Churchill as Prime Minister (1940–45). Pim saw Churchill on an almost daily basis and attended many overseas trips to set up temporary map rooms for the meetings with Franklin Delano Roosevelt and later Josef Stalin, including at Yalta in Crimea in 1945. At Cairo in November 1943, Churchill proudly showed the temporary map room to Jiang Jieshi (Chiang Kai-shek of the Republic of China) and his wife. Churchill's impressively detailed maps covered every theatre of war and were kept up to date with the latest Intelligence reports on the location of military units. Leading British officials also used maps. Sir Olaf Caroe, a policymaker in Delhi, was fond of the map of the world entitled 'Seven Theatres of Power'.

specific maps and diagrams in preparation for Operation Torch, the successful invasion of French North Africa in November 1942; for Operation Overlord, the invasion of Normandy in June 1944; and for an invasion of Fascist Spain, if it was judged necessary. Other works by Lobeck included 'Southeastern Europe: Strategic Map of Climatic Types' (1943) for the Army Map Service and, with others, *Military Maps and Air Photographs: Their Use and Interpretation* (1944). One of his co-authors in this, John K. Wright, had brought out in 1941 (with other co-authors) for the American Geographical Society, *The European Possessions in the Caribbean Area; A Compilation of Facts Concerning Their Population, Physical Geography, Resources, Industries, Trade, Government, and Strategic Importance*. Lobeck's *Geomorphology: An Introduction to the Study of Landscapes* (1939) was important in its use of scientific graphic representation, notably physiographic and block diagrams.

Aside from producing new maps, the need for information greatly enhanced the value of captured maps. In 1941 the Germans systematically looted Yugoslavia's Military Geographical Institute and also benefitted in the early stages of Operation Barbarossa, the invasion of the Soviet Union, from the capture of the Minsk map factory, followed by those in Kiev and Kharkov. Original Italian maps of 1940, with the addition of contours added by the South African Survey Company and aerial information supplied by the RAF, were the basis for the map of Tripoli in Libya by Britain's Geographical Section of the General Staff, published by the War Office in 1942, and then, in a revised version, by the United States' Army Map Service in 1943. The basis for a War Office map of the Tobruk defences on 12 January 1941, was an Italian map of 1936, with material, notably Italian positions and artillery, added by the British in an overprint. Published by a field survey company of the Royal Engineers, this map proved useful on 21 January when the British artillery helped prepare for the successful attack that led to Tobruk's surrender the following day. Earlier British versions had been produced by similar means weeks earlier. That 1936 Italian map of the area remained the base reference for Allied surveyors to produce a revised map in April 1942 (see page 34).

In response to the German invasion, the Soviet Union moved fast to complete the 1:100,000 map of the state west of the River Volga, a task that entailed 200 new sheets and that was completed by the end of 1941. Moreover, the Red Army developed a flexible production system in which maps were prepared and printed in order to meet immediate operational needs, which entailed the overprinting of maps with specific tactical information. Underpinning this activity in the Soviet Union was a more general attempt to improve the training for mapmaking and the production of maps, including the introduction of a uniform geodetic (coordinate) system.

Secrecy was at the heart of mapping for war, and, aside from producing their own maps, governments also restricted the distribution of existing and new maps that might help enemies. Thus, the United States restricted topographic maps produced by the United States Geological Survey (USGS) as well as the nautical charts created by the US Coast and Geodetic Survey (USC&GS). Claims that the Germans had used Soviet maps in their invasion in 1941 affected access to the latter in the Soviet Union. In some cases, the Germans had indeed employed them, and rumours circulated in the Soviet Union that the Germans had bought them in bulk prior to the attack. As a result, large-scale maps were banned from free circulation and removed from libraries. Moreover, information was restricted about industrial activity in those maps that did come into use. The OSS took steps to obtain the restricted Soviet maps.

Much espionage work was focused on gaining access to secret plans and maps. In the run up to Operation Overlord, the Allied invasion of Normandy, the Caen group of the French Resistance secured a copy of a map of the Atlantic Wall, detailing the German defences, while British agents provided reliable maps of the landing beaches for reconnaissance purposes. Espionage and subversion were also reliant on the existing maps, as with the use of Michelin maps by Britain when planning the dropping of agents and equipment into France.

Co-operative mapmaking

An important element of the strategic nature of mapmaking was provided by Anglo-American co-operation. In May 1942 the Loper-Hotline Agreement, negotiated by Herbert B. Loper, of the Intelligence Division at the Office of the Chief of Engineers in Washington, DC, divided military mapping and surveying of the world between the United States and Britain. Common technical features of the maps were also agreed. The United States' Army Map Service was given full responsibility for mapping the Americas, Australasia, the Pacific, Japan, the West Indies, and the North Atlantic; and the British were responsible for the rest. This allocation accorded with the systems the British already had in place for overseas production, most significantly centres in Egypt and India. As during the First World War, the Survey of India extended its activities beyond the British Empire. The Survey of India was given the task of mapping Iraq, Iran, Afghanistan, Burma (Myanmar), Thailand, French Indo-China, China, Malaya (Malaysia) and Sumatra, with most sections under direct military command, and there was an increase in air surveying, as a means to map hostile territory and to map at speed.

As part of a broader pattern of co-operation, the Geographical Section of the General Staff and the United States' Army Map Service exchanged map and geodetic material. The Western Allies initiated the World Aeronautical Chart, a 1:1,000,000 scale map series using the Lambert conformal conic projection, which provides maps on which straight lines approximate to a great-circle route between endpoints.

The British and Americans also used the maps produced already by the cartographic agencies of European states. Thus, in their topographic maps, the Geographical Section of the General Staff and the United States' Army Map Service homogenised different mathematical base data and geographical content from the numerous source maps, sorted out the complications of name spellings and brought the maps up to date to increase their reliability and usefulness.

Somewhat differently, during the German occupation (1940–44) France's National Geographic Institute printed maps for the German forces, but it also helped the British by providing maps of France and French North Africa, which in November 1942 was invaded by Anglo-American forces in Operation Torch.

An Italian perspective

Uncertainty over national strategy, and therefore over what needed mapping, was in large part an obvious product of a lack of clarity as to the likely character and strength of different countries' alliance systems. Indeed, this led to a great situation of flux in the war until the close of 1941. It is particular instructive to look at the strategic dimension of the war from the perspective of the second-rank powers, notably Italy under its Fascist dictator, Benito Mussolini (ruled 1922–43), who, like Hitler, did not understand how to read maps and was not interested in restrictions based on practicality. Prior to war, the Italians had backed a major naval build-up designed to contest the position of Britain and France in the Mediterranean. Moreover, whereas in 1934 Mussolini had moved troops to the Brenner Pass in response to an ultimately unsuccessful attempt by Austrian Nazis to seize control of Austria, by 1938 he was aligned with Hitler when the latter mounted a successful takeover of Austria. Mussolini shared both Hitler's contempt for the democracies and his opposition to Britain and France.

Mussolini was repeatedly regarded by Britain as a possible ally, but a major war had seemed possible over the Italian conquest of Ethiopia in 1935–36 (see pages 20–21), and Britain and France drew up plans, which included the closing of the Suez Canal and the bombing of northern Italian cities, notably the industrial centres of Milan and Turin. Allied maps were adequate for these purposes. In the event, there had been no conflict.

Italy did not enter the war until 10 June 1940. Furthermore, even after Germany declared war on the United States after Pearl Harbor, there was still uncertainty about the composition of the competing alliances. In particular, Soviet policy was difficult to fathom for all powers. There was speculation in 1942 and 1943, including in Britain, about the possibility of a separate German-Soviet peace. Until the Soviet Union attacked Japan in August 1945, it was also unclear whether the Soviets would ever do so.

Global outlook

Strategy was about means as well as tasks. As the war became a worldwide one, so it became more important to understand it accordingly. Aside from the interest in globes, aerial views and orthographic projections (a depiction as if from outer space), there were also concerns to improve cylindrical maps. Samuel Whittemore Boggs, the chief cartographer at the US State Department, commissioned Osborn Miller, the head of the

Department of Technical Training of the American Geographical Society, to do so. Miller amended the positioning of the poles used hitherto by the Mercator projection. This projection resulted in a similar view of the world to the Mercator, but with a reduced aerial distortion in polar regions. As a reminder that people and work are not in compartments, Miller, a British artillery officer in the First World War, was also an expert in the use of oblique air photos for making small-scale reconnaissance maps and he contributed the article 'Topographic Mapping from High Oblique Air Photographs' to the first *Manual of Photogrammetry* (1944).

Air power was not only a key requirement for mapping, it was also of strategic significance in terms of ground and sea control. The acquisition and protection of air bases reflected strategies carefully plotted in spatial terms, with particular interest in the range of aircraft located there, as well as their exposure to attack; and the bases helped to determine strategic options and operational means. Thus, in 1944, in response to Japanese advances in China, where American bases were overrun, the United States emphasised gaining control of islands, notably Saipan, that were within range for air raids to be launched against Japan.

So also in Europe, the Middle East and North Africa. General Thomas Blamey of the Australian Corps wrote in August 1941 about eastern Libya: 'Cyrenaica is regarded as most urgent problem of Middle East as control to [the city of] Benghazi would give [British] fleet freedom of movement as far as Malta and advance air bases to allow cover of sea operations.' The Anglo-Soviet occupation of Iran in 1941 was seen as a prerequisite if the British were to be able to mount air attacks on the important Baku oilfields in Azerbaijan if the Germans seized them from the Soviet Union, as they sought to do in 1942. Similarly, the capture of southern Italy in 1943 provided bases from which to conduct Allied strategic bombing, including, in 1944, heavy raids on the oil refineries at Ploesti in Romania, as well as on Munich. Aerial views and orthographic map projections helped explain these links focused on air power. Air bases were to remain a strategic priority in the first two decades of the Cold War, during which the importance of the mapping skills developed in the Second World War was to be proved repeatedly.

The range of the war led to the movement of air, sea and land units into areas with which they were very unfamiliar, for example the Germans into Egypt in 1941, and to demands for the (improved) mapping of these areas and, as an aspect of this improvement, for a degree of consistency in mapping. This was seen, for example, with the role of American air power against German submarines in the Atlantic. In tackling these new tasks, there were readily apparent contrasts in cartographic requirements and availability. Aircraft flying from American bases in North America, from American islands in the Caribbean, notably Puerto Rico, and from aircraft carriers, were operating in a very different context from those flying from new bases in Cuba, the Dominican Republic, Haiti, Panama and Brazil, the newness of which meant fresh mapping requirements.

American air bases in Brazil were also crucial to the long-distance movement of aircraft and shipment of arms to the Allies in the North African campaigns. Aircraft were flown from the United States via Brazil (and shipped from Britain) to Takoradi in the Gold Coast (Ghana) and then, via Lagos and Khartoum, on to near Cairo. Similarly, the air base at Keflavik in Iceland, built by the American military in 1942–43, was important to the American 'air bridge' to Britain via Newfoundland and Northern Ireland. This was part of a worldwide geostrategy, in which the United States Army Air Force's Air Transport Command (ATC) played a major role.

Strategy involved an effective use of information, both concerning one's own side and with reference to opponents. Thus, the location of opposing units was a key indication of apparent intentions. Like others, the British devoted much attention to this issue, relying on human Intelligence sources as well as signals Intelligence, notably intercepts. The mapping of such data was important. The Germans proved inadequate at this, as shown by their serious failure in 1941 to estimate accurately the number of Soviet divisions, which was the result of a German willingness to believe best-case guesswork.

The Allies eventually proved better than the Axis at understanding the areas in which they campaigned, and the resources that could be deployed; and in planning accordingly. Germany, Italy and Japan tended to improvise; Hitler's emphasis was on the socio-economic and political conditions he wished to see rather than what was occurring on the ground. In planning and campaigning, the Axis stress was often on the value of superior will, rather than on the realities of climate, terrain, communications and logistics. The constraints posed by the last three were ignored, for example, in the totally unsuccessful Japanese offensive against the British on the India–Burma (Myanmar) border in 1944. Such poverty of strategic understanding is difficult to capture in maps. Cutting Allied supply routes from India to China looked easier than it was.

The cartographic capabilities – from the availability of information through to the use to which it was put – of the United States were much greater than those of Japan. The American public was also offered an education in the significance of geography. Thus, 'China (seen from the Direction of Guam)', a vivid map by Richard Edes Harrison, published by *Fortune* in April 1941, addressed the reader directly:

> *'Here is nearly a full index for a geographer's primer – deserts, fertile flood plains, low hills, plateaus, precipitous mountains – practically everything except volcanoes. This geography has had a lot to do with China's history since July 7, 1937, when the Japanese invasion began: China's center of gravity shifted from the flat, vulnerable plains of the north and coast into the mountains of backward and neglected provinces. The migrations were many and spectacular: administration, education (see the small inset at the left), industry, and just plain people moved hundreds of miles. The army moved too, as the Japanese pushed it along the rivers, railways, and roads; and it still resists. And in ungarrisoned and unpatrolled islands within "occupied" territory; some of those who didn't go along carry on guerrilla warfare. It is not easy to supply a nation withdrawn far inland, but the blue lines and arrows show how essential goods are carried around the Japanese outposts, via the USSR and Burma, and are smuggled right through them, via southeast ports.'*

The strategic opinions offered to the public were not always practical. Thus, the *New York Journal-American's Pictorial Review* of 5 December 1943, explaining 'How Japan's Mongol Hordes Can Be Defeated' in the 'Largest Pincers Movement in Warfare's History', added to MacArthur's plan for an advance from New Guinea via the Philippines to Japan an account of the British advancing overland from India to recapture Southeast Asia before joining the Chinese to attack Japan, which was simply not credible.

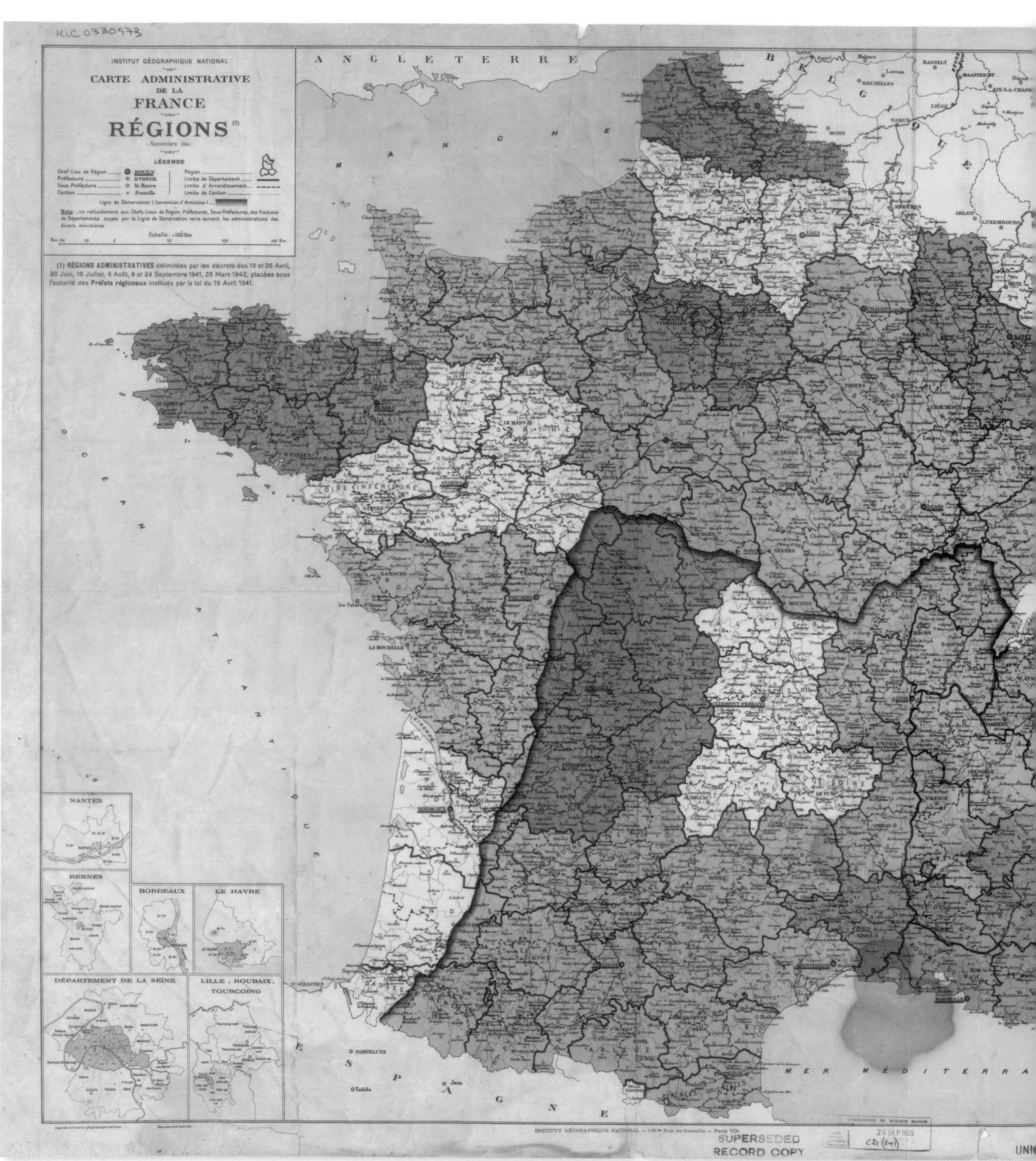
MLC 0330973
INSTITUT GÉOGRAPHIQUE NATIONAL
CARTE ADMINISTRATIVE
DE LA
FRANCE
RÉGIONS (1)
(Novembre 1941)
LÉGENDE
Chef-Lieu de Région ROUEN
Préfecture EVREUX
Sous-Préfecture le Havre
Canton Trouville
Région
Limite de Département
Limite d' Arrondissement
Limite de Canton
Ligne de Démarcation (Convention d'Armistice)
Nota: Le rattachement aux Chefs-Lieux de Région, Préfectures, Sous-Préfectures, des fractions de Départements coupés par la Ligne de Démarcation varie suivant les administrations des divers ministères.
Echelle: 1:400.000
Km. 50 25 0 50 100 150 Km.
(1) RÉGIONS ADMINISTRATIVES délimitées par les décrets des 19 et 26 Avril, 30 Juin, 18 Juillet, 4 Août, 9 et 24 Septembre 1941, 25 Mars 1942, placées sous l'autorité des Préfets régionaux institués par la loi du 19 Avril 1941.
ANGLETERRE
MANCHE
OCÉAN ATLANTIQUE
ESPAGNE
MER MÉDITERRA
BELGIQUE
NANTES
RENNES
BORDEAUX
LE HAVRE
DÉPARTEMENT DE LA SEINE
LILLE, ROUBAIX, TOURCOING
INSTITUT GÉOGRAPHIQUE NATIONAL — 136 Rue de Grenelle — Paris VII
SUPERSEDED
RECORD COPY
28 SEP 1985

FRANCE DIVIDED

Vichy regime, map of the administrative regions of France, April 1941.

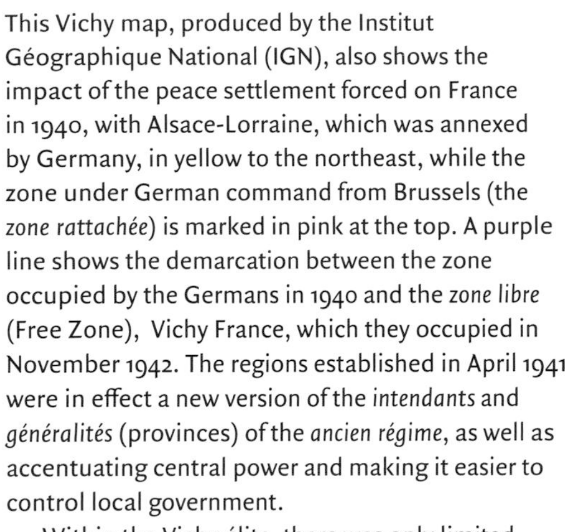

This Vichy map, produced by the Institut Géographique National (IGN), also shows the impact of the peace settlement forced on France in 1940, with Alsace-Lorraine, which was annexed by Germany, in yellow to the northeast, while the zone under German command from Brussels (the *zone rattachée*) is marked in pink at the top. A purple line shows the demarcation between the zone occupied by the Germans in 1940 and the *zone libre* (Free Zone), Vichy France, which they occupied in November 1942. The regions established in April 1941 were in effect a new version of the *intendants* and *généralités* (provinces) of the *ancien régime*, as well as accentuating central power and making it easier to control local government.

Within the Vichy élite, there was only limited support for Fascism, as opposed to a more broadly based conservative nationalism that was particularly open to Catholic activism. Yet, from the outset of Vichy France, there was a willingness to discriminate against Jews, and it did not require much German prompting, let alone pressure, to do so. The religious, cultural, political and social fault lines of the Dreyfus Affair re-emerged as Vichy strove to create an ostentatiously Christian (especially Catholic) France. With the hostility to liberalism, communism, socialism, Freemasonry, Judaism and Protestantism (although there were pro-Vichy Protestants) that was commonplace in right-wing circles during the Third Republic, Vichy also reflected the hostile conservative reaction to the Popular Front of the 1930s, and also drew on the influential Action Française, notably the late 1930s' anti-Semitic revival linked to opposition to Jewish refugees. Vichy presented the Third Republic, and particularly its politics, as decadent and weak, and as in large part responsible for the defeat of 1940. Anti-British themes were emphasised, especially with a cult of Joan of Arc, who had been canonised in 1920. Moreover, Vichy's account of the past was very much opposed to the French Revolution and to Enlightenment figures such as the Abbé Gregoire who had pressed for equality, notably on behalf of Jews and slaves.

Aside from deliberate destruction, such as the dynamiting in January 1943 of the Old Port of Marseille by the French police as part of Hitler's 'purification' of the cosmopolitan city, France was heavily pillaged for the benefit of Germany – indeed, it greatly helped the latter to maintain its war effort. The occupation costs in billions of francs rose from 81.6 in 1940 to 144.3 (1941), 156.7 (1942), 273.6 (1943) and 206.3 (1944); with the percentage proportion of GDP that represented being respectively 19.5, 36.8, 36.9, 55.5 and 27.9. A manipulated exchange rate also helped Germany to control the French economy: the mark-to-franc rate moved from 1 to 11 on 10 June 1940 to 1 to 20 on 25 June. Food was purchased or requisitioned for the occupying forces, and also sent to Germany. As a result, daily calorie intake in France fell by a half, and there was serious malnutrition. The occupation of the so-called Free Zone subsequently became harsher because a shortage of resources affected occupiers and the local population, leading to black-market activity and riots. Resistance activities became more intense, affected by the large-scale deportation of men for forced labour (although some individuals volunteered to work in Germany) and by the harshness of German repression, which also benefitted from support by collaborators.

BARBAROSSA *BLITZKRIEG*

German situation map, 2 July 1941.

On 22 June 1941, in Operation Barbarossa, German and Romanian forces (Finnish, Hungarian and Italian forces joined in subsequently) with 3,350 tanks invaded the Soviet Union, the biggest target yet for a mechanised assault. This situation map 10 days into the invasion shows the strength of the Soviet forces on the Southwestern Front and also the extent of the penetration of the German Army Group North. The Soviet Western Front was to demonstrate later in July that the advance of the German Army Group Centre would be more difficult than anticipated.

The German invasion plan reflected bold and untested assumptions about mobility, but necessarily so. The attack was to be concentrated between the Pripet Marshes and the Baltic, with much of the armour (two of the four panzer groups) under Army Group Centre, whose commander was Field Marshal von Bock. This unit was ordered to destroy opposing Soviet forces and then move from Smolensk to help Army Group North under Field Marshal von Leeb capture Leningrad (now St. Petersburg). This success was seen as a prelude to the advance on Moscow, which reflected Hitler's priorities rather than military advice, which had pressed for a concentration on Moscow. Meanwhile, Army Group South, under Field Marshal von Rundstedt, was to capture Kiev and then encircle Soviet forces in Ukraine, preventing these forces from falling back to defend the interior. In the next phase, forces from Leningrad and Smolensk were to drive on Moscow, while Army Group South advanced to Rostov in order to open the way to the Caucasus, both to seize the oilfields there and to threaten an advance into the Middle East.

The plan reflected an over-bold prospectus that drew heavily on assumptions about German armoured capability and likely Soviet responses. The Germans attempted to seize all the objectives simultaneously, a source of potential weakness that arose from the failure both to settle the core target of the operation and to devise a time-sequence of even limited plausibility. It was also assumed that the defeat of the Red Army near the frontier would lead to the enemy's collapse. On 31 July 1940, Hitler indicated to senior commanders that he was determined to shatter the Soviet Union with one blow.

Soviet forces were numerous, including 10,400 tanks (albeit of very mixed quality) in the western Soviet Union, but they were poorly prepared and deployed. The extent to which the Germans would be able to advance had been underestimated. Stalin's instructions – that units hold their positions and not retreat, and his encouragement of counter-attacks – ensured that the Soviets proved vulnerable in 1941 to German breakthrough and encirclement tactics, losing heavily as a result, including many tanks.

Aside from his responsibility for the condition of the Soviet military at the outset of the war, and particularly for the dire state of the command structure, Stalin had not appreciated that the Soviet Red Army would suffer from *blitzkrieg*, and certainly not to the degree that it did. The Winter War with Finland in 1939–40 had revealed Soviet deficiencies, but only on the attack. A full-scale assault by Germany was far more serious because Soviet preparations, strategy and doctrine were all inadequate. For example, the Soviet Union had relatively few anti-tank and anti-personnel mines in store in 1941 because they were seen to represent a defeatist focus on the defence.

The Germans were dazzled by the success of the frontier battles, their deep advances into the Soviet Union and the large number of prisoners; and these victories spurred Nazi euphoria and planning, including that of the Holocaust. The speed of the advance also impressed foreign observers. In part, this was attributable to German capabilities and in part to Soviet deficiencies. The Germans were more successful than the Soviets at the tactical level in linking firepower and mobility; operationally, they outmanoeuvred Soviet defenders and were able to impose their tempo on the flow of conflict.

The Soviets suffered from: a general lack of relevant command and fighting experience, certainly in comparison with the Germans; the absence of supply and maintenance systems to match the impressive quantity of weapons they had, including poor logistics and inadequate transport; and a General Staff that was no match for the Germans. Thus, the Soviets did not understand, as well as the Germans did, the operational implications of mechanised warfare. Yet, recovering rapidly from the initial confusion, many Soviet Red Army units, as was noted by the British Military Mission in Moscow, fought hard and effectively from the outset, as in the defence of Brest-Litovsk, inflicting greater losses than the Germans had anticipated. This was especially true of the Southwestern Front, which was strong in tanks. With Stalin anticipating an advance on the resources of Ukraine, most Soviet tanks were in this area and, despite confusion and frequent German air attacks, attempts were made to launch counter-attacks, showing Soviet forces would not be easily overcome.

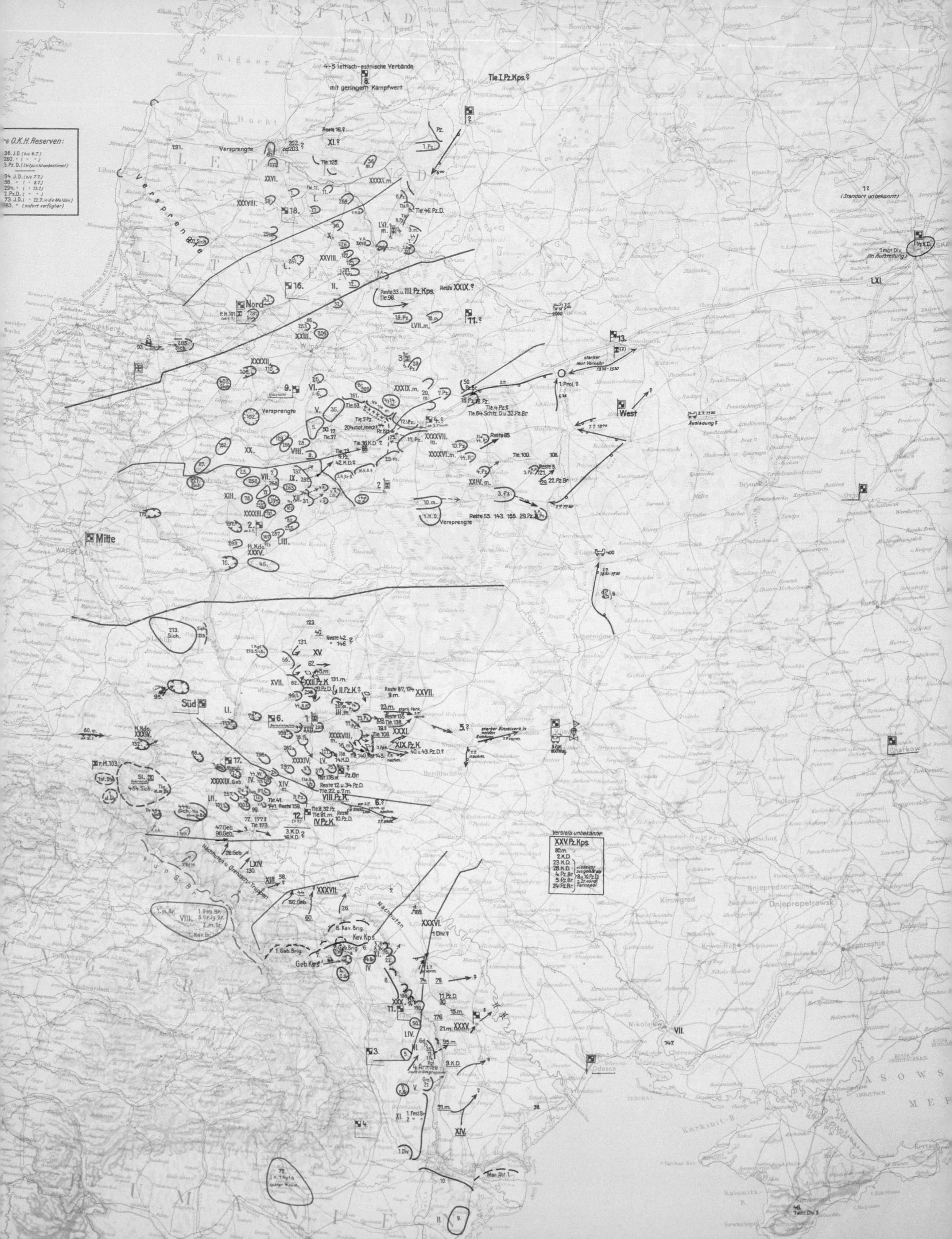

re O.K.H. Reserven:
96. J.D. (bis 6.7.)
260. " (" ")
5. Pz. D. (Zeitpunkt unbestimmt)
94. J.D. (bis 7.7.)
98. " (" 8.7.)
294. " (" 13.7.)
2. Pz.D. (" ")
73. J.D. (" 22.7. in die Moldau)
183. " (sofort verfügbar)
Rigaer Bucht
4-5 lettisch-estnische Verbände
8.
mit geringem Kampfwert
Tle. I. Pz. Kps.?
Versprengte
Nord
Mitte
Süd
West
13.
16.
18.
9.
4.
2.
6.
17.
11.
3.
Smolensk
Mogilew
Orel
Kiew
Charkow
Kursk
Warschau
Königsberg
Libau
Tschernigow
Berditschew
Kirowgrad
Dnjepropetrowsk
Saporoshje
Nikolajew
Cherson
Odessa
Sewastopol
Tle. 46. Pz. D.
Reste 33. u. III. Pz. Kps.
Tle. 98.
Reste XXIX.?
Reste 85.
Reste 55. 143. 155. 29. Pz.Br.
22. Pz. Br.
XXIV. m.
XXXXVII.
XX.
XIII.
XXXXIII.
XXXV.
Reste 42. 146.?
XV.
XVII.
XXII. Pz. K.
II. Pz. K.
Reste 87. 124.
XXVII.
XXXI.
XIX. Pz. K.
40. 43. Pz. D.?
VIII. Pz. K.
IV. Pz. K.
Reste 12. u. 34. Pz. D.
Reste 10. Pz. D.
Verbleib unbekannt:
XXV. Pz. Kps.
80. m.
2. K.D.
23. K.D.
28. K.D.
4. Pz. Br.
5. Pz. Br.
24. Pz. Br.
vielleicht zus.gefaßt als 8. u. 10. Pz. D. z. Zt. nördl. Tarnopol
Nachhuten u. Grenzschutz-Truppen
Nachhuten
XXXVII.
XXXVI.
Kav. Kps.
Geb. Kps.
1. Geb. Brig.
6. Kav. Brig.
XXXV.
XIV.
9. K.D.
4. Armee
LXI.
1. mot. Div. (in Aufstellung)
VII.
Mar. Abt. 1

ARMY GROUP NORTH

German situation map, 19 July 1941.

While attention was focused on the operations of Army Groups Centre and South, and notably Hitler's decision on 14 July to divert the panzer divisions of the former to affect an encirclement of Soviet forces in Ukraine, Army Group North continued to advance. The River Dvina line had been crossed easily and Soviet resistance in Estonia was not terribly strong, but this resistance stiffened as the Germans approached Novgorod to the east. Moreover, the Finns, eager to regain territory lost in the 1939–40 Winter War, had attacked the Soviets on 2 July. The Finns reached Koirinoye on Lake Ladoga on 16 July, although isolated Soviet forces fought on.

The closest German unit to Leningrad was the 41st Panzer Corps, which had advanced 650 miles (1,050 kilometres) in less than a month and had established bridgeheads over the River Luga, but there was no backup from the rest of Army Group North, which was slower-moving infantry.

The Finns provided both troops and a base from which, supported by the Germans, pressure could be exerted on Leningrad and the Soviet ports of Murmansk on the Barents Sea and Archangel on the White Sea (marked on the map as 'Weisses Meer'). However, the Finns fought the Soviets less well than they had in the Winter War of 1939–40, in part because they were less adept in attacking Soviet positions than they had been in defending Finland, but also because their morale was not so high. The Soviet defence proved particularly determined in the Karelian Isthmus and, rather than continuing the offensive, the Finns went onto the defensive in December after they had regained the territorial losses suffered in the Winter War. From then until 1944 the conflict was largely static on the Finnish front. The Soviet offensive that year led the Finns to change sides.

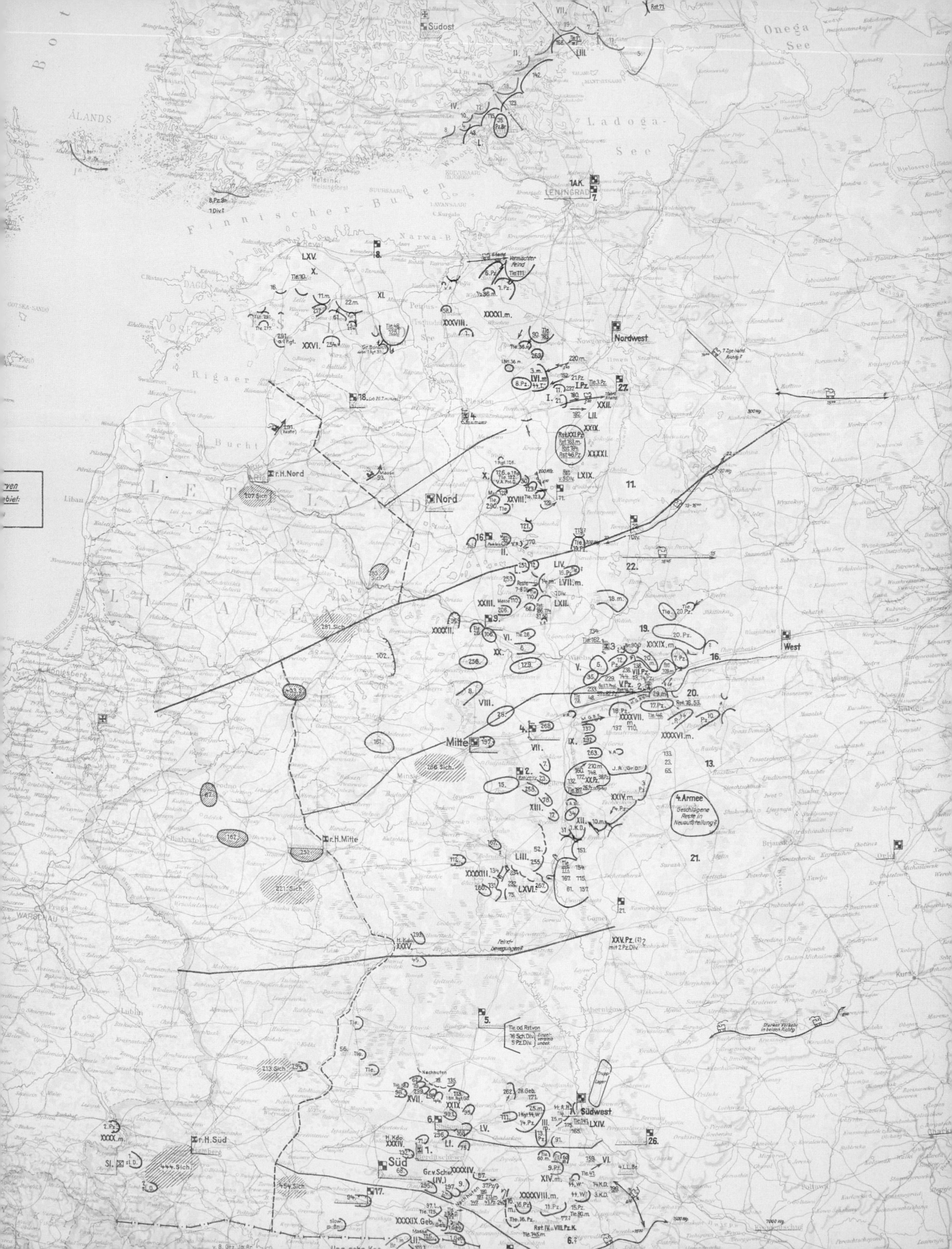

Südost
Onega See
Ladoga-See
LENINGRAD
Finnischer Busen
Helsinki
ÅLANDS
Narwa-B.
Vernichteter Feind
Nordwest
Rigaer Bucht
Nord
Gr.H.Nord
LETTLAND
LITAUEN
Königsberg
Mitte
Minsk
Grodno
Gr.H.Mitte
West
4. Armee Geschlagene Reste in Neuaufstellung?
Brjansk
Orel
Gomel
WARSCHAU
Lublin
XXV.Pz.(?) mit 2 Pz.Div.
Feindbewegungen?
Südwest
Süd
Gr.H.Süd
Kursk
Starker Verkehr in beiden Richtg.

LOSING MOMENTUM

German situation map, 28 November 1941.

This map produced a deceptive impression of German dominance, with German forces close to Moscow and Leningrad, including a German salient to the south of Moscow and a developing one to the north. But the map did not provide information on the strength of prepared Soviet defences near Moscow, nor on the weather; also the scale of the reinforcements the Soviets were receiving was not understood. Moreover, even had Moscow fallen, the Soviets planned to hold the line of the River Volga east of the city.

The German strategy in the Soviet Union was deficient on a number of levels. It was not a case of a good but poorly executed strategy, which leads to a focus on particular command decisions, such as that to turn troops southward to overrun Ukraine. Instead, the strategy was a bad one because it was not based on plausible time-achievable criteria and measures.

Winter freezing succeeded summer dust and autumn mud. Each posed serious difficulties, notably so to tank movements. As part of a broader failure of Intelligence and preparation, which included a serious underestimation of Soviet strength and resilience, the German High Command (OKH) had failed to make the necessary logistical arrangements to support an extended campaign. More generally, the serious deficiencies in the German Army, and their improvised solutions in combat conditions, had not been exposed by earlier opponents, but in late 1941 they became clear. Logistical support could not keep up with the advancing German forces because the gauge of the Soviet railway system had to be changed, destruction had been caused during the advance and there was a lack of trucks. For tanks there were particular problems with the supply of fuel and with numbers. Tank numbers had risen prior to the resumed German drive on Moscow launched in October, thanks to the arrival both of new tanks from the factories and of two divisions that had been refitted after their successful deployment in the Balkans earlier in the year. By November 1941, however, too many German tanks had been destroyed, damaged or were no longer fit for service. The poorly supplied German Army was in a very poor state, having taken heavy losses.

This situation affected the renewed German offensive, launched in mid-November, which sought to benefit from the valuable firmness that frost brought to the ground. In practice, however, German panzer divisions were unable to act as the intended operational centre of gravity because of the problems they faced, to which the High Command (OKH), and Hitler even more so, had failed to attach enough significance. The weather hit both combat operations and the crucial element of logistical support. The cold froze oil and lubricants and it killed many of the horses on which the army still relied for resupply. More significant was the failure to outline coherent and winnable strategic goals.

The Germans had mistakenly assumed that the Soviet Union would be a repetition of their success of 1940: it simply entailed transforming the deep penetration they had achieved in France in 1940 onto the enormous distances of the Soviet Union, and in a similar time sequence, despite the climate, weather, terrain and communication system all being far more challenging. But the tactical and operational means of 1939–40, in the shape of a combined arms attack (or *blitzkrieg*, 'lightning war') that had worked in a confined geographical space, could not work over the far greater spaces of the Soviet Union. Scale was a key dimension of strategy and, not least through imposing different time parameters, an important variable in it being implemented successfully. In addition, the Germans had developed an operational level doctrine in an improvised fashion. Alongside a seriously misplaced self-confidence (perhaps a product of racial arrogance toward Slavs) this ad hoc character helped set them up for serious problems when they attacked the Soviet Union and faced opposition in depth.

The German offensive ground to a halt in early December 1941, in the face of strong and effective Soviet resistance that was aided by the transfer of troops from Siberia and, less significantly, by British tanks that had been shipped in convoy to the Soviet White Sea ports.

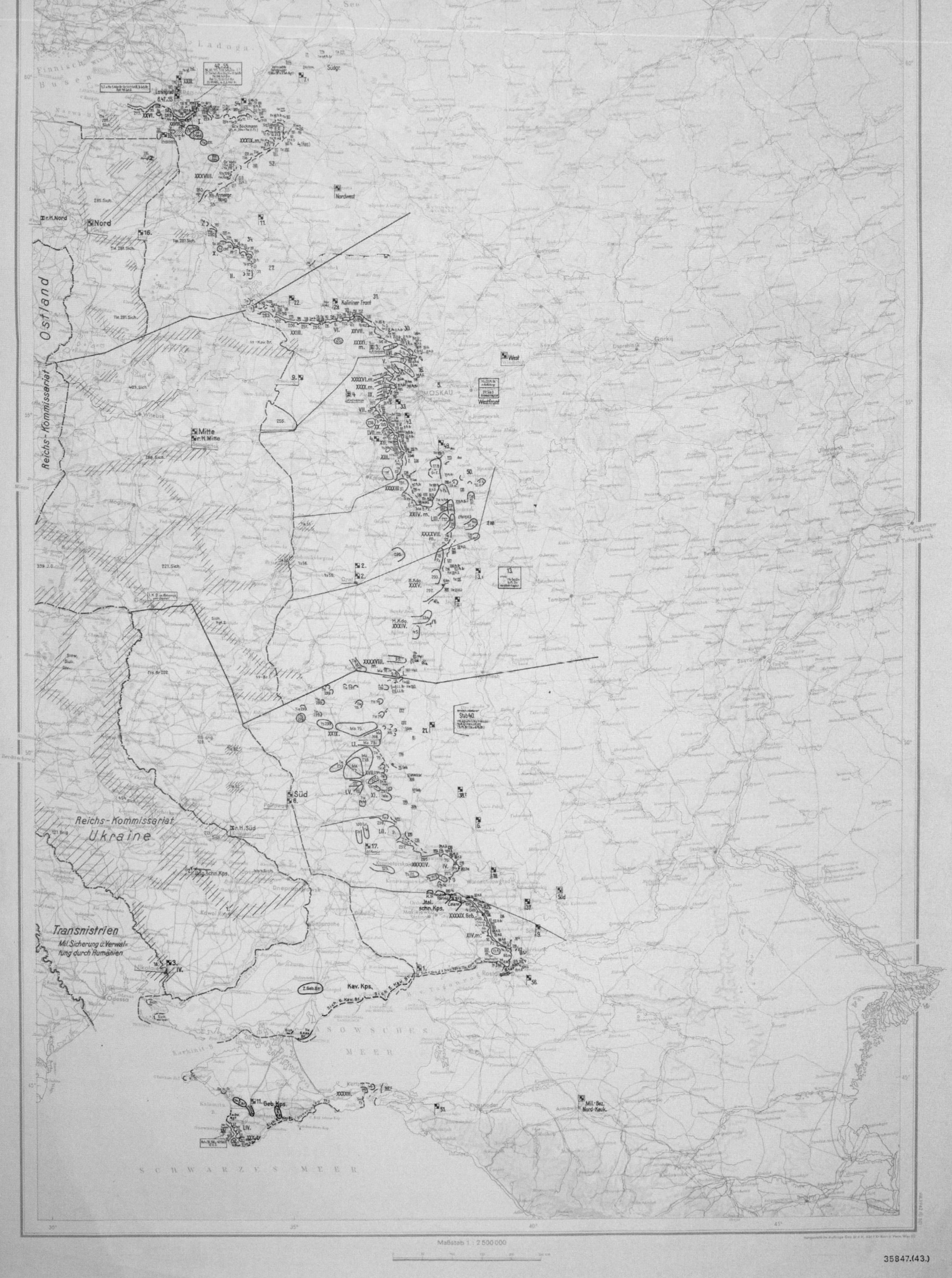

Onega See
Ladoga
Finnisch. Busen
Nordwest
Nord
16.
Reichs-Kommissariat
Ostland
Kalininer Front
22.
9.
West
MOSKAU
Westfront
Mitte
Gorkij
Kostroma
Tambow
Stab 40
Süd
17.
Reichs-Kommissariat
Ukraine
Transnistrien
Mil. Sicherung u. Verwaltung durch Rumänien
Odessa
Ital. schn. Kps.
XXXXIX. Geb.
Kav. Kps.
ASOWSCHES MEER
11. Geb. Kps.
Mil.-Bez. Nord-Kauk.
SCHWARZES MEER
Maßstab 1 : 2 500 000
35847.(43.)

KNOWING THE ENEMY

British Military Intelligence, 'Distribution of Axis Forces in the Balkans', December 1942.

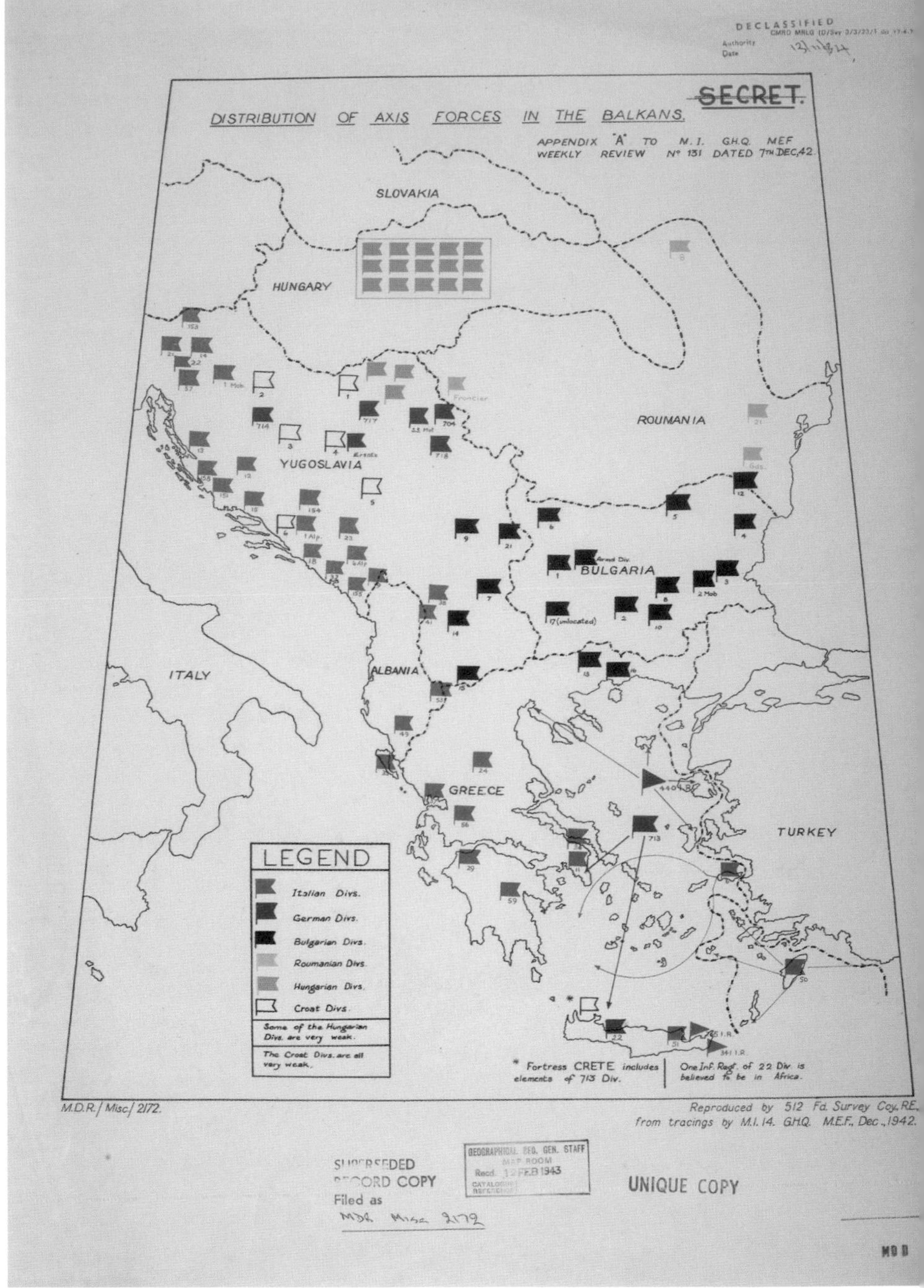

The order of battle of opposing forces was the prime Intelligence requirement for all combatants, and errors in this could be extremely serious, as the Germans discovered when they underestimated Soviet Red Army strength in 1941. Alongside the issues of numbers and location, with which Ultra intercepts could help, there was also the problematic one of quality, which was far harder to evaluate, just as the number of effective tanks was regularly overestimated when considering armoured divisions. The key, or legend, on this British Military Intelligence map notes the grave weakness of some Hungarian and all Croat divisions. Most of the Romanian Army was fighting in the Soviet Union, and that month it was to be hit very hard in the Soviet Stalingrad counter-offensive, while heavy Hungarian losses near Voronezh in early 1943 led to Hungary adopting a less bellicose position against the Soviet Union.

The location of German and Italian divisions was of particular interest to the British because they could otherwise be moved to oppose them in North Africa. The map revealed the significance of the resistance movements in Yugoslavia, divided between the communist Partisans and the royalist Četniks, in fixing Axis divisions. A German offensive had inflicted serious losses on the Partisans in late 1941 in Serbia and Montenegro, but Josip Tito had more success in Bosnia to which he moved his forces in 1942, from southeast to northwest Bosnia. At the same time, there were serious divisions between Tito and the Četniks under Draža Mihailovič – as well as between the Axis powers, in particular the German-created puppet 'Independent State of Croatia' (Nezavisna Država Hrvatska), which competed with Italy. Such confusion evaded mapmakers. In 1942, there were about 200,000 German troops in Yugoslavia, as well as 321,000 Italians, 170,000 Croats, 80,000 Hungarians, 70,000 Bulgarians and 57,000 Yugoslavian collaborators. The number of Cětniks may have been about 60,000 and of communist partisans about 100,000.

A PROBLEMATIC PENINSULA

United States' Army Map Service, Italy (South) Special Strategic Map, 1943.

Drawing in large part on Italian maps of 1935–41 at a scale of 1:1,500,000, and using the Lambert conformal conic projection, which preserves the shapes of objects at the expense of their relative sizes, this map offers very good guidance to the interplay of relief and communications, with two classes of roads and five of railways shown. Scales in miles and kilometres are offered. Such maps highlighted not only particular possibilities for amphibious attack, but also, through the terrain and rivers, clarified issues of breakout and exploitation, as with the mountains between the Gulf of Salerno, where there was a landing in 1943, and Naples. Anzio, the principal landing site in 1944, is marked south of Rome. Between (Monte) Cassino and Rome, the Allies had an uninterrupted chain of mountains on their right side, and many difficult hills between them and Rome. There was only one relatively good road. Moreover, the plain near the Tyrrhenian Sea was the Agro Pontine or Pontine Marshes. Mussolini had drained the area, but soon after the Anzio landing the Germans had destroyed the many small dams, which rendered the ground difficult for Allied tanks.

From 3 September 1943, the Allies invaded mainland Italy. Air power was seen as the way to stop German counter-attacks, but air power would not ensure success in mounting offensives. The density of German forces on the relatively narrow east–west defensive lines hampered Allied advances.

At the same time, in Italy there was an absence of relevant Allied doctrine and effective planning. This was particularly the case after the Gustav Line was broken in May 1944. The pursuit of the Germans was insufficiently close and vigorous, in marked contrast to the Soviet style. In August, Field Marshal Sir Alan Brooke, the Chief of the Imperial General Staff, wrote about General Sir Harold Alexander, the 15th Army Group commander:

> *'I am rather disappointed that Alex did not make a more definite attempt to smash Kesselring's forces up whilst they were south of the Apennines. He has planned a battle on the Apennine position and seems to be deliberately driving the Germans back onto that position instead of breaking them up in the more favourable country. I cannot feel that this policy of small pushes all along the line and driving the Boche [Germans] like partridges can be right. I should have liked to see one concentrated attack, with sufficient depth to it, put in at a suitable spot with a view to breaking through and smashing up German divisions by swinging with right and left. However, it is a bit late for that now ... very hard to get old Alex to grasp the real requirements of any strategic situation.'*

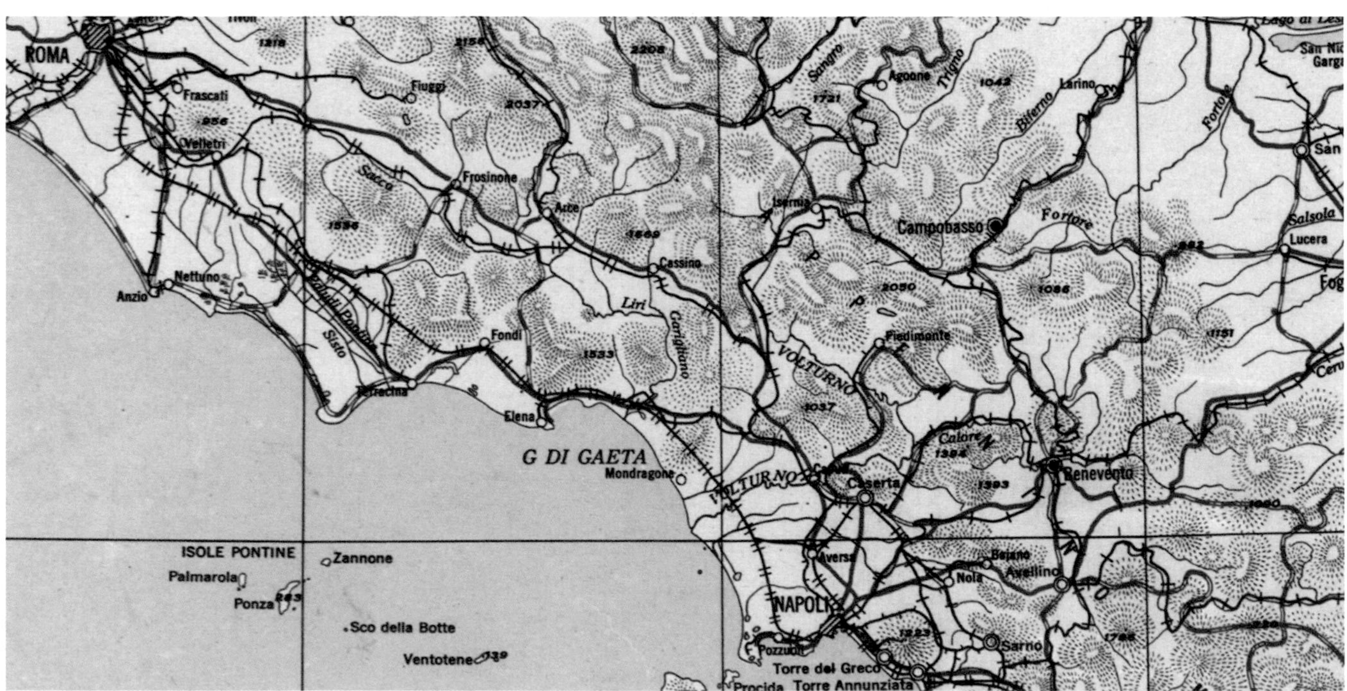

Prepared under the direction of the Chief of Engineers, U. S. Army, Washington, D. C.
Compiled by the Army Map Service from the following sources:
Europe, 1:1,000,000, GSGS 2758, Palermo NJ 33, 1935; Rome NK 33, 1935;
Sofia NK 34, 1938; Athens NJ 34, 1941.
Italy and Neighboring Countries, 1:1,500,000, GSGS 4176 ...

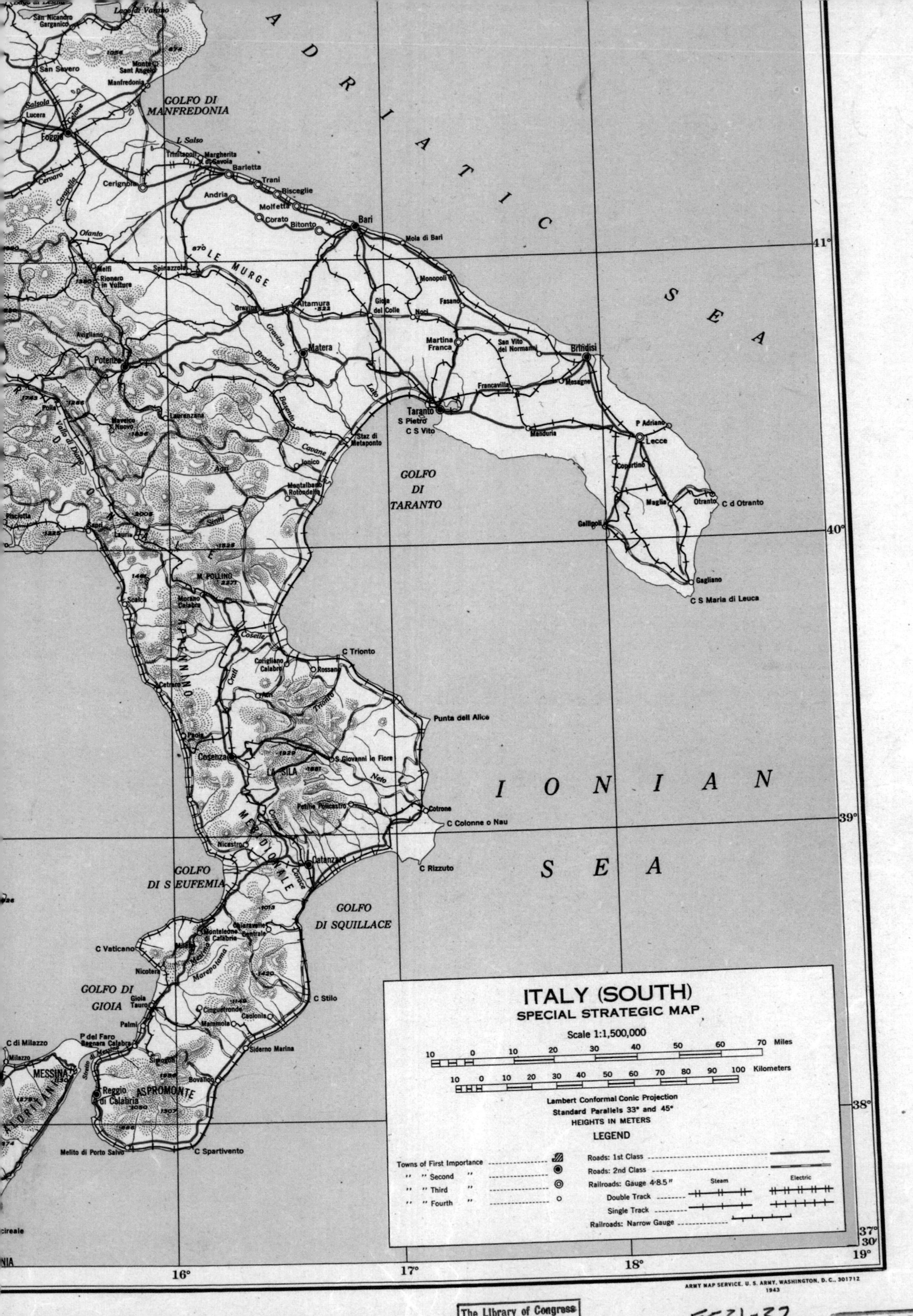

ITALY (SOUTH)
SPECIAL STRATEGIC MAP
Scale 1:1,500,000
10 0 10 20 30 40 50 60 70 Miles
10 0 10 20 30 40 50 60 70 80 90 100 Kilometers
Lambert Conformal Conic Projection
Standard Parallels 33° and 45°
HEIGHTS IN METERS
LEGEND
Towns of First Importance
" " Second "
" " Third "
" " Fourth "
Roads: 1st Class
Roads: 2nd Class
Railroads: Gauge 4'8.5"
Steam
Electric
Double Track
Single Track
Railroads: Narrow Gauge
ADRIATIC SEA
IONIAN SEA
GOLFO DI MANFREDONIA
GOLFO DI TARANTO
GOLFO DI S EUFEMIA
GOLFO DI SQUILLACE
GOLFO DI GIOIA
LE MURGE
LA SILA
ASPROMONTE
M POLLINO
San Nicandro Garganico
Lago di Varano
San Severo
Monte Sant Angelo
Manfredonia
Lucera
Foggia
L Salso
Trinitapoli
Margherita di Savoia
Barletta
Trani
Bisceglie
Cerignola
Andria
Molfetta
Corato
Bitonto
Bari
Mola di Bari
Ofanto
Melfi
Rionero in Vulture
Spinazzola
Monopoli
Gravina
Altamura
Gioia del Colle
Noci
Fasano
Avigliano
Potenza
Matera
Martina Franca
San Vito dei Normanni
Brindisi
Francavilla
Mesagne
Taranto
S Pietro
C S Vito
Manduria
P Adriano
Lecce
Copertino
Staz di Metaponto
Ionico
Montalbano Rotondella
Maglie
Otranto
C d Otranto
Gallipoli
Gagliano
C S Maria di Leuca
Polla
Lauria
Scalea
Morano Calabro
C Trionto
Corigliano Calabro
Rossano
Cetraro
Punta dell Alice
Paola
Cosenza
S Giovanni in Fiore
Neto
Petilia Policastro
Cotrone
C Colonne o Nau
Nicastro
Catanzaro
C Rizzuto
C Vaticano
Monteleone di Calabria
Chiaravalle Centrale
Nicotera
Gioia Tauro
C Stilo
Cinquefrondi
Caulonia
Mammola
Palmi
P del Faro
Bagnara Calabra
C di Milazzo
Milazzo
Siderno Marina
MESSINA
Bovalino
Reggio di Calabria
Melito di Porto Salvo
C Spartivento
41°
40°
39°
38°
37° 30'
16°
17°
18°
19°
ARMY MAP SERVICE, U. S. ARMY, WASHINGTON, D. C., 301712
1943

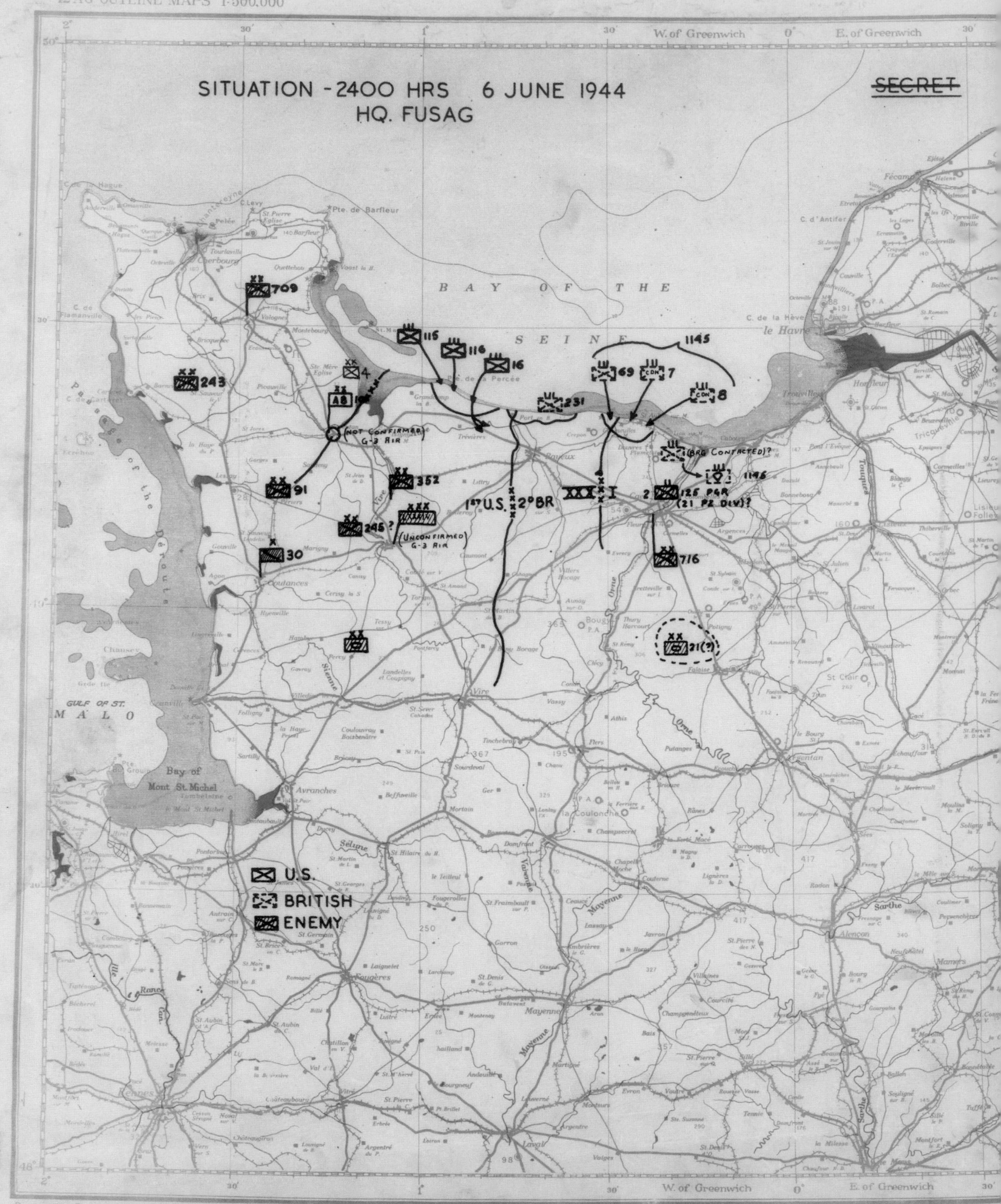

12 AG OUTLINE MAPS 1:500,000
SITUATION - 2400 HRS 6 JUNE 1944
HQ. FUSAG
SECRET
BAY OF THE SEINE
GULF OF ST. MALO
Bay of Mont St. Michel
(NOT CONFIRMED G-3 AIR)
(UNCONFIRMED G-3 AIR)
(BRG CONTACTED)?
125 PGR (21 PZ DIV)?
1ST U.S.
2ND BR
U.S.
BRITISH
ENEMY
W. of Greenwich
E. of Greenwich
Prepared by Engineer Section 12 AG 1944 from G.S.G.S. 4072 1:500000 Air Maps - Revised from G.S.G.S. 4438 Communications Maps 1943 & T.I.S. Report March 1944
Scale 1:500,000

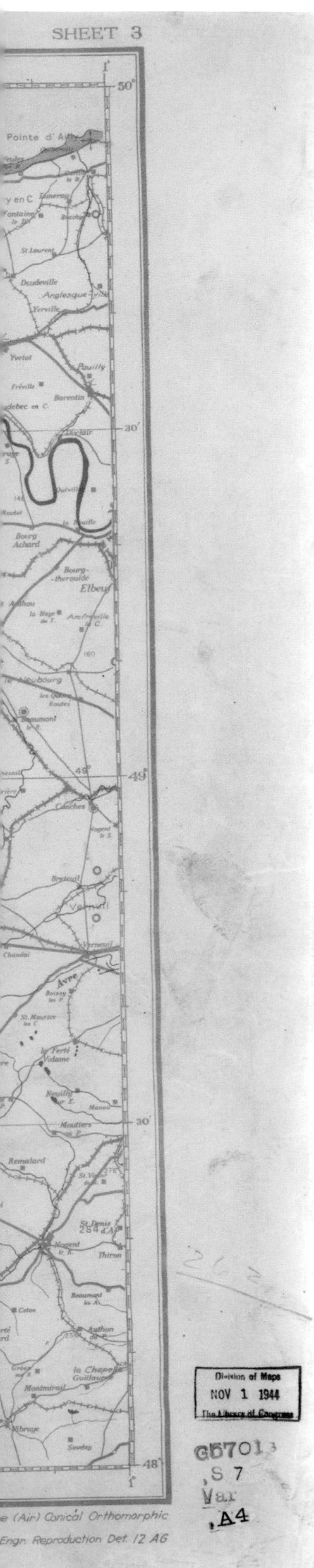

BREAKING OUT FROM THE BEACHHEAD

Engineer Section of US 12th Army Group, daily situation map, 6 June 1944.

The sheet shows the position of the 12th Army Group and known German units on just one day; a similar sheet was created for each day from D-Day (6 June 1944) until 26 July 1945. The sheets were accompanied by a 'G-3 Report' giving detailed information on troop positions. Relief is shown by spot heights.

This map presents the information at the close of D-Day and is particularly instructive on the location of German units. Allied planners were greatly concerned that the German panzer divisions in France would drive into the beachheads before they could become established and supported by sufficient anti-tank guns and armour. The German commanders, however, were divided about where the Allied attack was likely to fall, and about how best to respond to it. There was particular disagreement over whether to move their panzer divisions close to the coast, so that the Allies could be attacked before they could consolidate their position, or massed as a strategic reserve, which was the advice of General Geyr von Schweppenberg. Rommel, fearing the danger of any consolidated Allied beachhead, wanted to defeat the invasion at the waterline. The eventual decision, made by Hitler who had taken direct control, was for the panzer divisions to remain inland, but their ability to act as a strategic reserve was lessened by the decision not to mass them and by the destructive impact achieved by Allied air power. The tensions and uncertainties of the German command structure accentuated major failings in its Intelligence and planning.

The 21st Panzer Division, the sole German armoured division in the area, and a poorly commanded unit, did not counter-attack until the early afternoon. German tanks then approached the area between Juno and Sword Beaches, but they were blocked. Far from having division-sized manning and equipment, the 21st had 112 Mark IVs and some old French tanks, and it lacked the feared Tigers and Panthers. Most panzer divisions by this stage were seriously below complement, and this made (and makes) the use of tank divisions on maps, as if they were all of the same size, seriously misleading. Elements of the division that counter-attacked toward the sea were unnerved by seeing follow-on Allied glider forces landing near their position and withdrew for fear of being outflanked. In all, 156,115 men landed on 6 June and with far fewer casualties than had been feared: a casualty rate of 10.8 per cent (3,686 men) of those landed at Omaha made it by far the most costly beach. In part, this reflected the element of surprise, but the invaders also benefitted from a tremendous overprovision of air and naval support.

DEFENCE IN DEPTH

US 12th Army Group, situation map for 12.00 on 2 October 1944.

This map shows stiffening German defences as the Allies approached the German frontier and, in particular, concentrations of German troops near Arnhem, near Aachen, and in both Alsace and Lorraine. The reference to two German units as 'temporarily unlocated' underscored the problems encountered in producing such maps. Intelligence intercepts provided material, as did prisoners and, in France, Resistance sources, but these became less valuable at this juncture, and prisoners were in any case not helpful for units located in the rear. The German troops on the coast were at fortified positions: Calais, Dunkirk and the Scheldt islands, under assault from Allied forces. These German forces hampered logistical support for advancing Allied units, and thus provided the Germans with defence-in-depth while also tying down Allied units. However, the troops thus deployed by the Germans could be blockaded, which required fewer Allied units, and they were also absent from the main German order of battle. Hitler's determination not to withdraw German units from what he presented as fortresses repeatedly proved to be an effective loss of units, which were thus 'fixed'.

Aachen was the epicentre of German defence against the US First Army, which renewed the assault to the north of the city on 2 October, although it made only slow progress. The fortifications the Germans called the 'West Wall' were also known by both sides as the Siegfried Line. They had been upgraded and, in conjunction with the terrain and a hard-fought defence, proved effective. Further north, launched on 17 September, the British attempt to take Arnhem failed and the survivors were withdrawn across the Rhine on 25 September.

What does not clearly appear in maps is worth unpicking for this map registers two Allied failures. First, in showing a coherent German defence as well as no significant encirclement of large German forces in the field, the map reflects the Anglo-American failure to destroy the German Army after the Battle of Normandy. The emphasis on a broad-front advance ensured that the gaining of territory, rather than the destruction of German forces by more risky manoeuvring deep operations, was the option followed, but this proved part of a failure to engage adequately with the operational level of war, and more particularly with the synchronising or sequencing of operations. Concentration of force was not applied and there was no equivalent to the German counter-attack in the Battle of the Bulge. The second failure was the difficulty, despite aerial superiority, of preventing or adequately noting German troop movements and dispositions. This deficiency not only helped doom the Arnhem offensive, but also led to a failure to anticipate the Bulge offensive.

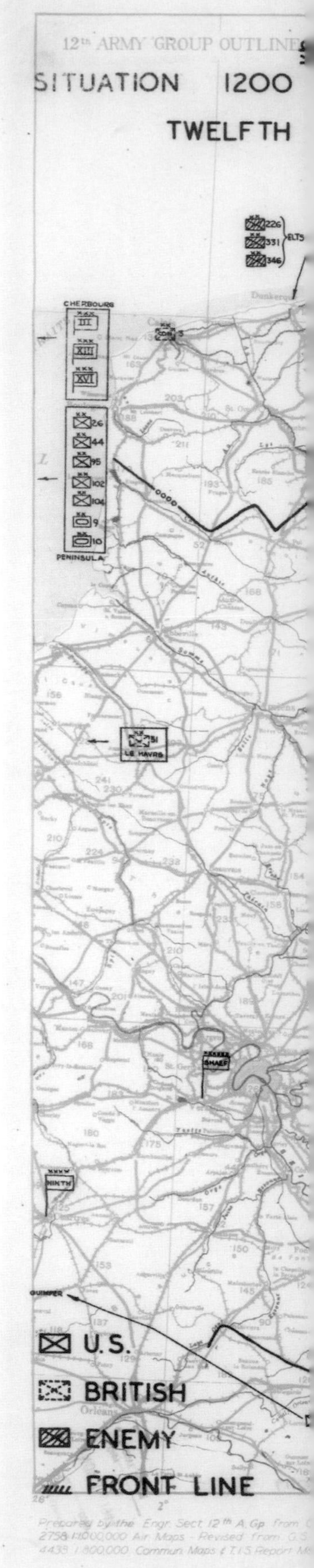

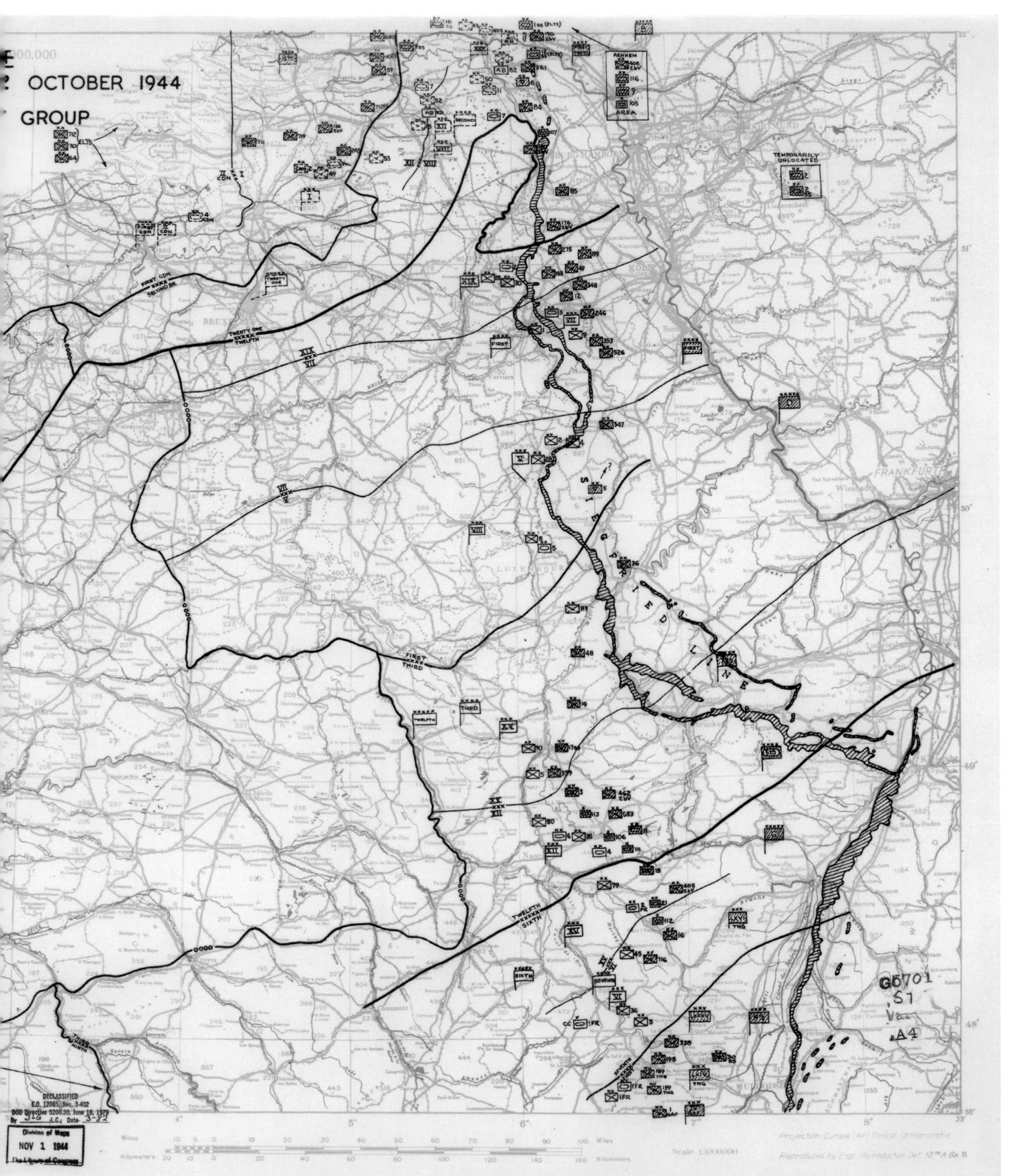
OCTOBER 1944
GROUP
TEMPORARILY UNLOCATED
ARNHEM AREA
FIRST CDN
SECOND BR.
TWENTY ONE
TWELFTH
FIRST
THIRD
TWELFTH
SIXTH
SEVENTH
FRANKFURT
SIEGFRIED LINE
G5701 S1 V4 .A4
DECLASSIFIED
E.O. 12065, Sec. 3-402
DOD Directive 5200.30, June 18, 1979
Division of Maps
NOV 1 1944
The Library of Congress
Miles
Kilometers
Scale 1:1000000
Projection: Europe (Air) Conical Orthomorphic
Reproduced by Engr. Reproduction Det. 12th A.Gp 11

UTTER DESTRUCTION

US Strategic Bombing Survey, Hiroshima before and after the bombing and 'Atomic Bomb Damage: Nagasaki', 1945.

These images show Hiroshima, the area around ground zero, photographed before and after bombing with 1,000-foot circles added; and the strategic bombing survey of damage to Nagasaki, with circles radiating out from 'ZERO', with a record of structural damage classified as follows: by fire only, by fire and blast, and by blast only. The least-severe fourth category is 'superficial damage by blast only'. The legend also identifies fire breaks and whether the area was a densely or sparsely populated one.

The creation of the atom bombs indicated the exceptionally deadly capabilities of the world's leading industrial society. This was the product not only of the application of science, but also of the powerful industrial and technological capability of the United States, and the willingness to spend about US$2 billion in rapidly creating a large nuclear industry.

At the Potsdam Conference, the Allied leaders issued the Potsdam Declaration on 26 July 1945, demanding unconditional surrender, as well as the occupation of Japan, Japan's loss of its overseas possessions and the establishment of democracy. The threatened alternative was 'prompt and utter destruction', but on 27 July the Japanese government decided to ignore the declaration, which they saw as a political ultimatum. President Harry S. Truman wrote, 'My object is to save as many American lives as possible'.

The already very deadly air assault on Japan – perhaps 90,000 died when Tokyo was incendiary bombed on 9 March 1945 – culminated in the dropping of atom bombs on 6 August and 9 August on Hiroshima and Nagasaki respectively, the bombs landing very close to the aim points. As a result, probably over 280,000 people died, either at once or eventually through radiation poisoning. This transformed the situation by demonstrating that Japanese forces could not protect the homeland. On 14 August, Japan agreed to surrender unconditionally. Particularly destructive products of industrial warfare, the atom bombs were employed to achieve the total war goal of unconditional surrender without having to resort to the fight to the finish that would have followed an American invasion.

The heavy Allied losses in capturing the islands of Iwo Jima and Okinawa earlier in the year suggested that the use of these bombs was necessary in order both to overcome a suicidal determination by the Japanese to fight on, and to obtain unconditional

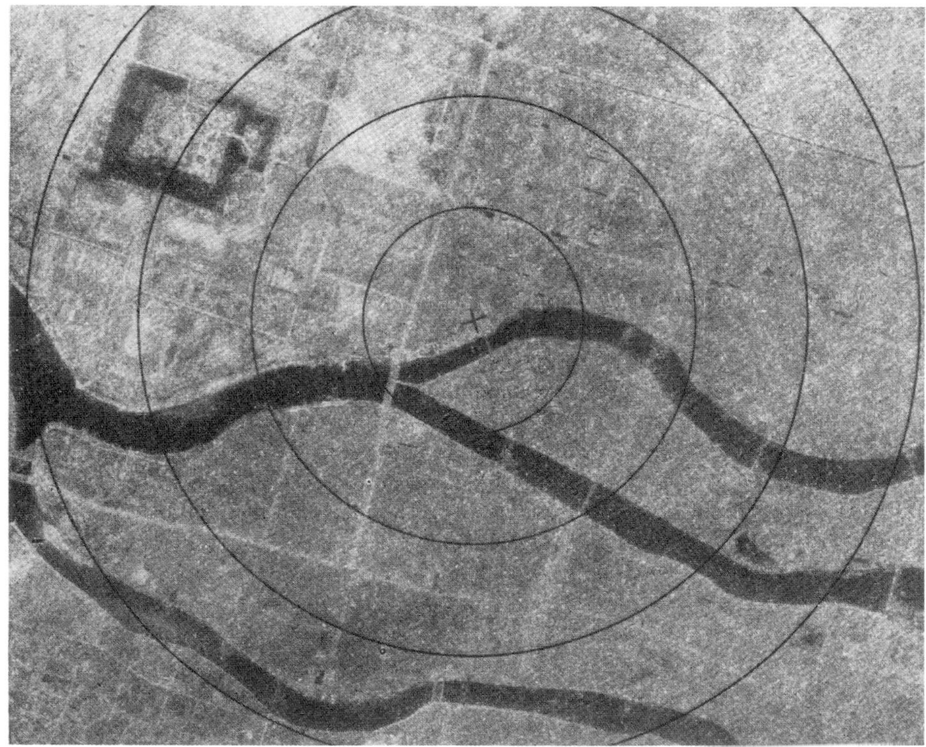

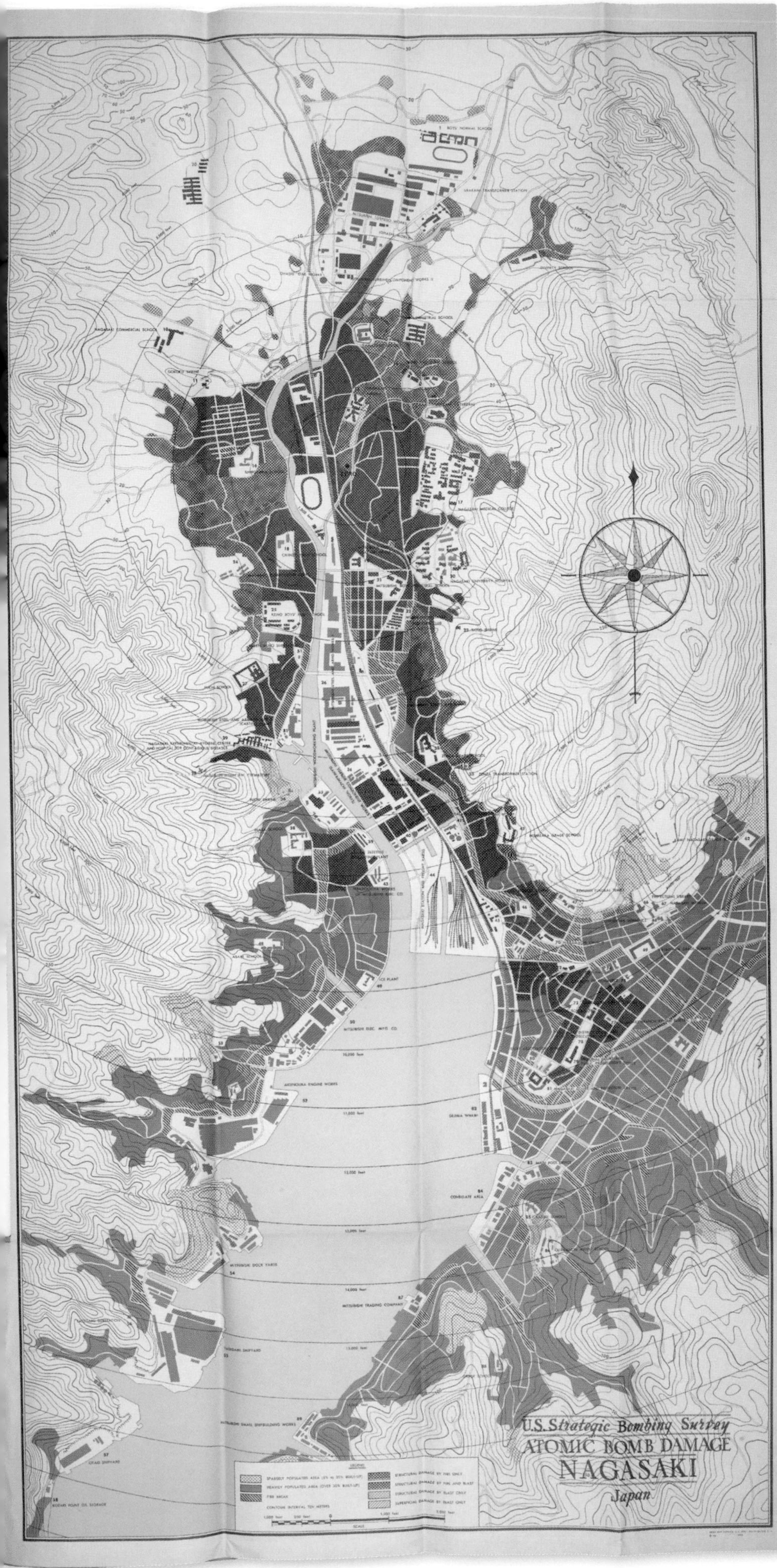

surrender. The dropping of the second bomb, on Nagasaki, had a greater impact on the Japanese government than the use of the first, clearly indicating an apparently inexorable process of devastation. The combined shock of the two bombs undoubtedly led the Japanese to surrender, although the Soviet invasion of Manchuria on 9 August was also a significant factor. The army had refused to believe what had happened and still wanted to fight after Hiroshima. The limited ability of the Americans to deploy more bombs speedily was not appreciated: no other bomb was available on 9 August, although one would have been about a week later. However, planning ahead, the Americans were already considering the use of atom bombs in tactical support of the envisaged landing on the island of Kyushu, which would have been the first of the main Japanese islands to be invaded.

Had the war continued, civilian casualties would have been immense. Aside from the direct and indirect consequences of invasion, the continuation of the conventional bombing campaign would have been very costly, both directly and indirectly: incendiary bombings from March 1945 onwards of Japan's major cities, which contained many wooden structures, caused much greater devastation than the atom bombing. Had the war lasted into 1946, the destruction of Japan's rail system by bombing would have led to famine because it would have been impossible to move food supplies.

B

Operational

3.

'The nation with the best photo reconnaissance will win the next war...'

Werner von Fritsch, 1938.

If it is accepted that aerial photography, notably by Do 17Ps and He 111Hs, did help the dramatic and successful German invasion of the Low Countries and France in 1940, then this declaration by Germany's Chief of Staff a decade earlier was prescient.

During the war, the need to coordinate operations on air and land, air and sea, and land and sea resulted in the increased complexity of many maps, so as to help with planning in three dimensions. This was especially true for bombing and tactical ground support from the air, and for airborne attacks by parachutists and gliderborne troops. Moreover, the development – explicit and implicit – of what has been termed the operational dimension of war, that between strategy and tactics, was significant. This encouraged demands for mapping at a certain scale, and contributed to standardised scales for all maps to make them comparable. The operational dimension of offensives – especially in the Soviet Red Army in 1943–45, notably Operation Bagration in 1944, which took Soviet forces westward across Belarus to near Warsaw, while inflicting very heavy losses on the Germans – hinged on an ability to seize and retain the initiative, including by bringing forward supporting units, which was more likely to result in the capability to outmanoeuvre opponents by advancing in depth rather than adopting the linear approach of advancing along the entire front. This ability, in turn, depended on staff officers being well informed about the distances and locations involved. An understanding of the geospatial dimension and terrain of the battlefield, and appropriate maps, were crucial to the German mission command system or *auftragstactik*.

For air warfare the precise knowledge that could be reproduced on maps was a matter of conditions as well as locations. Weather data, while being kept secret, was mapped, and meteorological mapping and analysis became increasingly important and sophisticated, notably with the depiction of air masses and the fronts between them, with particular symbols showing the type of front. There was important standardisation by the Americans from 1942, in part in order to help prediction, while by 1944 the Allies were employing radar and radio direction finders to track the weather, and hydrogen balloons were used to indicate wind speed and direction.

Nonstop flights became far longer in the war, with issues for navigation, mapping and fuel supplies. Alongside institutional pressure, notably from the United States' Air Transport Command, these issues spurred the development of more comprehensive aeronautical maps, notably the Western Allies' 1:1,000,000 scale World Aeronautical Chart map series. Long-range bombing, such as that of Japan by the Americans in 1944–45, necessitated particularly accurate maps because it pushed the limit of the bombers' flying range and the United States' Office of Military Intelligence acquired and reproduced maps of Japan. The US Coast and Geodetic Survey's map of Japan, which was issued in April 1944, provided an azimuthal equidistant projection (all points on the map are at proportionally correct distances from the centre point) that depicted straight line routes converging on Tokyo.

Radio navigation in the Second World War was far more difficult due to an emphasis on the night-time bombing of defended targets, an approach for which the use of landmarks as navigation tools was not possible. There was a lack, over enemy territory, of ground-based radio aids, notably directional beacons providing bearings. As a result, hyperbolic navigation systems developed. These used receivers on aircraft and ships to fix location using the difference in the timing of radio waves received from land-based beacon transmitters. Thus, location could be plotted from synchronised signals emitted from a considerable distance. For example, the Germans used the Knickebein system to guide bombers to the target.

Moreover, to help bombers, existing printed maps were acquired by all powers, and supplemented by various surveillance activities. To guide their bombers to targets in Britain from 1940, the Germans used British Ordnance Survey maps (see pages 66–67) enhanced with photographic information from aerial reconnaissance. More generally, as part of the process in which the war saw increased preparation for a total struggle, aeronautical charts for aircraft were improved not only technically but also by becoming increasingly sophisticated, and were tailored to different tasks and speeds. In the United States, five basic, standardised charts, each with a distinctive scale, were developed during the war: for flight planning, cruising, descent, approach, and landing and taxiing. Target perspective charts provided the view from the cockpit. Radio navigation charts presented information from

INSTRUMENT FITTERS install a Type F.24 aerial camera, the mainstay of RAF photography, into a North American Mustang Mark IA of No. 35 (Reconnaissance) Wing, RAF 2nd Tactical Air Force, based at Gatwick, Sussex, early 1944.

AERIAL RECONNAISSANCE

Aerial photography improved significantly during the war, thanks to better cameras that offered enhanced magnification. Long-range reconnaissance missions by aircraft flying at high altitude, notably Do 17Ps and He 111Hs, preceded the German attack on the Soviet Union in 1941 and the Soviets failed to prevent this.

A chart was captured from a midget submarine attacking Pearl Harbor on 7 December 1941 and it showed that the Japanese, who had some very good reconnaissance aircraft, could have reasonable prior knowledge about locations. Another example recovered from a crashed Japanese aircraft was also not without crude value. The Philippines were also reconnoitred before the Japanese landings in December 1941, while in May 1942 the Japanese produced a sketch map of Port Moresby in New Guinea that was drawn from aerial photographs.

As the doctrine as well as the technology of aerial reconnaissance advanced, each affected the other. Britain and the United States produced simpler and much less heavy cameras than their German counterparts, which could therefore be carried in large numbers by smaller and faster aircraft, such as the British Supermarine Spitfire and the North American Mustang. The British eventually used the de Havilland Mosquito PR 34, which was less vulnerable to ground fire and aerial interception than other aircraft because of its speed and service ceiling. The Mosquito also had an impressive range, which proved an important attribute for aerial reconnaissance.

Cameras mounted in fast fighters helped to provide a greater number of images of ground conditions, and regular overflights charted the progress of change, whether that was new factories or bomb damage to facilities. This process was further helped by the wider use of colour film as a consequence of Kodak's 1942 development of Ektachrome reversal film. These technological enhancements made a doctrinal transformation possible, in the shape of a deliberate focus on repeated photographs of the same site, which thereby provided more information on the extent to which the enemy was using and changing a site.

Equally significantly, photo-interpreters greatly improved their analysis of the resulting film. They had to respond to dispersion and disguise procedures in order to ensure that targeting remained appropriate, and because of their focus on long-range strategic bombing the British and Americans put more of an emphasis on photographic analysis than did the Germans or Japanese.

Air photos were not only crucial for planning and assessing bombing missions, but also important for operations on land and shallow water, by providing information on topography, beach gradients and water depths, including over coral, which was a key point for the Americans in the Pacific. Aerial photography was generally organised in terms of index maps, such as that of the road from Mongmit to Myitsone in Burma, which was produced by Britain's Photographic Reconnaissance Force to aid the successful advance into Burma in 1945.

The scale and sophistication of Allied aerial photography was such that there would have been adequate American mapping for the large-scale invasion of Japan planned for late 1945. In contrast, other commitments and the total loss of the air war ensured that the Axis powers progressively suffered from a lack of adequate up-to-date aerial photography in order to help plan and carry out their land and air operations.

transmission stations, while radar recognition charts – to facilitate target recognition – were produced by Britain's Air Ministry, for example of the North German coast.

Bombing itself greatly damaged the existing production of maps, as well as the current map repositories, which were an important resource. Thus, the [British] Ordnance Survey headquarters on London Road in Southampton was badly hit by German air raids in late 1940, with the destruction of many valuable documents, including height survey records and drawings for the standard one-inch map series. German map production and repositories, in turn, were eventually hit far harder. For example, the Allied bombing of Leipzig in 1944 destroyed Velhagen & Klasing and Wagner & Debes, both of which were major map publishing companies.

A variant for air power was the use of a spatial reference system to guide long-range missiles, most significantly the German V-rockets that did not have a pilot who could guide them. In planning routes for the V1, which was launched from 13 June 1944, the Germans came to appreciate that there were discrepancies between British and Continental data sets, although V1 (like V2) targeting was so imprecise that this was not a crucial problem. Due to the lack of a reliable guidance system, the missiles, although destructive, could not be aimed accurately. The use of missiles reflected the inability of the Germans to sustain air attacks on Britain, as well as Hitler's fascination with new technology and the idea that it could bring a quantum leap forward in military capability and satisfy the prospect of retaliation for the British bombing of Germany.

Use of existing maps

In addition to aerial reconnaissance (see box, above), existing maps were, of course, used. For example, Britain had published from 1934 a 1:100,000 topographical map of Palestine. During the war, the plates were handed over to British Army units serving there, for updating and printing for military purposes. In Egypt in 1940–42, albeit with the attack coming from the west and not, as in the First World War, from Turks to the east, the British benefitted from being able to operate in an area they had controlled and mapped prior to the war, with the Survey of Egypt producing valuable data. During the war, British military surveyors in Egypt used aircraft, and aerial photography was important there, as elsewhere, for mapping behind enemy lines.

More generally, a range of existing map sources was employed. Having seized Norwegian maps to help in their 1940 conquest there, the Germans who invaded Belgium and France later in 1940 had copies of the 1938 editions of Michelin maps, while, less impressively, British aircraft over Norway in 1940 used a *Norway, Sweden, and Denmark* Baedeker handbook, which had been revised in 1912. German preparations for an invasion of Britain in 1940 included large-scale town plans marked with strategic locations, and copies of Ordnance Survey maps, with overprints highlighting sites which were targets, the shoreline marked in terms of cliff, sand, steep and flat, and photographs to help with selected invasion beaches. Field Marshal Erwin Rommel's annotated map of El Alamein on 23 October 1942, which was captured by the British, was good (see page 79).

In their campaigns in North Africa in 1940–43, the British adapted Italian and French maps for Libya and Tunisia respectively, as with the maps for the Mareth Line in Tunisia, which the advancing British overcame in March 1943. The Japanese used French and Dutch maps when moving into French Indo-China and the Dutch East Indies in 1940–42.

New maps

Operation Overlord, the invasion of Normandy in 1944, was preceded by the precise study and mapping of climatic conditions because these directly affected the troops' ability to land. As well as predicting average weather conditions, there was also work on five-day forecasting. The invasion was preceded by more than two years of aerial photography, and the resulting maps were highly detailed. British COPPs (Combined Operations Pilotage Parties) collected, mostly at night, information on hydrographic and beach conditions. The resulting maps had to include tidal levels and coastal defences. The nature of the German gun positions were carefully graded, for example: 'Fixed coast gun, in open position / Medium battery, in open position / Heavy battery in casemate / Medium, fixed coast howitzer / Light, mobile gun or gun-howitzer / Anti-tank gun, less than 50 mm / Light, machine gun' and so on.

Aerial photography was one of the sources for Operation Plunder, the successful Allied crossing of the Lower Rhine on 23/24 March 1945 (see pages 98–99). The preparatory maps revealed impressive reconnaissance, including the notes on the crossing sites, the mapping of their allocation and approaches (to avoid confusion among the shipping), as well as the mapping of the range of German anti-aircraft fire. The mapping was highly progressive, such that it was deemed necessary to destroy previous versions, a process made possible by the scale of production. By this point in the war, mappers were moving away from the use of existing sources because air reconnaissance and newly created maps meant they were able to upgrade older maps.

In the Soviet Union, the geodetic survey service switched to focus its aim on providing the military with maps and lists of coordinates, and to carrying out surveys where required. As most of the large-scale Soviet mapping envisaged in 1938 had not yet been done, immediate need was the priority. Similarly, between 1941 and 1944, the German military mapped Europe, western Asia and North Africa, at the scale of 1:1,000,000, which was the same scale as a new Soviet state map on which work was begun in 1941. In turn, the British had to produce new maps for Southeast Asia, a task made more difficult by the major impact on aerial reconnaissance of the seasonal extent of low cloud cover, notably in Burma (Myanmar). There was the same problem for the Americans and Australians in New Guinea.

Mapping was affected by the operational flow and needs of the war, and by changes in the scale and structure of mapmaking at every level. With time, the Allies built up their capacity, in part as the Soviet Union recovered from the disruption caused by the German invasion and the move eastward of production facilities to escape the German advance. Later, when the Soviets advanced westward beyond their pre-war borders, so they had to redraw local maps to create ones that matched Soviet specifications. Their annexations in 1939 and 1940 had already posed an issue for Soviet mapmakers, but this became even more serious once they moved further westward. This was a process that matched the earlier one faced by the Germans as they advanced in 1939–42, but one in which the Soviets displayed greater thoroughness and to which they devoted more resources.

The United States created the Army Map Service by merging the Engineer Reproduction Plant, the Map Library and the Cartographic Section of the War Department's General Staff; built new facilities, such as those at Brookmont, Maryland, for the Army Map Service, and the Aeronautical Chart Plant at St. Louis, Missouri; and recruited far more staff, including many women. In 1942, American activity was in part ad hoc because bold plans drawn up in 1941 for new mapping were superseded by immediate demands, not least under the pressure of Japanese advances and attacks. As a result, existing maps were rapidly and simply revised, notably by adding a grid. In contrast, in 1943 fresh maps were compiled, often on a smaller scale and incorporating different and new material.

In addition to maps produced in the United States, there was much activity in the field. Each US Army division had a Topographical Battalion able to carry out field surveys to produce maps and to print them, which aided the process of the rapid dissemination of appropriate maps to platoon-level units. The British had lorry-mounted map printers in France in 1939–40; lost at Dunkirk, they were replaced. Rapid mapping in large quantities meant that it was readily possible to record progress. Thus, in the North-West Europe theatre in 1944–45, the US 12th Army Group produced daily situation reports, showing its position, and that of adjacent Allied forces and German units (see pages 52–53). Information was overprinted on the base maps. Each sheet was accompanied by a report giving detailed information on troop positions. Other forces also developed and/or recovered capability, the Canadians benefitting from British and American methods, while the Free French created a mapping agency in 1943 thanks to American help.

Maps were also produced by resistance movements in Europe, which were based on existing topographic maps that were supplemented by their observations. Sometimes, they produced completely novel maps for the purpose of some particular action. In the case of Yugoslavia, there was a highly developed cartographic service consisting of photographers, drafters and surveyors, organised as geodetic sections of particular corps. Their map production was known as 'Partisan cartography'.

The Pacific provided a clear example of the impact on mapping of the flow of operations. The Land Survey Department of the Japanese General Staff Headquarters produced not only maps of Japanese territories, but also *gaihozu*, 'maps of elsewhere'. These drew on Japanese survey squads and existing, reproduced foreign maps. Thanks to these and other sources, Japan could draw on good maps for some areas. Thus, the Japanese aeronautical maps of China were good, while there was also good mapping by the Japanese of parts of the Soviet Union. A 1944 map of Lake Baikal in Siberia included information about the thickness of the ice at different times of the year, and highlighted storage sites and other locations of military materiel.

However, maps of the Pacific produced before the war for Japanese forces were heavily derivative of the limited colonial mapping by the Western imperial powers attacked by Japan in December 1941: Britain, the Dutch and the United States. Once the Japanese had conquered widely, they prepared more maps in order to support fresh offensives and to protect against Allied counter-attacks, notably in Burma, New Guinea and the Solomon Islands. Produced in a hurry and often without much, or any, input from aerial reconnaissance, the Japanese maps were inadequate as operational tools, let alone for tactical purposes given the precision the latter required. However, on those islands further from the Allies – for example, the Philippines, which the Americans did not invade until late 1944 – the Japanese had more time for mapping. Nevertheless, their maps still lacked the degree of cartographic accuracy and information seen in their Allied counterparts. Cartography did not have high standing in the Japanese Army, but in terms of preparing defensive positions, the situational awareness was fit for purpose.

For each power, there were 'Cinderella' areas that received less attention, for example Burma for the British. Nevertheless, in 1943 the Geographical Section of the General Staff produced a new map of Burma. In 1944 it followed up with a series of maps of the region showing airfields in Japanese-occupied territory. In preparation for the recapture of Rangoon in 1945, the British map of 1944 was reproduced on new sheet lines and with minor corrections; while toward the end of the war, as the British planned the recapture of Singapore, the survey of Singapore, first made by the Federated Malay States Surveying Department in 1938, was updated by the War Office in 1944 (see pages 12–13) and then reissued in 1945.

In a wartime situation of shortage the allocation of manpower and aircraft, in particular, was important to new mapping, such as in North Africa where the British only had sufficient air reconnaissance after the deployment in late 1941 of the aircraft of the South African Survey Company once the conquest of Italian East Africa (Eritrea, Abyssinia and Italian Somaliland), in which it had been involved, had been completed. In North Africa, a key challenge was the need to cover the Sahara Desert, because the Italian mapping of Libya, while good, was largely restricted to the coastal strip. This need was increased by Allied attempts to advance into the interior and also to turn the flank of Axis coastal positions. There was also a need for more detailed local maps. In Southeast Asia and the Dutch East Indies (Indonesia), where distances were far greater, Allied mapping, for example of Borneo, depended on advances that led to the capture of airfields from which aerial reconnaissance could be made, in this case in the Moluccas.

Alongside the production of completely new maps, there was the reprinting and updating of others in order to meet operational needs. Thus, the Anglo-Soviet occupation of Iran in 1941 led to the upgrading of existing maps. For example, the map of the Bam area, compiled and published originally under the direction of the Surveyor-General of India in 1912, and revised to 1941, was published by the War Office in 1942 and reprinted by the United States' Army Map Service in 1942. A note drew attention to issues with the map, notably the significance of water supplies:

> *'The word "Perennial" against water features indicates that they generally contain water but the accuracy of the information regarding perennial water on this map is not guaranteed and must not be accepted without further investigation. Hill features are shown by vertical hachuring except in the extreme S W corner where they are shown by form-lines at approximate vertical intervals of 250 feet.'*

New maps of parts of Iran were compiled and reproduced by the India Field Survey Company in 1943. In preparation for Operation Torch, the Allied invasion in November 1942, maps of French North Africa were collected and reprinted.

Assessing effectiveness

Gauging operational effectiveness is never easy. Dependent on combined arms operations, the Western Allies proved especially good at creating and using maps of a number of types, and this was particularly the case for invasions for which preparations were extensive in order to be able to overcome the advantages enjoyed by the defenders. Thus, there was almost a degree of overprovision of force for the invasion of Sicily in 1943 and for both D-Day and Operation Dragoon in 1944; but this reflected understandable anxieties about the consequences of failure.

Aside from the very extensive and frequently updated maps of the Allied landing sites, there was much provision for the subsequent campaign. This included not only conflict but also the supply of facilities. Produced in 1943, a 'Geo-Topographical and Airfield Map,' for use in selecting sites in North-West Europe for rapid airfield construction, depicted existing airfields alongside a colour-based grid: 'Blue rulings (horizontal) refer to density of possible sites in an area, judged on levelness and closeness of country,' while 'Red rulings (vertical) refer to soil types which govern permeability and drainage'. The categories were: 'Granular Soils (permeable); Loam Soils (fairly permeable); Clay soils (impermeable); and Marsh (waterlogged).' A July 1943 map of theoretical beach and port capacities covered very different criteria. For the beaches, there was a valuable distinction drawn between adequate and incomplete Intelligence, while waters sheltered from prevailing winds was one of the criteria for beaches. Weather was also part of the equation.

A standard account of the war was to argue that Germany and (to an extent) Japan had militaries with superior fighting quality, but that they were overwhelmed by the greater resources deployed by the Allies, notably the United States, but also the Soviet Union, with the resources of the latter including its size and the winter conditions. As such, there was a pivot moment in which resources overtook fighting quality, the key turning point generally seen as early December 1941, when the Germans were being blocked in the Battle of Moscow and also declared war on the United States. This account, however, suffers from an exaggeration of German fighting quality and because it underplays the numerous serious flaws in German warmaking. In particular, the effectiveness of the German *blitzkrieg* ('lightning war') attacking methods in 1939–41 was exaggerated by contemporary and later commentators, both German and other, under the spell cast by the sheer shock and drama of the German offensives, which were in such striking contrast to the nature and speed of conflict in the First World War, particularly in the case of aircraft.

As a result, there was an overrating of the impact of military ideas and methods that, in practice, represented more of an improvisation than the fruition of a coherent doctrine; or an evolution rather than a military revolution. Rather than focusing, as is so often done, on the use of tanks and ground-attack aircraft, in practice the Germans benefitted not from means but from methods. In particular it was the German Army's doctrine, training and leadership – and, notably, the stress placed in mission-command upon flexibility, personal initiative and action. Germany's opponents could not match these elements, either individually or in combination. This was what made Germany's advances especially effective in 1939–41 and that is difficult to capture on maps.

That success in 1939–41 also owed much to poor strategic and operational decisions by Germany's opponents, notably in terms of defensive planning and the allocation of reserves, especially as a consequence of having to defend an overlong perimeter. This was the case in Poland (1939), on the Western Front (1940), and in both Yugoslavia (1941) and Greece (1941). Alongside the undoubted flaws of opponents, German forces displayed serious shortcomings in their capabilities in 1940, and in practice the margin between success and failure was closer than is generally appreciated, whereas the potential of weaponry and logistics based on the internal combustion engine, the tank, the lorry and aircraft was less dramatic than talk of *blitzkrieg* might suggest. Artillery remained a key factor, as was to be seen on all the military fronts, and it was the major killer among weapon systems.

More seriously, Germany's invasion of the Soviet Union in June 1941 (Operation Barbarossa, see pages 42–47) revealed major, and ultimately fatal, flaws in German strategy, operational practice and fighting methods. From the outset, Soviet forces fought better than the Germans had anticipated, and the Red Army was larger than expected. In addition, the substantial amount of Soviet territory conquered, and the millions of troops killed or captured by November 1941, were at the cost of heavier-than-expected German losses. With its emphasis on defence and its stress on artillery, Soviet doctrine proved increasingly effective once the serious initial shock and surprise of the German attack had been absorbed, an effectiveness that was enhanced by the immense size of the Soviet Union. The Red Army learned to cope with both German tactics and German operational methods: anti-tank guns proved crucial in weakening German tank attacks, while establishing defences in depth hindered the German exploitation of breakthroughs.

Recent scholarship has led to a re-evaluation of earlier views. The failure of German offensives in 1941, 1942 and 1943 are now less frequently attributed to the factors cited by German generals and others – notably the size of the Soviet Union, disparities in resources, the harshness of the winter (which affected both sides) and maladroit command interventions by Hitler, important as they all were. Instead, the scholarly emphasis has come to be on the quality of the Soviet defence; although, in addition, fresh insights have been gained into German deficiencies, such as a failure to engage with logistics and a brutal mishandling of the civilian population.

Although the United States entered the war in December 1941, there was no deployment of American forces on the European mainland until the invasion of Italy in 1943 and none in major strength sufficient to threaten German hegemony until the invasion of Normandy in June 1944. As a result, the Germans on land (although not in the air or at sea) were able to concentrate on the Soviets. Despite this, and despite relaunching the offensive on the Eastern Front in 1942 and, on a smaller scale, in 1943, they were unable to do so successfully.

The stability of the Soviet regime was also a key factor. The political dimension was more significant in deciding the outcome of the 1941 campaign than the military issue of German operational failure. As, albeit very differently, with China in 1937 and Britain in 1940, there was no Soviet military or political collapse comparable to that in France in 1940, Yugoslavia in 1941 or Italy at the Allies' hands in 1943.

The fall of Mussolini

The interplay of factors involved in success and failure is shown in the interaction of operational with strategic and geopolitical factors which led to the fall of Mussolini in 1943. The crisis of the Italian military system became that of Fascism in part because the limits of both were similar, and Mussolini could not, and would not, confront this. His determination that Italians would fight, and that to compromise was to degrade Fascism, ignored realities.

The Allied invasion of Sicily precipitated a growing crisis of confidence in Mussolini among the Fascist leadership, as well as among other Italian leaders, especially the king, Victor Emmanuel III. After the meeting of the Fascist Grand Council on 24 July 1943, Mussolini was arrested at the king's behest on 25 July, and Marshal Pietro Badoglio, a former Chief of Staff, formed a government, although he was an individual consistently without ability. The Italian Army and the Church backed this government, and it excluded the Fascists. An armistice with the Allies, the Armistice of Cassibile, was signed by the new Italian government on 3 September 1943, and was announced five days later. However, the Germans rapidly seized most of Italy, with the new Italian government, which had fled from Rome to Brindisi, unable to organise effective opposition. The Italian Army was not given appropriate instructions, although much of the fleet was able to sail to join the British in Malta, albeit under German air attack. The Italian Army effectively disintegrated, again something that did not lend itself to mapping. Units were dissolved by their commanders, to avoid coming into conflict with the German Army, or were disbanded by private initiative.

VICE ADMIRAL HENRY KENT HEWITT, US Navy, and Lieutenant General Mark W. Clark, USA, onboard the troop ship USS *Ancon* (AGC-4) during Operation Avalanche at Salerno, Italy, September 1943. Hewitt's determined leadership was important to Allied success at Salerno. The unexpected strength of the German counter-attacks led the Allies to bring up more naval firepower but some warships were damaged by German radio-controlled bombs. Clark was the commander of the US Fifth Army, which liberated Rome in June 1944, but he was criticised for prioritising that objective instead of encircling the German forces as planned. As a result, the Germans withdrew to take up new positions behind defensive lines further north.

The advance in Italy

The main Allied invasion force was landed on 9 September 1943, in the Gulf of Salerno, south of Naples. The Allies pressed on to enter Naples on 1 October. The Germans had earlier withdrawn in the face of the Allied advance and of a popular rising in Naples that they could not crush, a revolt that indicated the advantages enjoyed by insurgents with good 'mental maps' of the urban landscape. The Allies were left in control of southern Italy, but the Germans had established a strong line across the peninsula, making plentiful use of the mountainous terrain to create effective defensive positions, from which they were to fight ably and cause heavy casualties for the attacking Allies. Mapping came into its own at this point for both sides.

The war in Italy was a hard-fought one, and an Allied landing at Anzio in January 1944, designed to open up the conflict, failed to produce the expected outcomes. If they had read their maps correctly, the planners would have realised that the beaches at Anzio were surrounded by heights that had to be taken quickly, because otherwise the German defenders could move in, take the high ground and pummel the troops on the beaches, as at Gallipoli in 1915. Acting in accordance with the maps required aggressiveness in taking the high ground, but poor leadership meant that Rome did not fall until June, and the Germans still held northern Italy at the end of the 1944 campaigning season, and did so with a relatively modest amount of troops. In part, Hitler held on because he wanted to limit Allied air attacks on southern Germany from Italian bases, an issue indicated by maps that recorded bombing ranges.

There was no Allied breakout into northern Italy until April 1945. This development helped revive the resistance movements in Italy, which had been bloodily suppressed by the Germans the previous autumn and, in particular, winter. Anti-partisan operations had specific requirements in mapping, notably those of areas of partisan activity, terrain and communications. In 1943–45, there was in Italy, at the same time, a patriotic war against the Germans and a civil war between Fascists and anti-Fascists. The resistance itself was seriously divided between the communists and the non-communists. The former, who dominated the actual fighting, very much took directions from Moscow. Again, this situation is very difficult to capture in mapping.

The Italian perspective can readily be repeated for other states, such as Finland, France, Romania and China. They had their particular, and sometimes divided, trajectories as active participants in the war, and they were also areas in which conflict took place. The complex geopolitics of the war was not separate to its operational flows and intensity, and an understanding of terrain and geography was critical to operational success.

STADTPLAN VON PLYMOUTH

German General Staff, map of Plymouth, 1941.

Produced by the Germans, the 'Stadtplan von Plymouth' identifies many target sites, including fortifications, railway stations, the airport, the harbour, barracks, bridges, factories (by type) and a tunnel. The Germans printed their information over a British Ordnance Survey map. This was but one example in a pattern of maps of many cities.

These maps were intended to help both air attack and also any invasion force. This provision might seem surprising in 1941 because the Germans were then focusing preparations on attacking the Soviet Union. To a degree, this reflected a lack of German strategic clarity, although preparations for a range of possibilities were normal for all powers.

Sequential warmaking played a major role in German thinking. Faced by what appeared to be a spread of Soviet power in Eastern Europe, which might facilitate a Soviet attack and/or make a German attack on the Soviet Union more difficult, Hitler was focused on war with the Soviets. After that, it seemed pertinent to mount an attack on Britain. Defeating the Soviet Union would provide Germany with an opportunity to build up the air and naval forces necessary for war with Britain and, even, possibly to renew planning for an invasion by land forces. What became the intractable nature of the war with Britain led to a view that defeating the Soviet Union would become a way to weaken Britain: as with Napoleon in 1812, Moscow was seen as a stage on the way to London. Germany's bold plans for a naval build-up, notably Plan Z (a 1939 plan for a pre-1914 style naval competition to match Britain by 1948), suggested that in time it might be possible to attain the naval superiority that would permit an invasion.

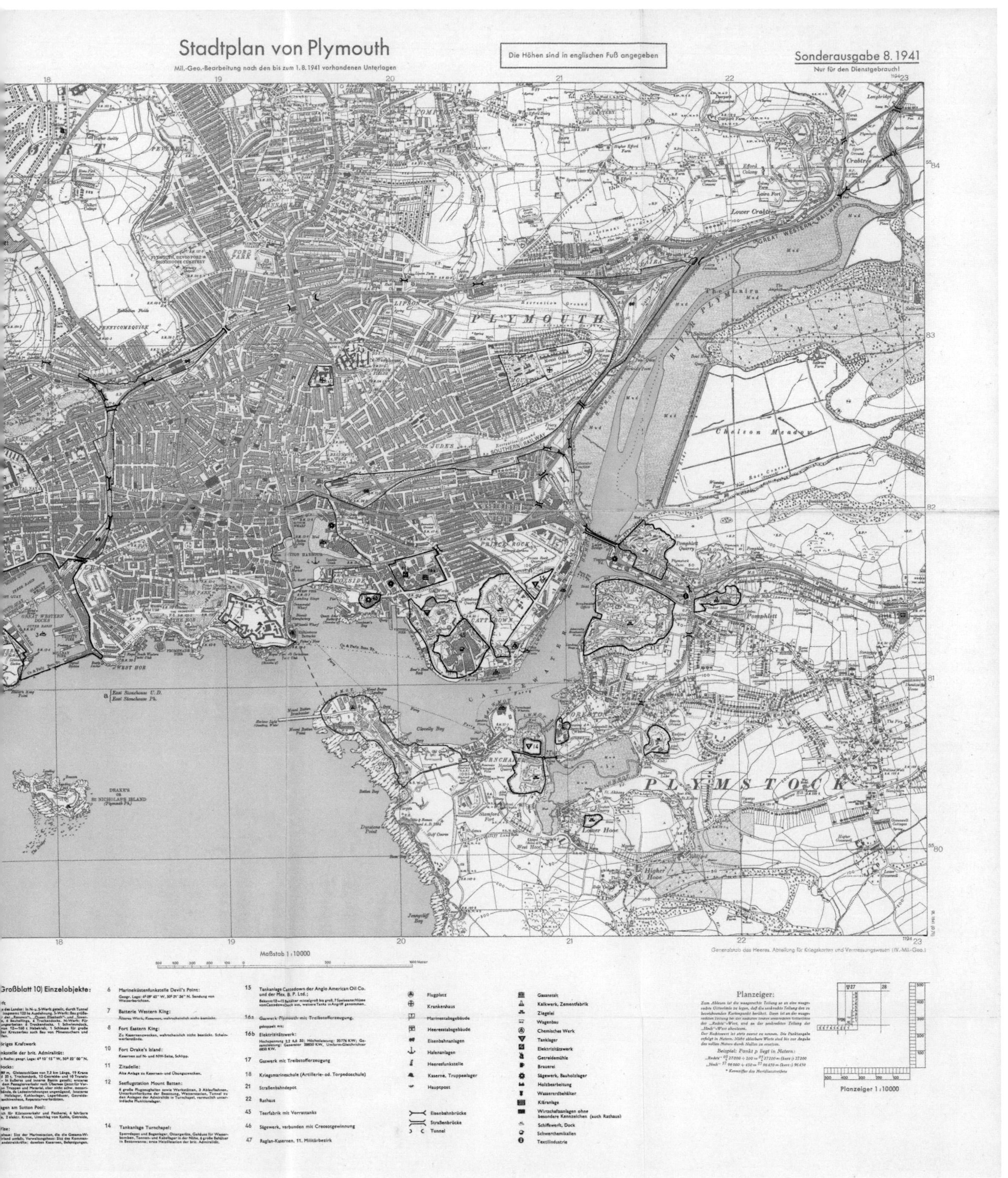
Stadtplan von Plymouth
Mil.-Geo.-Bearbeitung nach den bis zum 1.8.1941 vorhandenen Unterlagen
Die Höhen sind in englischen Fuß angegeben
Sonderausgabe 8.1941
Nur für den Dienstgebrauch!
PLYMOUTH
PLYMSTOCK
Chelson Meadow
Pomphlett
Lower Crabtree
Crabtree
Cattewater
Drake's or St. Nicholas's Island
Maßstab 1:10000
Planzeiger
Planzeiger 1:10000
Generalstab des Heeres, Abteilung für Kriegskarten und Vermessungswesen (IV. Mil.-Geo.)

DESTROYING THE DOCKS

German aerial photograph of Liverpool, Britain's key Atlantic port, November 1940.

The extensive harbour facilities along both banks of the Mersey can be readily discerned. The German annotation identifies three sections in red: Liverpool's northern (a) and southern (b) dock systems, with more than 30 named docks, as well as the dock system and shipmaking facilities at Birkenhead (c) on the Wirral Peninsula. The port of Liverpool's northern system, from Gladstone to Princes, consisted of more than 20 docks, while the southern system, from Canning to Herculaneum, had a dozen. The entire, interconnected system extended along approximately eight miles (12 kilometres) of waterfront and because it was enclosed and isolated from the tides it enabled ship movements 24 hours a day. Cammell Laird shipyard produced nearly 200 commercial and military vessels during the war.

From 7 September 1940, the Luftwaffe bombed London heavily and repeatedly, with the docks on the River Thames being hit hard. This was what became a German strategy to starve Britain into surrender, in part by destroying the docks through which food was imported. Liverpool, Plymouth and Southampton were also to be heavily attacked for that reason, as well as to deny the Royal Navy bases. Liverpool had already been attacked on the nights of 28–31 August and 4–6 September 1940, but was attacked more heavily from mid-November and the attacks continued until January 1942, when other ports were added, including Avonmouth, Cardiff and Portsmouth. On Merseyside nearly 4,000 civilians were killed. The location where the German aerial photograph bears the label 'Bootle' is actually Wallasey. Bootle was directly opposite and the location for many of the northern docks, such as Langton, Brocklebank, Canada and Huskisson, where vital war materials such as timber and grain were imported from North America.

Reconnaissance was significant from the outset of air power in war. Instruments for mechanically plotting from aerial photography were developed in 1908, while a flight over part of Italy by Wilbur Wright in 1909 appears to have been the first on which photographs were taken. Reconnaissance aircraft were particularly important in the First World War, but this value, and the vulnerability of the aircraft as they maintained straight and level courses to take their photographs, made them targets, and considerable effort was spent in chasing, if not destroying, these aircraft to prevent this aerial information from being gathered.

The Luftwaffe lacked the capability to match its destructive purpose, hampered by limitations with their aircraft and a lack of well-trained pilots. The bombers' load capacity was inadequate for the delivery of an effective strategic bombing campaign. Night bombing and the length of the winter night ensured that the assault was damaging but the English ports continued operating. Moreover, the Blitz – the assault on national morale that the Luftwaffe had advanced as a reason for its significance – in practice lacked strategic impact. The German hope that the British people would realise their plight, overthrow Churchill and make peace, proved a serious misreading of British politics and public opinion. Instead, Intelligence reports in Britain suggested that the bombing had led to signs of 'increasing hatred of Germany', as well as to demands for 'numerous reprisals'. The idea that Hitler had to be defeated and removed was strengthened.

The attack on Britain was the last time in which a tri-service strategy was to the fore for Germany. The attempt to use submarines to close Britain's maritime links also failed. There was much focus on submarine activity on Liverpool's Western Approaches, but most vessels continued to arrive in the port safely.

WATERLOO
with
SEAFORTH
LITHERLAND
LIVERPOOL
Leuchtturm
BOOTLE
Egremont
BIRKENHEAD
KANAL
a
b
c

CONQUERING CRETE

Luftwaffe, map of operations against Crete, 1941.

The island of Crete was held by a lightly armed garrison of 35,000 British Commonwealth and Greek personnel, most of whom had just hastily retreated from Greece. The topography of Crete is difficult to defend, being 160 miles (nearly 260 kilometres) long and varying in width between eight and 35 miles (13–56 kilometres). The interior is mountainous, with poor north–south communications. The main road artery runs close to the north coast and connects Suda Bay with the towns of Maleme (unmarked in the map, but west of Canea near 'Platanias'), Canea (marked 'Chania'), Retimo ('Rethimnon') and Heraklion ('Iraklion'). British supply bases were situated to the south in Egypt, but Suda Bay lies on Crete's north coast, where the airfields had also been built. Maps like this one, used by the Luftwaffe for the attack on the island in 1941, provided important information that had to be supplemented by up-to-the-minute aerial reconnaissance of an opponent's positions and moves

On 20 May 1941 Operation Mekur began, with German parachute and glider troops under Major-General Kurt Student landing in 70 gliders and from 550 aircraft to seize the airfields at Maleme, Retimo and Heraklion. By 31 May Crete had fallen. Despite being warned of German plans by intelligence from Ultra (deciphered German intercepts) material, the British defence was poorly prepared and command proved inadequate: much of the garrison was short of equipment, especially artillery, and lacking in air support. The German success in seizing and holding Maleme airfield provided a bridgehead for reinforcements by means of transport aircraft. The airborne aspect of the attack succeeded and led to German victory over a poorly commanded resistance. In invading Norway in 1940, German airborne assaults had also been crucial to success.

In contrast, British naval superiority, which had already enabled the evacuation of 43,000 troops from mainland Greece, meant the Germans were unable to support their Crete operation by moving troops by sea. Two German convoys doing so were successfully intercepted. Nevertheless, German air attacks hit the British attempt to reinforce, supply and, eventually, evacuate Crete by sea, yet again demonstrating the vulnerability of warships to dive bombing, and notably so in inshore waters where the room for manoeuvre was limited. Many warships, including three cruisers and six destroyers, were sunk, and despite the evacuation some troops were left behind and captured.

The conquest of Crete was the first time a major objective had been taken by massed airborne assault, and the entire operation raised important questions regarding the utility of air power. Admiral Cunningham, the British naval commander in the Mediterranean, observed: 'You cannot conduct military operations in modern warfare without air forces which will allow you at least to establish temporary air superiority.' The British feared that the Germans might press on to attack Cyprus, but Crete did not prove to be a strategic launching pad for the Germans. However, German losses (5,567 dead) were so great that Hitler ordered that no similar operation should be mounted in the future, limiting the range of future capability.

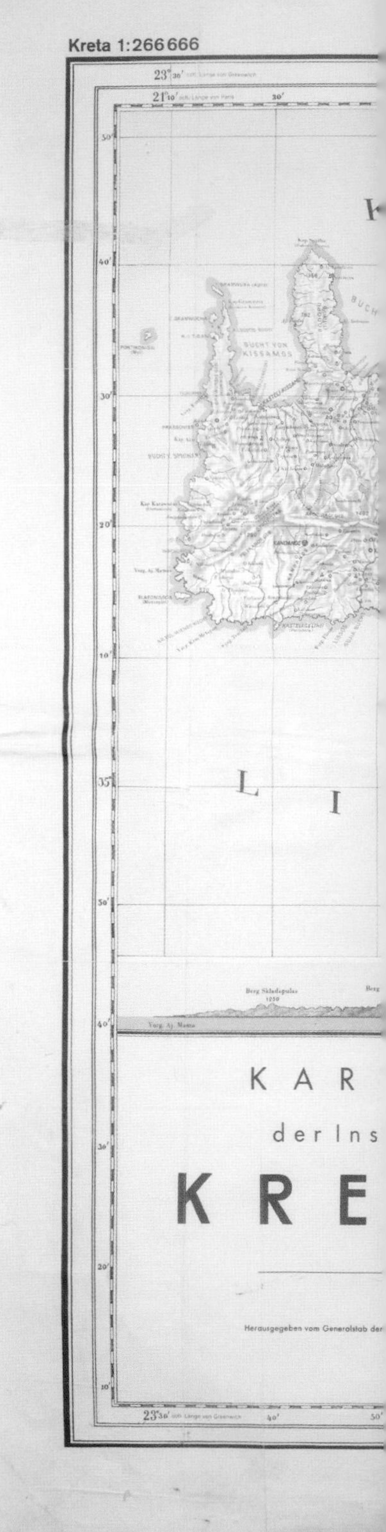

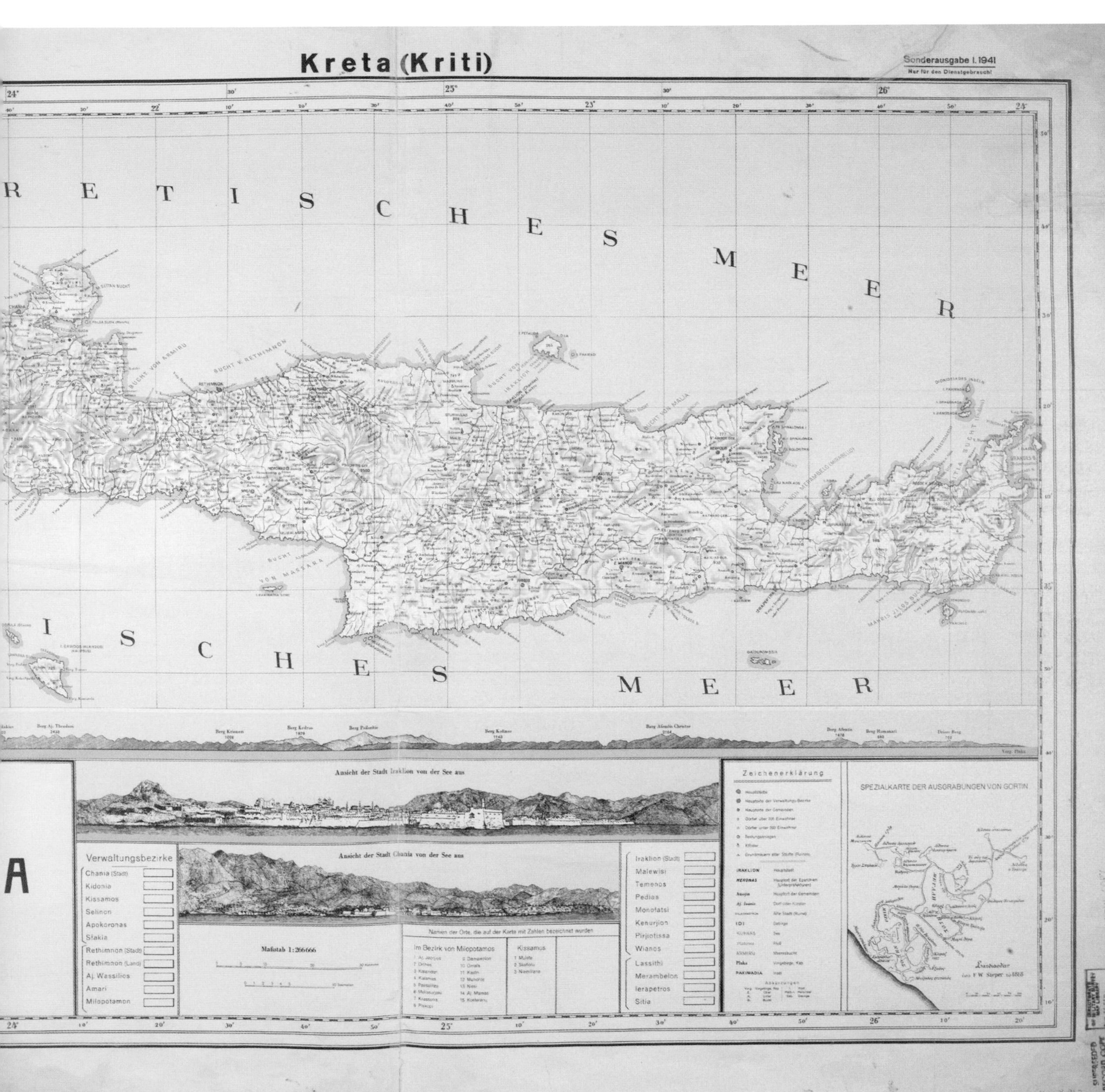

Kreta (Kriti)
Sonderausgabe I. 1941
Nur für den Dienstgebrauch!
R E T I S C H E S M E E R
I S C H E S M E E R
Ansicht der Stadt Iraklion von der See aus
Ansicht der Stadt Chania von der See aus
Verwaltungsbezirke
Chania (Stadt)
Kidonia
Kissamos
Selinon
Apokoronas
Sfakia
Rethimnon (Stadt)
Rethimnon (Land)
Aj. Wassilios
Amari
Milopotamon
Maßstab 1:266666
Namen der Orte, die auf der Karte mit Zahlen bezeichnet wurden
Im Bezirk von Milopotamos
Kissamos
Iraklion (Stadt)
Malewisi
Temenos
Pedias
Monofatsi
Kenurjion
Pirjiotissa
Wianos
Lassithi
Merambelon
Ierapetros
Sitia
Zeichenerklärung
SPEZIALKARTE DER AUSGRABUNGEN VON GORTIN
MAPS.X.942

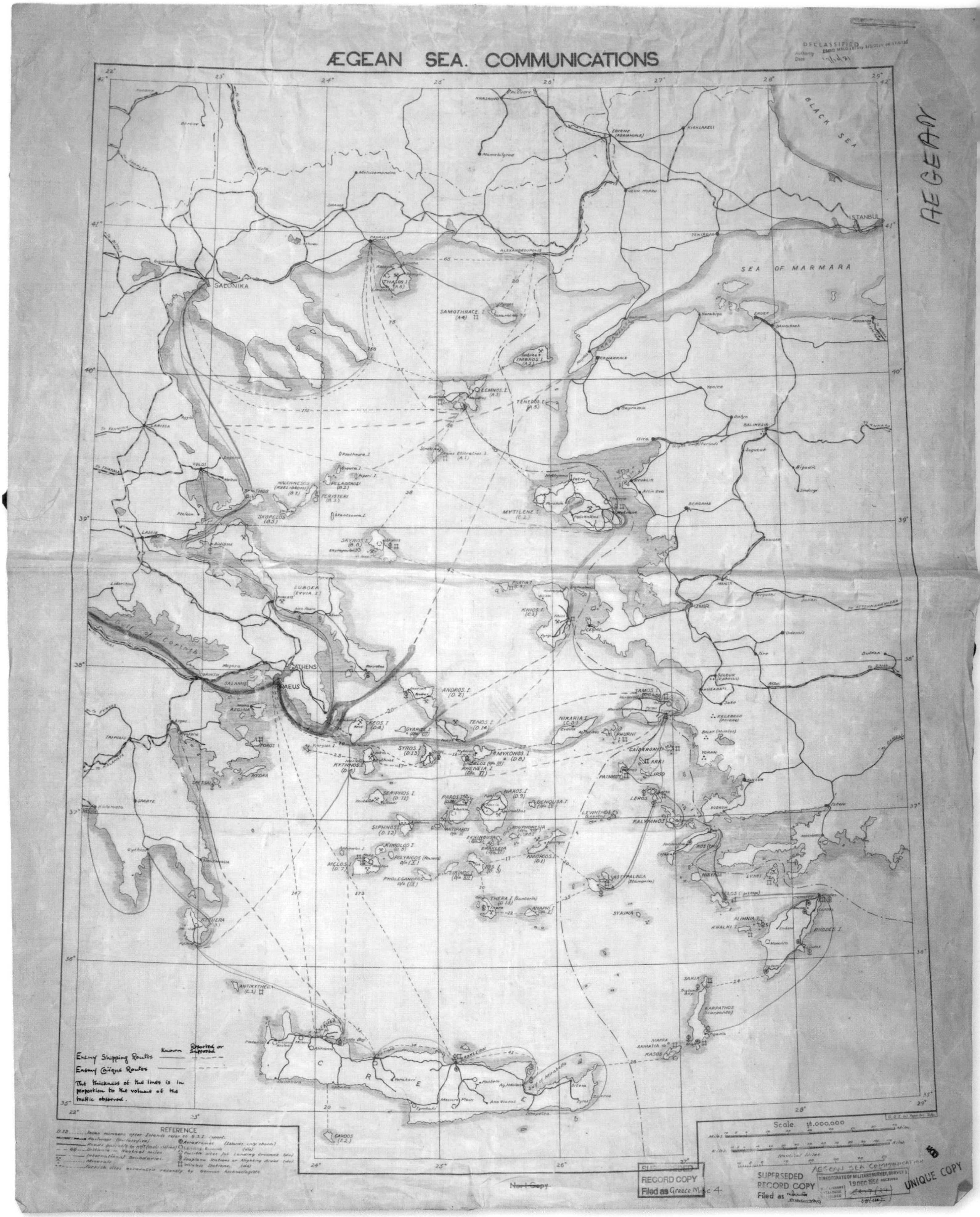

ÆGEAN SEA. COMMUNICATIONS

AEGEAN

BLACK SEA

SEA OF MARMARA

ISTANBUL

SALONIKA

KAVALLA

ALEXANDROUPOLIS

THASOS I.

SAMOTHRACE I. (A4)

IMBROS I.

LEMNOS I.

TENEDOS I. (A.5)

LARISSA

SKOPELOS (B.3)

SKYROS I.

MYTILENE I. (C.2)

EUBOEA (EVVIA I.)

KHIOS I. (C1)

IZMIR

Gulf of Corinth

ATHENS

PIRAEUS

SALAMIS

AEGINA

HYDRA

ANDROS I. (D.2)

TENOS I.

SAMOS I.

NIKARIA I.

KEOS I.

SYROS I. (D.13)

MYKONOS I.

KYTHNOS I.

SERIPHOS I. (D.11)

PAROS I.

NAXOS I. (D.9)

SIPHNOS I. (D.12)

KIMOLOS I.

MELOS I.

PHOLEGANDROS I.

AMORGOS I.

THERA I. (Santorin)

ANAPHI I.

PATMOS

LEROS

KALYMNOS

ASTYPALAEA

KOS

RHODES I.

KHALKI I.

ALIMNIA I.

SYRINA

SARIA

KARPATHOS (Scarpanto)

KASOS

KYTHERA

ANTIKYTHERA (E.1)

HERAKLEION

C R E T E

Gulf of Mirabella

GAVDOS (E.2)

Known — Reported or Supposed

Enemy Shipping Routes

Enemy Caïque Routes

The thickness of the lines is in proportion to the volume of the traffic observed.

REFERENCE

D.12 .... Index numbers after Islands refer to G.S.I. report.

Railways (Unclassified)

Roads passable to M/T (Unclassified)

— — 40 — — Distance in Nautical miles

International Boundaries.

Minerals

Turkish sites excavated recently by German Archaeologists

Aerodromes (Islands only shown)

Landing Grounds (do)

Possible sites for Landing Grounds (do)

Seaplane Stations or Alighting Areas (do)

Wireless Stations. (do)

Scale. 1/1.000.000

Miles

Kilos.

Nautical Miles.

THE AXIS NETWORK IN THE AEGEAN

British Military Intelligence, 'Aegean Sea Communications', 1941.

With the thicknesses of the lines for the routes said to be in proportion to the volume of the traffic observed, this British Military Intelligence map seeks to understand the dynamics of the Axis system in the Aegean, modern consideration of which tends to focus solely on the German conquest of Crete in 1941. In practice, the situation was far more complex, with the Italians in control of the Dodecanese Islands, most prominently Rhodes, from which they also attacked Crete in May 1941. Moreover, the Germans advanced in the Aegean by using ships they seized, capturing the islands there – notably Thasos, Samothrace, Lemnos, Lesbos and Chios – against minimal resistance, especially in the latter stages of the campaign.

Once the Axis forces were established in the Aegean, the British became interested in it as a potential area to be attacked, by air and/or sea, as well as one where Greek Resistance activity could be supported. This map presumably drew on information provided by the Greek Resistance as well as from British agents based in neutral Turkey. The map identifies railways, roads, aerodromes, seaplane stations, wireless stations and landing grounds, as well as potential sites for landings; it also notes locations where there are minerals.

'Concern' about a possible invasion of Greece, an impression deliberately encouraged by Allied deception designed to distract attention from preparations against Sicily, led the Germans to reinforce their forces there in June and July 1943. The invasion did not come, but the strength of the German position helped ensure that the British were unsuccessful in the Aegean when they tried to exploit the armistice with the new Italian government after it was announced on 8 September. This attempt reflected Churchill's longstanding interest in the region, dating to the First World War and the Gallipoli Campaign of 1915, and his hope in the Second World War of bringing Turkey into the conflict. The Americans sensibly advised against the plan. In the event, a swift German response and the inability of the British to secure air superiority led to a serious failure in the Dodecanese Islands campaign, with the loss of newly inserted British garrisons on Cos (3–4 October) and Leros (12–16 November) as well as of supporting warships. This failure indicated the continued importance of air power, and formed a striking contrast with American successes in the Pacific: there, island targets were isolated before they were attacked. In the event, Turkey entered the war only in its closing weeks, making scant difference.

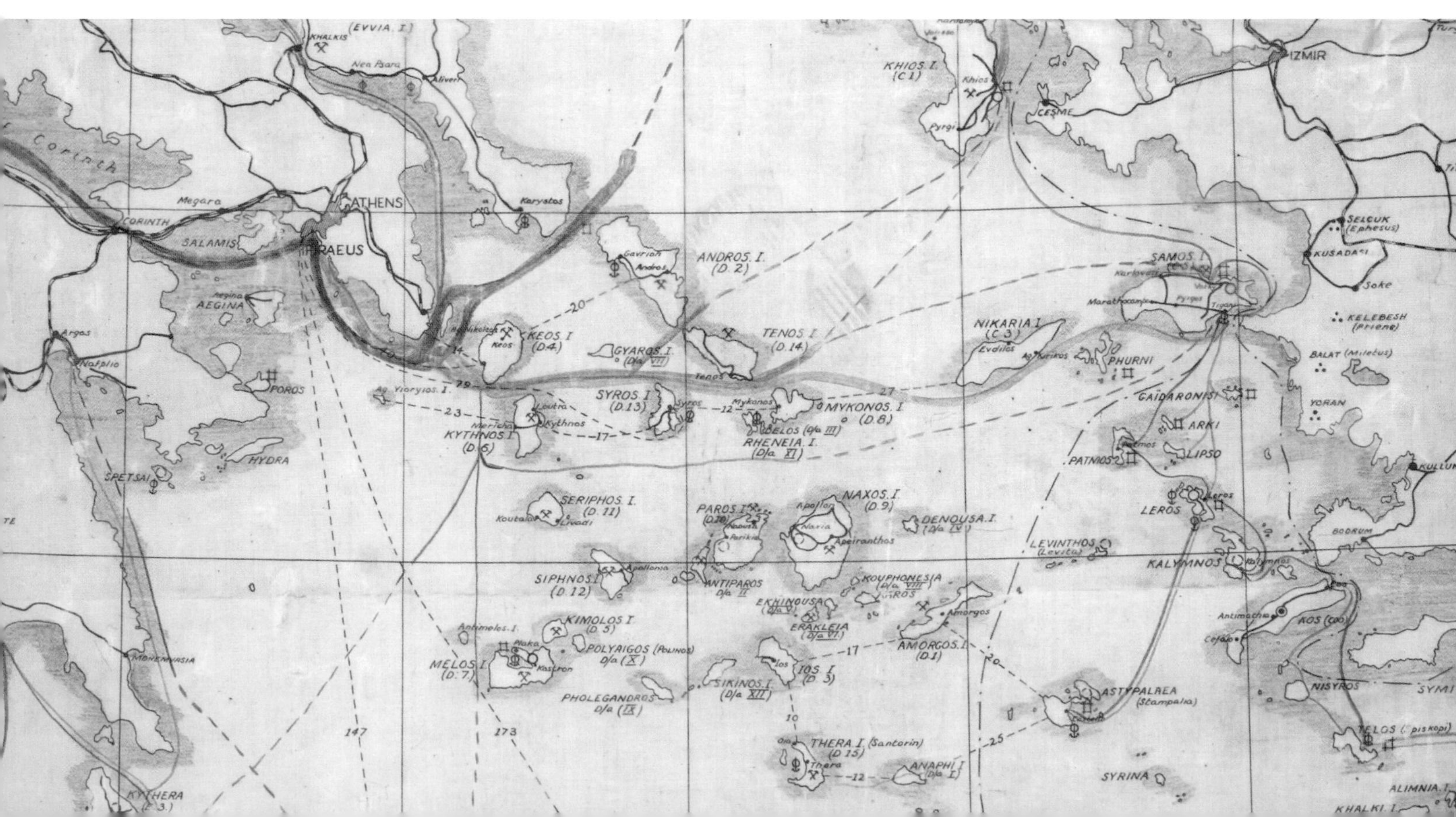

FAILURE ON THE VOLGA

German General Staff, map of the southern part of Stalingrad, 1942.

This map shows the Soviet defences at Stalingrad, including artillery positions on the other side of the River Volga (marked 'Wolga'). Strategic failure in the shape of the simultaneous pursuit of very different goals had greatly handicapped German operations against the Soviet Union in 1941, and this was repeated in 1942. Operation Blue focused on an advance into the Caucasus in order to capture the Soviet oilfields and then put pressure on Allied interests in the Near and Middle East. The oilfields were seen as a preparation for the lengthy struggle that the United States' entry into the war appeared to make inevitable, and Hitler expanded the original objective in order to seize all the oilfields, including Baku. Despite Germany's synthetic oil production domestically and its control of the limited production in Austria, Hungary, Poland and, most significantly, Romania, Germany was very short of oil.

The 1st Panzer Army was instructed as part of Army Group A to overrun the oil-producing areas, and the 40th Panzer Corps crossed the River Don on 20 July 1942, creating a bridgehead at Nikolaevskaya, before advancing south rapidly across the Kalmyk Steppe and beating off a Soviet counter-attack by two tank brigades at Martynovska. Stavropol was captured on 3 August while the most westerly oilfields round Maikop were captured on 9 August, by the 3rd Panzer Corps and 57th Panzer Corps, and the Germans advanced to within 19 miles (30 kilometres) of Grozny, but logistical problems mounted as supply lines lengthened, while resistance was also a problem. The practicalities of an operational space the size of the Soviet Union was not adequately grasped through maps using scales that made plotting moves too easy. In addition, German economic experts warned about the problems involved in deploying captured resources. Oil could not be as readily moved to the Reich as some assumed, or, indeed, within the conquered zone. Hitler understood the need to gain the Black Sea coast in order to be able to move the oil from the most westerly oilfields, but not the problems that would subsequently occur in moving sufficient oil. Moreover, the Soviets had become more effective in destroying sites threatened with capture, and oil installations were particularly easy to wreck. Those of Maikop were destroyed as were those, in a precautionary measure, at Grozny, while preparations began at Baku.

Hitler's ambition was even bolder, as the advance of Army Group South to the Caucasus was seen as a preparation for the destruction of the British position in the Middle East, which it was hoped would compensate for Britain's strength at sea and in the air. Victory on land over the Soviets and then the British was a strategy configured to the nature of the German military and to its planning system, one designed to counteract the problems caused by the German lack of a unified command structure and joint staff. Lebensraum, thus, would become an immediate as well as a long-term strategic goal.

Meanwhile, German planning became increasingly focused on what had been a flank position, the city of Stalingrad on the Volga, as Hitler became convinced that it had to be captured.

Despite a massive commitment of resources, the Germans were fought to a standstill at Stalingrad, which had been turned by their air and artillery attacks into an intractable urban wasteland that, in practice, offered major advantages to the Soviet defenders. The fighting there became attritional, and the German force was 'fixed'. Soviet losses in combat were heavy, but they helped to stop the Germans. The Soviets had mass, some good commanders and effective artillery – and the Red Army shot about 15,000 soldiers in order to stiffen morale.

On the pattern of the Germans attacking Verdun in 1916, Hitler hoped that Stalin would commit his forces to hold the city that bore his name and thus enable the Germans to destroy them, thereby achieving a result of strategic consequence that would avoid the need to wage comparable struggles against other cities. But this problematic approach undermined the overstretched Caucasus operation and was matched by poor command decisions at Stalingrad, which included a failure both to drive against the flanks of the Soviet position in the city, in an attempt to cut it off from the river, and to focus on the same targets throughout.

Maps of Stalingrad gave German commanders a misleading impression of their ability to strike through the city. In practice, individual urban complexes – especially the Tractor, Red October and Barricades factories, and the city's iconic grain silo and elevator – proved formidable, well-defended obstacles, their situation enhanced by the use of reinforced concrete for the construction of industrial plants and other installations.

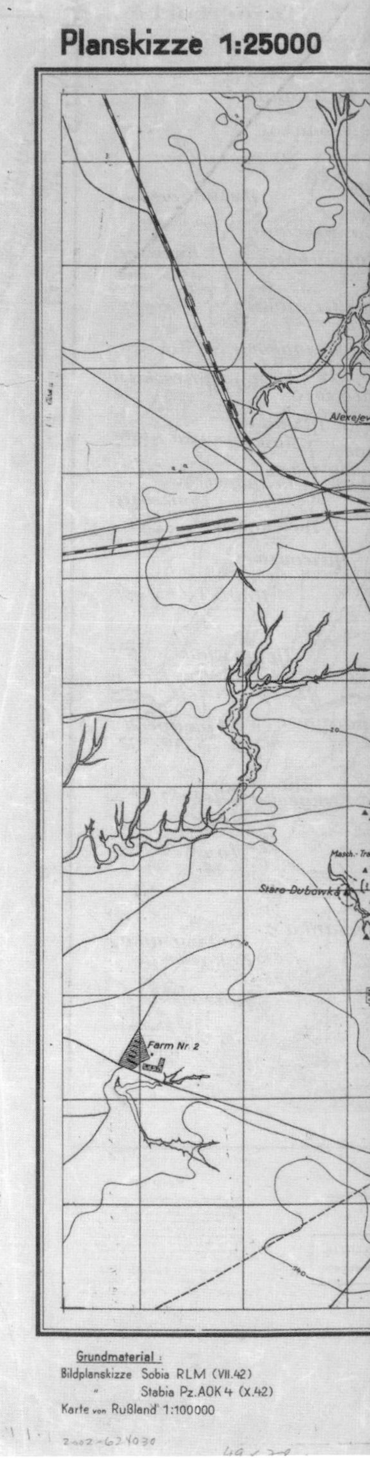

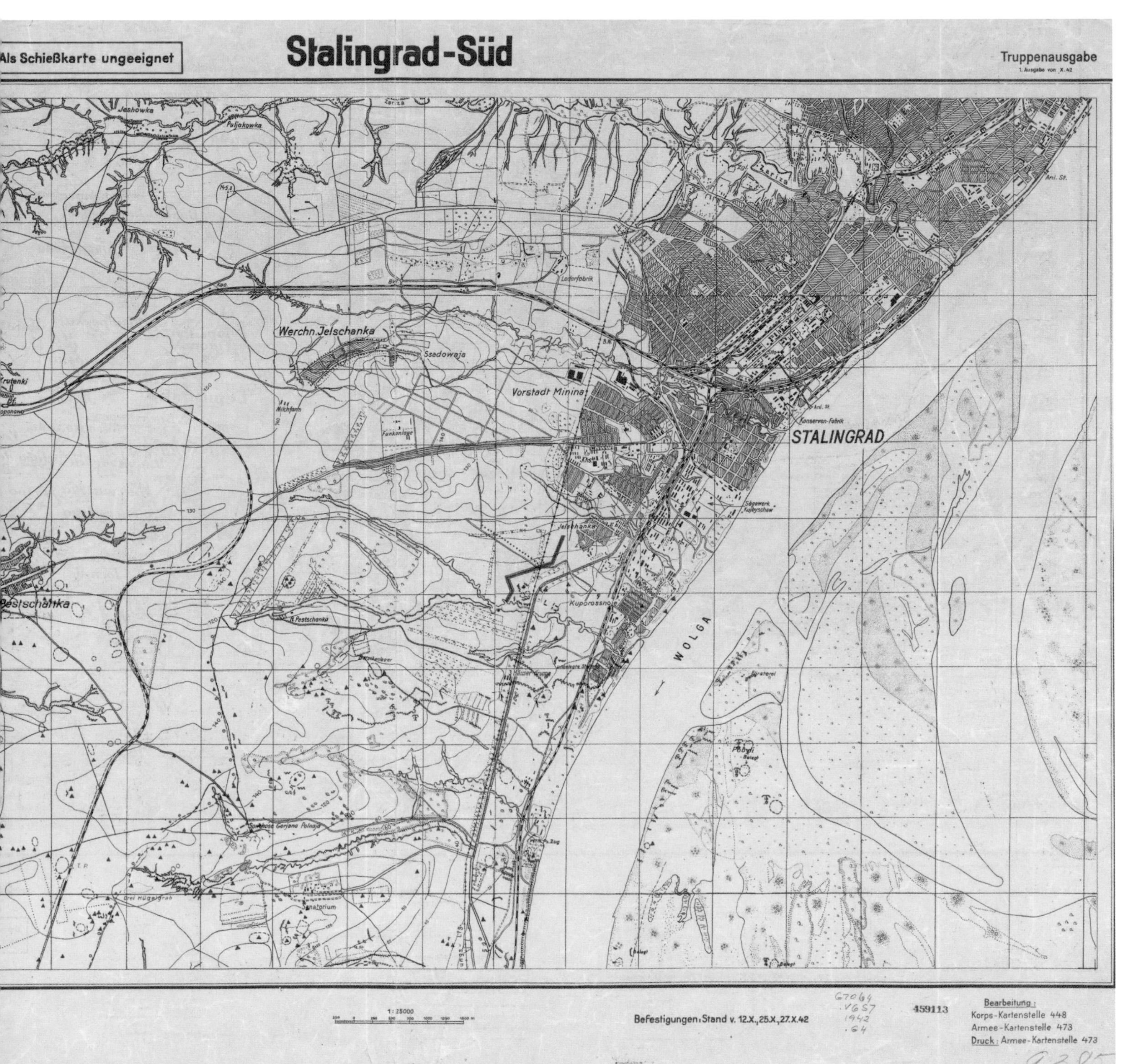
Als Schießkarte ungeeignet
Stalingrad-Süd
Truppenausgabe
Jeshowka
Puljakowka
Lederfabrik
Werchn. Jelschanka
Ssadowaja
Vorstadt Minina
Konserven-Fabrik
STALINGRAD
Jelschanka
Kuporossnoj
WOLGA
Pestschanka
Sanatorium
1:25000
Befestigungen, Stand v. 12.X., 25.X., 27.X.42
459113
Bearbeitung:
Korps-Kartenstelle 448
Armee-Kartenstelle 473
Druck: Armee-Kartenstelle 473

TREACHEROUS TERRAIN

Royal Australian Survey Corps, map of the area around Buna in New Guinea, 1942.

In July 1942 the Japanese, advancing from New Britain, had landed at Buna, Gona and Sananananda on the northeastern coast of Papua New Guinea with the intention of pushing southward across the peninsula to Port Moresby. This map was produced by the Royal Australian Survey Corps (RASC) from a variety of sources and is from the four-miles-to-the-inch 'Strategical Series' for New Guinea. It has an additional reliability diagram, and a note that the map is 'not the result of Military Surveys'. Instead, responding to the military directive of 13 July that photomaps be produced of the Buna and Kokoda areas, the surveyors had rapidly gathered sketch maps, civilian oil surveys, missionary maps and hydrographic charts to try to compile a resource, however inaccurate, that might be of some use during the defence of Port Moresby. As intelligence from aerial photography became available it was used to revise the series.

The coastal area was most reliable and the height was shown by hachures, not contours. In addition, the Royal Australian Navy's Hydrographic Service produced relevant charts.

Amphibious operations were to be crucial in attacking the Japanese on the northern coast of New Guinea, thus avoiding the problems of advancing overland and supplying troops through intractable terrain. However, in the autumn of 1942 the focus was on the Japanese base at Buna, and this was to be attacked both by American troops who had been brought by sea or airlifted, and by Australian and American troops that had advanced overland. The deep-sea route to the east and north of the Trobriand or Kiriwina Islands was overly exposed to Japanese aircraft from Rabaul, and it was therefore necessary to consider inshore routes, which was done by the Hydrographic Service. The ships, two requisitioned trawlers, survived a Japanese bombing attack.

In November, the advancing Australians focused on Gona and Sanananda Point, and the Americans on Buna and Cape Endaiadere. The capture of these positions was seen as necessary in order to consolidate the failure of the Japanese advance on Port Moresby overland via the Kokoda Trail, and prevent it being repeated.

In turn, the American and Australian advances also faced heavy losses, notably from disease (especially malaria and fevers), as well as serious logistical issues. Buna village and Buna Mission were attacked from 19 November, but only fell on 14 December 1942 and 2 January 1943 respectively. There were also problems over command and fighting quality. General Thomas Blamey, the Commander-in-Chief of the Australian Military Forces and Commander of Allied Land Forces under General Douglas MacArthur, Commander-in-Chief, Southwest Pacific Area, reported:

> *'I had hoped that our strategical plans would have been crowned with complete and rapid success in the tactical field. It was completely successful strategically in as much as we brought an American division on to Buna and an Australian division on to Gona simultaneously. But in the tactical field ... it was a very sorry story ... the American troops cannot be classified as attack troops ... [and] from the moment they met opposition sat down and have hardly gone forward a yard. The action, too, has revealed a very alarming state of weakness in their staff system and in their war psychology ... the American forces, which have been expanded even more rapidly than our own were in the first years of the war, will not attain any high standard of training or war spirit for many months to come.'*

He also noted the difficulty of the terrain: 'Throughout the region the tangled undergrowth, broken ground and ever prevailing damp made movement off the track difficult and arduous in the extreme.... Beyond Nauro the track is inches deep in mud while whole leeches abound everywhere.' After the fall of Buna Mission, and Cape Endaiadere the same day, it took several more weeks of bitter fighting to mop up and secure the beachhead, where there were countless bunkers of Japanese troops, many of whom fought to the death.

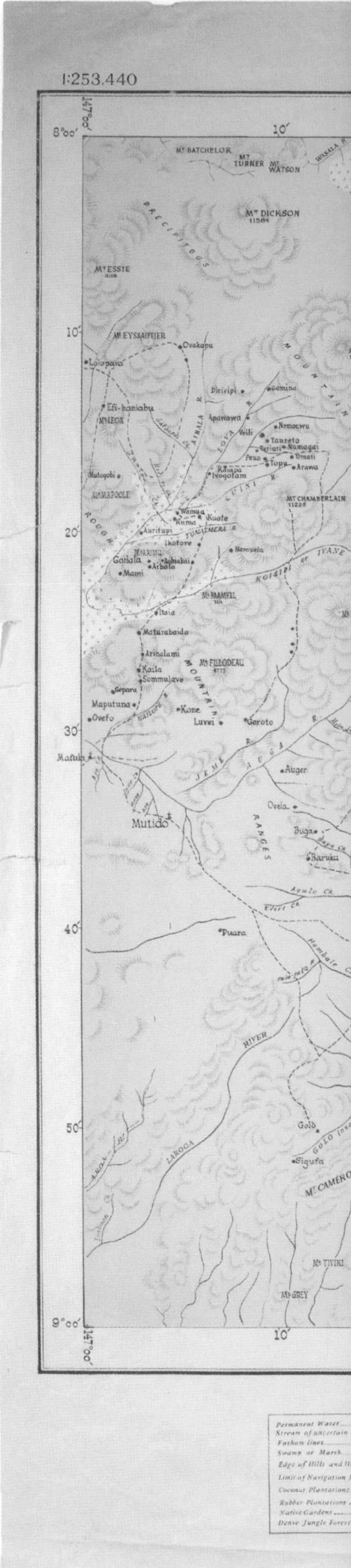

BUNA
NEW GUINEA
REFER TO THIS MAP AS 4 MILE STRATEGICAL SERIES N.G. BUNA FIRST EDITION
PACIFIC
OCEAN
HOLNICOTE BAY
MAMBARE BAY
KOKODA
BUNA
Scale 4 Miles to 1 Inch
INDEX TO ADJOINING SHEETS
RELIABILITY DIAGRAM
This Map is Compiled from Numerous Plans and Notes, from various sources, and is not the result of Military Surveys.
THIS MAP IS AN OFFICIAL DOCUMENT. IF FOUND, IT MUST BE HANDED INTO THE NEAREST MILITARY HEADQUARTERS OR POLICE STATION.
89885 (27.)

MOMENTUM IN THE DESERT

A British copy of Field Marshal Erwin Rommel's annotated map of El Alamein and a counter-battery chart, 23 October 1942.

These differing views of the same situation in part reflect the contrasting requirements of defence and attack. The map referred to as Rommel's personal map (opposite) was one of defence, but the 'C.B. Chart' (drawn up later by the Royal Engineers in January 1943) is given dynamism by being part of a plan to attack the Axis forces on the night of 23–24 October, which was when the Allied offensive began. Covering the northern part of the battlefield, the map (see page 80) presents both sides' FDLs (Forward Defended Localities), locations of flash spotting ('F.S.') posts and sound ranging ('S.R.') bases, the advance routes of the individual Allied divisions and the artillery situation, for the key element here is counter-battery ('C.B.') fire in order to suppress Axis artillery. In this, the Allies proved successful.

Field Marshal Rommel, the commander of the Afrika Korps, the German force sent to North Africa in February 1941 to help Germany's Italian allies, certainly thought that he would receive more troops and material, and that Hitler had approved offensive plans beyond those of the official Oberkommando des Heeres (OKH), or High Command, orders to him. Uninformed about the Operation Barbarossa plan for an invasion of the Soviet Union, Rommel mistakenly believed that he would receive more divisions as soon as the German campaign against Greece, launched in April 1941, was over. Rommel's offensive vigour and drive therefore appears justified from what he knew about the strategic situation.

This point underlines the degree to which it could help for the commanders of major theatres to be involved in the strategic situation, in order for them to be able to shape their operational plans accordingly. Rommel was not, because as far as the OKH was concerned, he had been given his orders. But from the incomplete knowledge of Germany's strategic plans, and possibly due to Hitler's encouragement, Rommel was convinced that he could deal Britain a serious, maybe fatal, blow by acting against his orders and attacking. Rommel's selfish personality also played a role. On the Allied side, commanders who acted contrary to their orders, tended to be replaced, or were given the necessary reason to change their mind. In contrast, although Rommel might be an extreme example, he was an instance of the process by which Hitler, inherently an opportunist, allowed the war to expand unnecessarily.

Rapidly driving the British back into Egypt in early April 1941, Rommel proved a bold and effective commander of the German forces, while British tank–infantry–artillery co-operation was inadequate, and, in tank-on-tank combat, the outgunned British tanks were poor. Using the British technique, as applied against the Italians, of outflanking their less mobile opponent with their tank units through the desert, the Germans proved more successful in combined arms capability and tactics, and benefitted from the intelligence gained by radio intercepts. Moreover, air support for the British was not well organised in this period of the war. German successes challenged Allied morale, which helped focus the wider significance of tank combat in North Africa. In the *Sunday Pictorial*, a British newspaper, of 27 April 1941, J.F.C. Fuller commented on the Germans driving back the British forces in North Africa: 'Like a ladder in a girl's stocking, our splendid desert campaign is running backwards up our strategical leg from its ankle to its knee.' The British forces in North Africa had been seriously depleted by the requirement to send troops to Greece and Syria.

British weaknesses made Rommel's task a lot easier without detracting from his bold and imaginative tactical use of his armour. Several logistical innovations made the use of German tanks more effective: tank transporters were employed to take their vehicles to the battle area, which made their maintenance a lot easier; and they also had a highly effective tank recovery and repair organisation, which ensured that many tanks that had been disabled were quickly returned to frontline service. The British lagged the Germans in both of these areas, and the majority of British tanks were notoriously unreliable, which compounded the situation.

The German advance into Egypt in 1942 was checked at El Alamein, about 60 miles (96 kilometres) west of Alexandria, on 1–4 July and at Alam Halfa from 30 August through to 7 September. In part, this reflected the impact of a lack of fuel for any German deep-flanking operation, which for example immobilised the 21st Panzer Division on 30 June when Rommel had originally planned to attack. The British benefitted from new six-pounder anti-tank guns, notably when resisting attack at Deirel Shein on 1 July. These were a major improvement on the two-pounders.

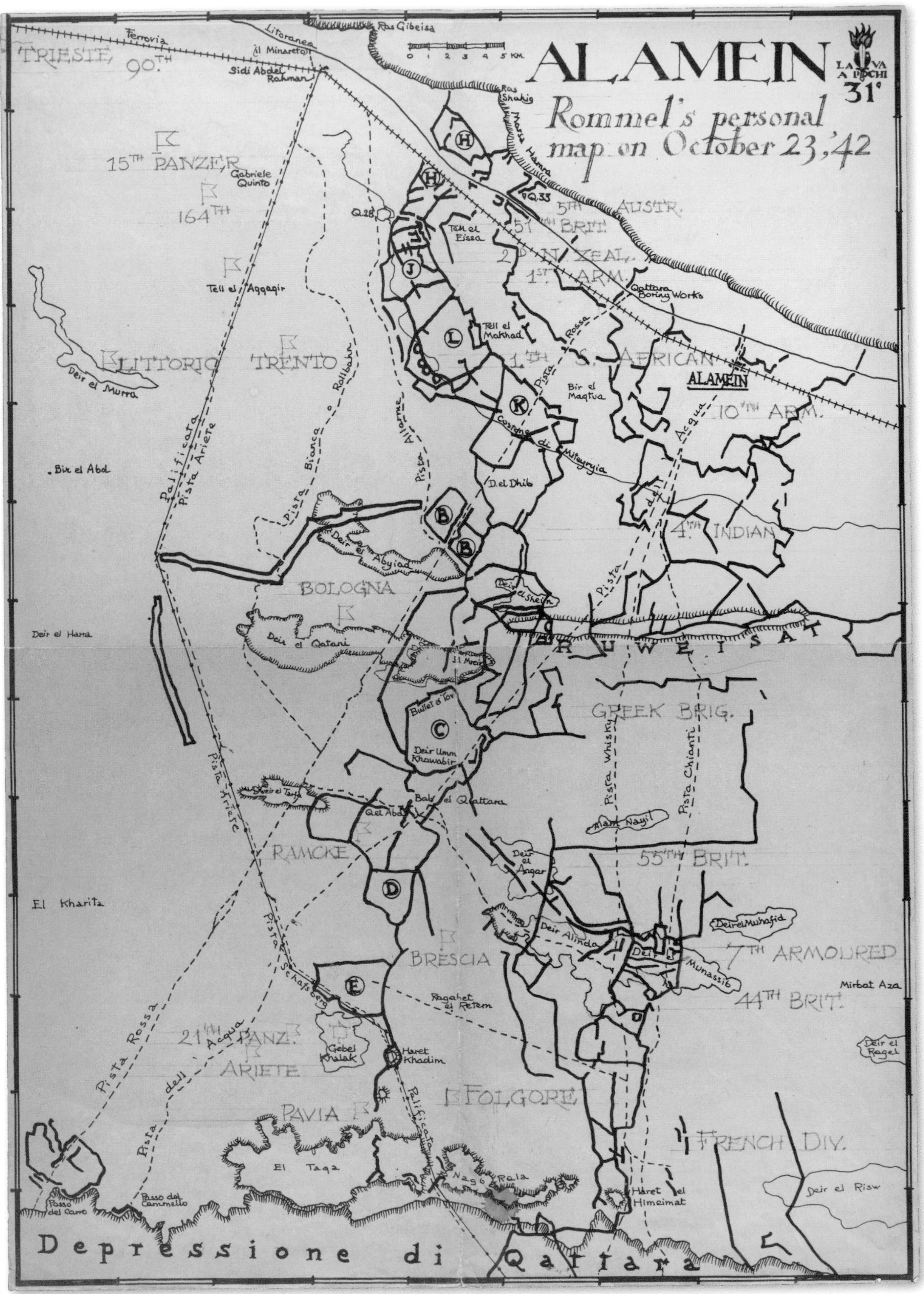

ALAMEIN
LA VA A POCHI
31°
Rommel's personal map on October 23,'42
0 1 2 3 4 5 KM.
Ferrovia
Litoranea
Ras Gibeisa
il Minareto
Sidi Abdel Rahman
TRIESTE 90TH
Ras Shahig
15TH PANZER
Gabriele Quinto
164TH
Q.28
Q.33
5TH AUSTR.
51ST BRIT.
Tell el Eissa
2D N. ZEAL.
1ST ARM.
Tell el Aqqaqir
Qattara Boring Works
Tell el Makhad
LITTORIO
TRENTO
1ST S. AFRICAN
Deir el Murra
Pista Rossa
Bir el Maqtua
ALAMEIN
10TH ARM.
Costone di Miteyriya
Rollbahn
Pista Bianca
Pista Alleme
Palificata
Pista Ariete
Bir el Abd
D. el Dhib
4TH INDIAN
Deir el Abyiad
BOLOGNA
Deir el Shein
Pista di Acqua
Pista
Deir el Harra
Deir el Qatani
RUWEISAT
31 Mrei
GREEK BRIG.
Bir el Tov
Deir Umm Khawabir
Pista Whisky
Pista Chianti
Pista Ariete
Bab el Qattara
Qat Abd
Alam Nayil
RAMCKE
Deir el Angar
55TH BRIT.
El Kharita
Deir el Muhafid
Deir Alinda
Deir Munassib
BRESCIA
7TH ARMOURED
Mirbat Aza
Pista Schaffers
Ragahet el Retem
44TH BRIT.
Pista Rossa
21ST PANZ.
Pista dell' Acqua
Gebel Khalak
Haret Khadim
Deir el Ragel
ARIETE
FOLGORE
Palificata
PAVIA
FRENCH DIV.
El Taqa
Nag Rala
Haret el Himeimat
Deir el Risw
Passo del Carro
Passo del Cammello
Depressione di Qattara

SECR

EL ALAME
C. B. CHAR

- FIELD
- MEDIUM
- HEAVY
- A.A.& D.D.
- No. of TROOPS
- F.S. POSTS
- S.R. BASE
- OWN FDL
- ENEMY FDL

NIGHT 23/24O

ARABS GULF

90 LT. DIV.

THREE ITALO-GERMAN ARMOURED BATTLE GPS

REGT 164 DIV.

REGT TRENTO

9 AUST. DIV.

51 (H) DIV.

REGT 164 DIV.

N.Z. DIV.

REGT TRENTO

1 S.A. DIV.

REGT 164 DIV.

REGT BOLOGNA

VON DER HEYDTE

4 IND. DIV.

7 MED 27

7 MED 426

64 MED 211

2/8 RAA

2/12 RAA

2/7 RAA

64 MED 212

127 FD

126 FD

128 FD

Mersa el Hamra

Ras el Shaqiq

El Alamein Station

Ruweisat Ridge

Deir el Qatani

M.D.R. Misc./2245

Drawn & reproduced by 524 Pal. C. Fd. Survey Coy. from tracings by Counter Battery Pool R.A., R.A. Base

MOD

In addition, British operational command and tactics had now improved. There was a readiness to engage in mobile warfare, making effective strikes in combination with the holding of defensive positions. This was an aspect of a more general qualitative transition in the British Army, as it came to be better prepared to take the offensive against the Germans. At the same time, Lieutenant-General Montgomery judged the British and Allied forces not yet ready for a successful offensive. Their tactical grasp of combined arms combat was limited. On 15 July, the British armour did not act to protect the New Zealand troops that had seized Ruweisat Ridge and then had to face, with few anti-tank guns, German tank counter-attacks. The commander of the 5th New Zealand Infantry Brigade noted the disorientating impact of the German tank movements, and the need, in response: '[to carry] Wrigley's grenades (sticky bombs) .... Towards morning the tanks seemed to form up in lines on either side of the main axis of our advance ... enabled them to use cross-fire.'

At Alam Halfa, later in the summer, Montgomery relied on anti-tank guns, a technique learned from Rommel, and inflicted serious losses on the attacking German armour. In contrast, the British tanks took defensive positions and were not launched in a follow-up attack. Earlier, attacking in Operation Splendour on 22 July, the 23rd Armoured Brigade had incurred heavy losses. Poor armour–infantry coordination and inadequate tanks were serious problems for the British. As a result, Montgomery refused, despite intense pressure from Churchill, to attack until the Eighth Army was ready and had built up the adequate reserve judged necessary for sustaining any attack.

The British launched a full-scale attack in the final battle of El Alamein from 23 October to 4 November 1942. They faced prepared positions, defended by extensive minefields and well-located anti-tank guns, and supported by armour. Greater British familiarity with combined arms tactics than hitherto was important, but it was only one of a range of factors contributing to British success, which included: skilful generalship; the availability of deciphered intelligence about German movements; larger numbers of men, artillery and tanks; better tanks; improved morale; effective use of artillery; air superiority and support; and attacks on Rommel's vital fuel supplies from Italy. Rommel also deployed his artillery and reserves poorly, and mishandled his Italian allies, as did most German commanders. Rommel, more generally, suffered from the understandable focus being given to German resources on the Eastern Front.

The shift in tank warfare was shown on 2 November when a German counter-attack led to heavy German tank losses. These broke the German–Italian armour, destroying the majority of the German tanks and most of the Italian units. By 3 November, Rommel had only 187 tanks left, and of these 155 were small Italian ones that were relatively ineffective – and certainly so against opposing Allied tanks. At the end of an attritional struggle, one in which, as in 1918, superior British artillery had been crucial, Rommel felt obliged to order a general withdrawal, leaving the 132nd Ariete Armoured Division in the rearguard. This Italian force, positioned south of 21st Panzer Division and backing up the 17th Pavia Infantry Division and 185th Folgore Paratroop Division, acted to cover the Axis disengagement. The Ariete, outgunned by the Allies, made a last stand on 3/4 November but was destroyed around Tell El Aqqaqir.

Although he had failed to recognise limitations adequately at the outset, including in the British armour, Montgomery had read the terrain ably and, alongside his adaptability and flexibility, his sequential blows eventually succeeded, not least by forcing Rommel to commit his forces, thus facilitating the decisive British blow. Montgomery's 'corps de chasse' provided the flexibility to change the main point of attack during the battle, and also the strength to maintain the momentum of the attack after the break with the Axis position. However, the initial progress had been slow, and Montgomery's ability to read the battlefield should not be exaggerated. Moreover, focused on the immediate battle, he proved poor at planning for the exploitation phase of the battle, although in part his target was removed by the rapid flight of the Germans, combined with the traffic congestion affecting the larger British forces.

ESSENTIAL WATER DATA

US Army Corps of Engineers, water map of Siena region, October 1943.

These maps offered data printed onto a standard base map and were based on a study by the United States Geological Survey (USGS) in September 1943 and compiled in October by the US Army Corps of Engineers in collaboration with the USGS. Water supplies were crucial to military operations, not only for supplying troops, but also for supporting equipment. The data, notably that offering the usual range from low to high water, and the ordinary maximum discharge of rivers, was crucial in terms of operational and tactical moves, notably with reference to the fordability of waters, a factor to both attackers and defenders, not least because it could force a focus on existing bridges and on bridging points where pontoon bridges could be constructed.

That these maps were provided reflects the rapidly developing situation in Italy. The conference at Washington, DC, in May 1943 had agreed on an invasion of Italy, but the pace of operations was speeded up in early September as the Germans speedily and successfully responded to Italy changing sides: an armistice with the new Italian government was signed on 3 September and announced five days later. Allied forces invaded Calabria from nearby Sicily on 3 September, meeting no resistance, while on 9 September an amphibious force seized Taranto and the main force was landed further north in the Gulf of Salerno, south of Naples. An invasion further north still was regarded as more risky due to greater exposure to German counter-attacks and airpower, as well as being at a greater distance from Allied bases.

At that stage, it was assumed that there would be a rapid advance northward. This proved totally mistaken, and thanks to a combination of poor intelligence, a flawed strategy and a degree of mission creep the Allies soon found themselves involved in a very different campaign, which involved attritional battles as they tried to break through the Gustav Line of defence. The situation was compounded by the terrain: steeply indented river valleys running off the Apennine spine provided superb defensive terrain for the Germans who proved very adroit at exploiting the battlefield opportunities. Rainfall contributed to the impasse in the autumn of 1943 and the subsequent winter, and this accentuated Allied interest in surface water conditions and the impact of the weather.

In the event the Allies did not advance into Tuscany, the area shown in this map, until after the

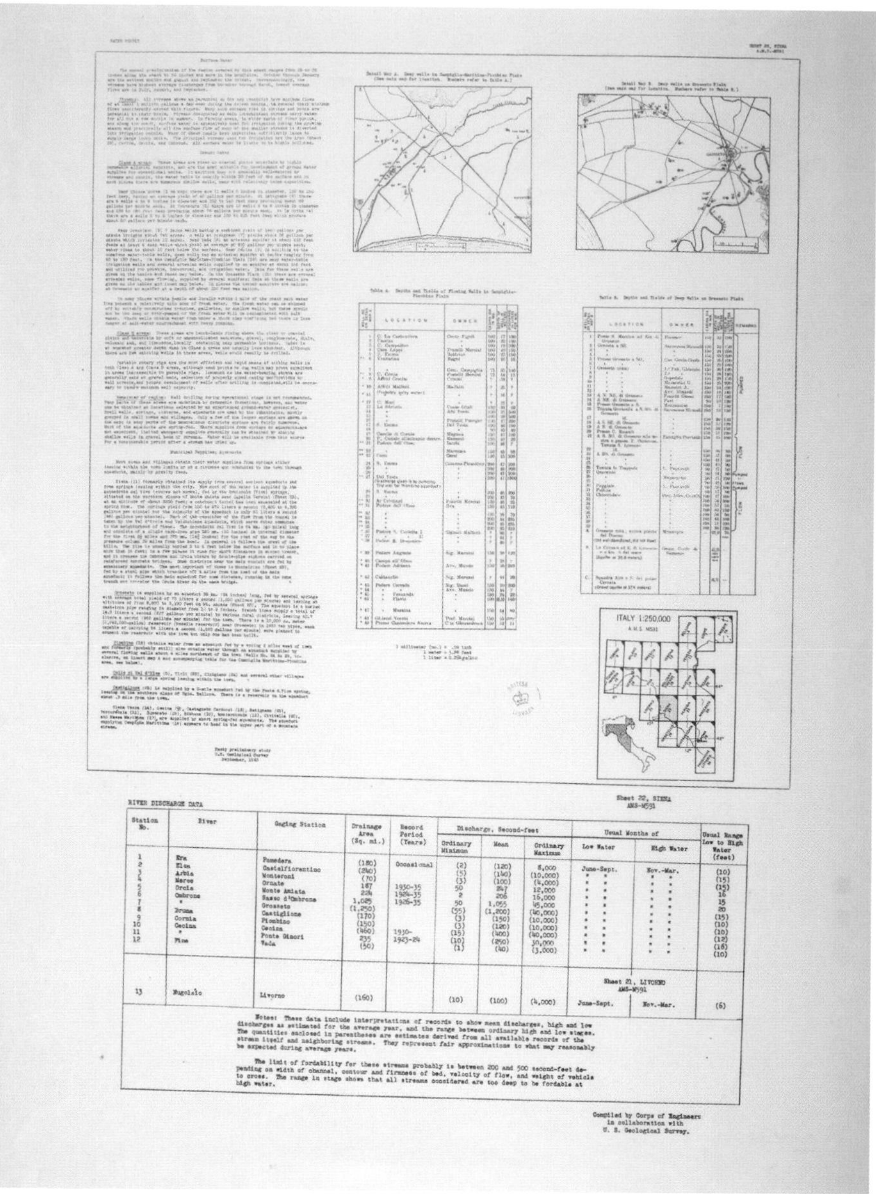

Gustav Line was broke in May 1944. The summer of 1944 saw the Allies progress into Tuscany, which, partly due to a focus on capturing Rome, failed to envelop the retreating Germans and they were able to consolidate a new line of defence. Piombino (see '18' on the map) on the coast west of Siena was captured on 26 June. On 5 July, Siena fell to North African troops of the French Expeditionary Corps, who were welcomed by a jubilant crowd. Heavy autumnal rain was, however, to swell rivers that September, again frustrating the Allied attacks, which were hindered by coming up against the Gothic Line. Overall, the Allies sustained high casualties in Italy and they were incurred for only limited strategic benefit.

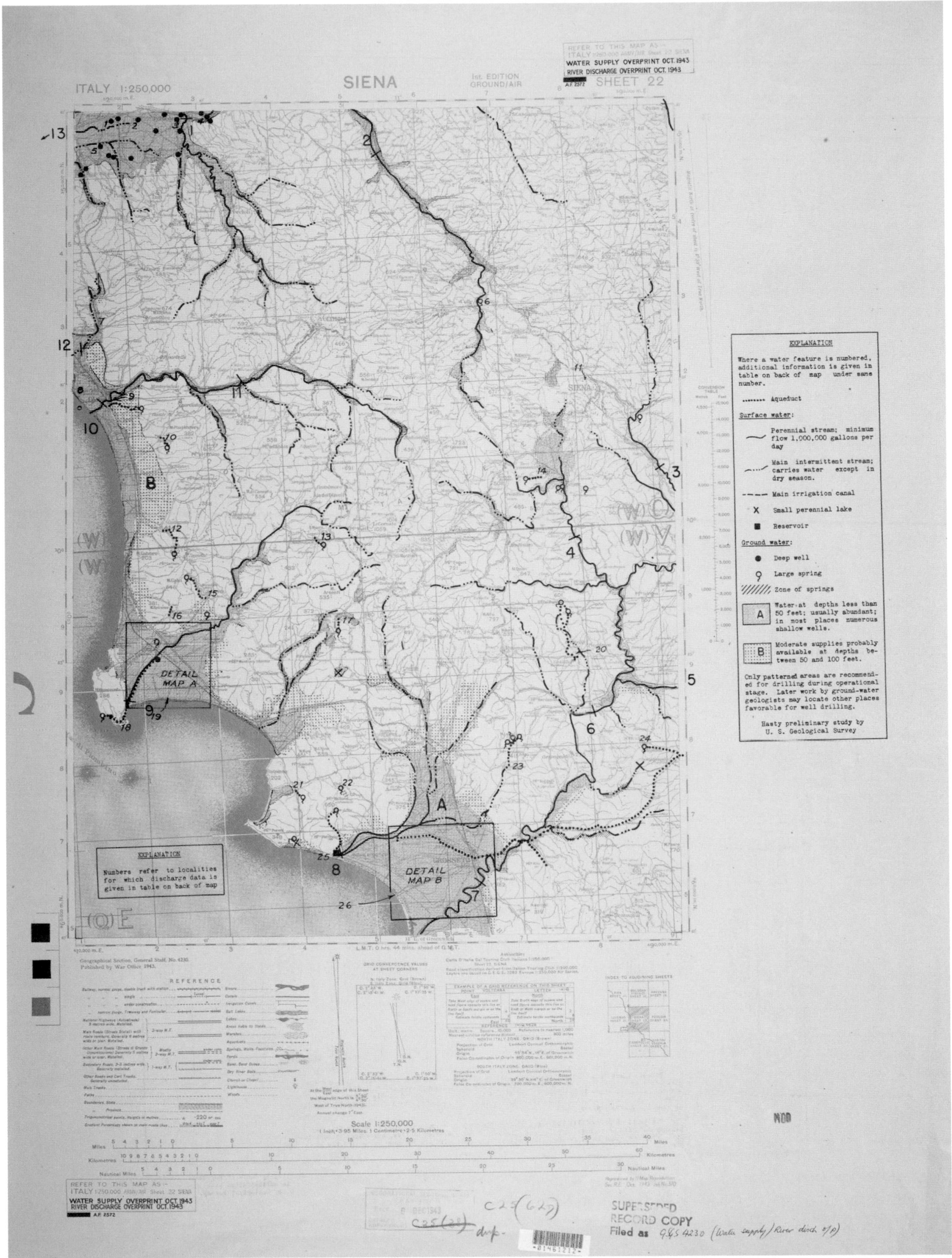

WATER SUPPLY OVERPRINT OCT. 1943
RIVER DISCHARGE OVERPRINT OCT. 1943
ITALY 1:250,000
SIENA
1st. EDITION
GROUND/AIR
SHEET 22
EXPLANATION
Where a water feature is numbered, additional information is given in table on back of map under same number.
Aqueduct
Surface water:
Perennial stream; minimum flow 1,000,000 gallons per day
Main intermittent stream; carries water except in dry season.
Main irrigation canal
Small perennial lake
Reservoir
Ground water:
Deep well
Large spring
Zone of springs
Water at depths less than 50 feet; usually abundant; in most places numerous shallow wells.
Moderate supplies probably available at depths between 50 and 100 feet.
Only patterned areas are recommended for drilling during operational stage. Later work by ground-water geologists may locate other places favorable for well drilling.
Hasty preliminary study by U. S. Geological Survey
DETAIL MAP A
DETAIL MAP B
EXPLANATION
Numbers refer to localities for which discharge data is given in table on back of map
Scale 1:250,000
WATER SUPPLY OVERPRINT OCT. 1943
RIVER DISCHARGE OVERPRINT OCT. 1943
SUPERSEDED
RECORD COPY
Filed as

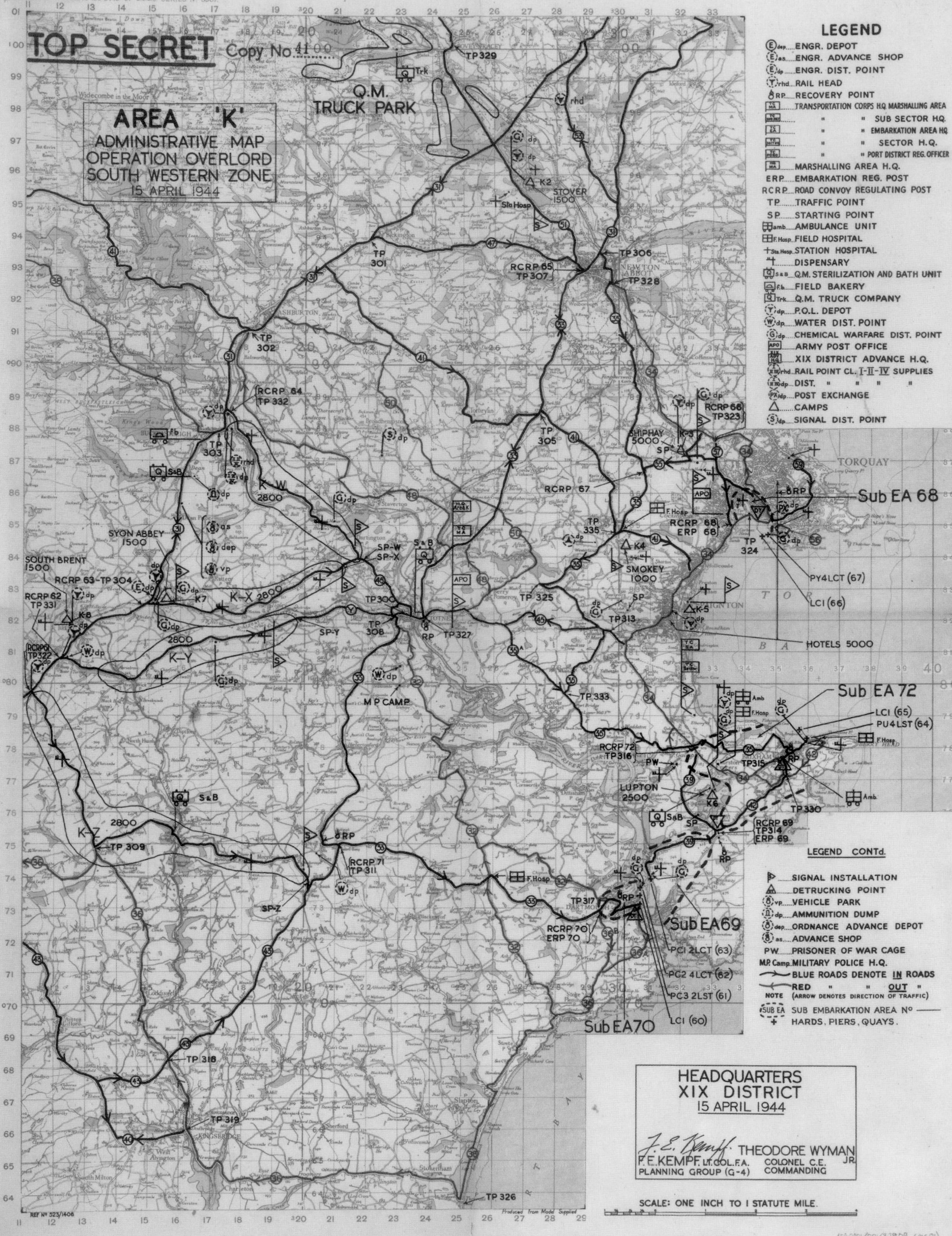

PTS. OF 1" SHEETS 138 & 145 OF G.S.G.S. SERIES No. 3907.
TOP SECRET
Copy No. 41
AREA 'K'
ADMINISTRATIVE MAP
OPERATION OVERLORD
SOUTH WESTERN ZONE
15 APRIL 1944
Q.M. TRUCK PARK
LEGEND
ENGR. DEPOT
ENGR. ADVANCE SHOP
ENGR. DIST. POINT
RAIL HEAD
RECOVERY POINT
TRANSPORTATION CORPS H.Q. MARSHALLING AREA
" SUB SECTOR H.Q.
" EMBARKATION AREA H.Q.
" SECTOR H.Q.
" PORT DISTRICT REG. OFFICER
MARSHALLING AREA H.Q.
ERP EMBARKATION REG. POST
RCRP ROAD CONVOY REGULATING POST
TP TRAFFIC POINT
SP STARTING POINT
AMBULANCE UNIT
FIELD HOSPITAL
STATION HOSPITAL
DISPENSARY
Q.M. STERILIZATION AND BATH UNIT
FIELD BAKERY
Q.M. TRUCK COMPANY
P.O.L. DEPOT
WATER DIST. POINT
CHEMICAL WARFARE DIST. POINT
ARMY POST OFFICE
XIX DISTRICT ADVANCE H.Q.
RAIL POINT CL. I-II-IV SUPPLIES
DIST. " " "
POST EXCHANGE
CAMPS
SIGNAL DIST. POINT
LEGEND CONTd.
SIGNAL INSTALLATION
DETRUCKING POINT
VEHICLE PARK
AMMUNITION DUMP
ORDNANCE ADVANCE DEPOT
ADVANCE SHOP
PW PRISONER OF WAR CAGE
M.P. Camp MILITARY POLICE H.Q.
BLUE ROADS DENOTE IN ROADS
RED " " OUT "
NOTE (ARROW DENOTES DIRECTION OF TRAFFIC)
SUB EMBARKATION AREA No.
HARDS, PIERS, QUAYS.
STOVER 1500
TP 306
RCRP 65 TP 307
TP 328
TP 301
TP 302
RCRP 64 TP 332
TP 303
K-W 2800
SYON ABBEY 1500
SOUTH BRENT 1500
RCRP 63-TP 304
RCRP 62 TP 331
K6
K7
K-X 2800
K-Y 2800
SP-W
SP-X
SP-Y
TP 300
TP 308
TP 327
M P CAMP
TP 305
SHIPHAY 5000
RCRP 66 TP 323
RCRP 67
TP 335
SMOKEY 1000
TP 325
TP 313
TORQUAY
Sub EA 68
RCRP 68 ERP 68
TP 324
PY4LCT (67)
LCI (66)
HOTELS 5000
TP 333
Sub EA 72
LCI (65)
PU4LST (64)
RCRP 72 TP 316
TP 315
LUPTON 2500
TP 330
RCRP 69 TP 314 ERP 69
RCRP 71 TP 311
K-Z 2800
TP 309
SP-Z
TP 317
RCRP 70 ERP 70
Sub EA 69
PC1 2LCT (63)
PC2 4LCT (62)
PC3 2LST (61)
LCI (60)
Sub EA 70
TP 318
TP 319
TP 326
HEADQUARTERS
XIX DISTRICT
15 APRIL 1944
F. E. Kempf
F. E. KEMPF LT. COL. F.A.
PLANNING GROUP (G-4)
THEODORE WYMAN JR.
COLONEL C.E.
COMMANDING
SCALE: ONE INCH TO 1 STATUTE MILE.
REF No. 523/1406
Produced from Model Supplied

PREPARING FOR THE ALLIED INVASION

Headquarters XIX District, Area 'K' Administrative Map, Operation Overlord, South Western Zone, 15 April 1944.

The extensive American preparations in southwestern England for Operation Overlord, the 1944 invasion of Normandy, are shown in this map, including roads that were coloured blue or red to identify them as 'in' and 'out' in order to ensure traffic flow. Allied air and naval superiority meant that German aerial reconnaissance of preparations was very limited, although Slapton Sands on the bottom of the map marked the site of a disastrous attack on an American landing exercise by nine German E-boats. In November 1943 villagers in Strete, East Allington, Stokenham, Blackawton and Slapton were told that the area from the eastern end of the bay at Blackpool Bay south to Beesands must be evacuated and they were ordered to relocate by 21 December. Montgomery had proposed the beaches in the area as being perfect for practising the planned landings. Unfortunately, on 27/28 April 1944 when a night practice was conducted in Exercise Tiger the Germans detected radio traffic and stumbled across the convoy waiting in Lyme Bay. More than 700 Americans died in the ensuing night-time chaos when several landing ships were torpedoed, and the following day several hundred more died from a 'friendly fire' artillery and naval bombardment in the live-firing exercise when they landed on the beach.

As with the 1943 invasion of Sicily, for which there were decoy exercises that suggested attacks on Sardinia or Greece, an Allied decoy exercise was important to the success of the invasion. In addition to hiding these preparations in southwestern England, there had been a concerted attempt underway since 1942 to persuade the Germans that the invasion would be mounted from southeastern England, which offered a shorter sea crossing to France and a shorter overland route to Germany. Normandy, in contrast, was easier to reach from invasion ports further west, notably Portsmouth and Plymouth. At the heart of this map is the River Dart in southern Devon, the location for the Royal Naval College at Dartmouth (which in December 1943 became the US Naval Advanced Amphibious Base) and one of the areas in which shipping was made ready for the invasion. The assault force for Utah Beach, the most westerly of the Normandy landing beaches, the US 4th Infantry Division, was prepared in Devon and sailed from departure points along the Devon coast. The resources present in the legend of this map indicate the folly of assuming that it should have been possible to invade in 1943 because it took time, effort and massive resources to prepare such an invasion force.

NAVIGATING OMAHA

Commander Task Force 122, maps of Omaha Beach, 21 April 1944.

These maps of the Normandy coast at St. Laurent-sur-Mer and Colleville-sur-Mer (right), and at Vierville-sur-Mer (overleaf), better known to the Allies as Omaha Beach (East and West, respectively), bear the classification 'BIGOT'. The origin of the acronym BIGOT is uncertain but some believe it to stand for British Invasion of German Occupied Territory. 'BIGOT' was the highest level of military security – above 'Top Secret'. Following the rehearsal disaster at Slapton Sands (see page 85), Eisenhower ensured that the bodies of all 10 missing men who had 'BIGOT'-level clearance were accounted for and had therefore not been captured, with the risk that information would leak.

Designed to help navigation, the maps include panoramic shoreline sketches that depict the beach as seen from water-level height, approximately 2,000 yards (1,800 metres) offshore, with the sea at about 16 feet (five metres) above low water. There are warnings to use all soundings and beach contours with care, and that building landmarks might be destroyed, making terrain features more reliable for visual navigation. The reverse side offers detailed beach gradients as well as a sunlight and moonlight table and information on inshore currents and tidal stages (see page 89).

The American tank attack faced problems. There were two American tank battalions (743rd and 741st), each of 48 tanks. The 741st Battalion, landing with the US 1st Infantry Division, lost 29 of its 32 swimming Duplex-Drive Shermans because they were launched too far offshore – 5,000 yards (over 4,500 metres), and in a sea with six-foot (1.8-metre) waves, and the crews therefore drowned. The rest of the 741st landed dry-shod, as did most of the 743rd, operating further west with the US 29th Infantry Division

The Germans had two 88mm guns at Omaha Beach. They were in fixed bunkers, and not mobile. One of the surviving tanks destroyed the emplacement on the western end of Omaha Beach. There were also mines and anti-tank ditches on the beach, and the latter had to be cleared by M4 tankdozers: 16 were scheduled to land in the early assault, but only six got ashore and five of those were knocked out. Once the beach was cleared, tanks were able to move inland from Omaha to help clear the town of Colleville-sur-Mer.

However, aside from Hitler's exaggeration of the scale and effectiveness of the Atlantic Wall defences, this strategy still left the initiative to the Allies,

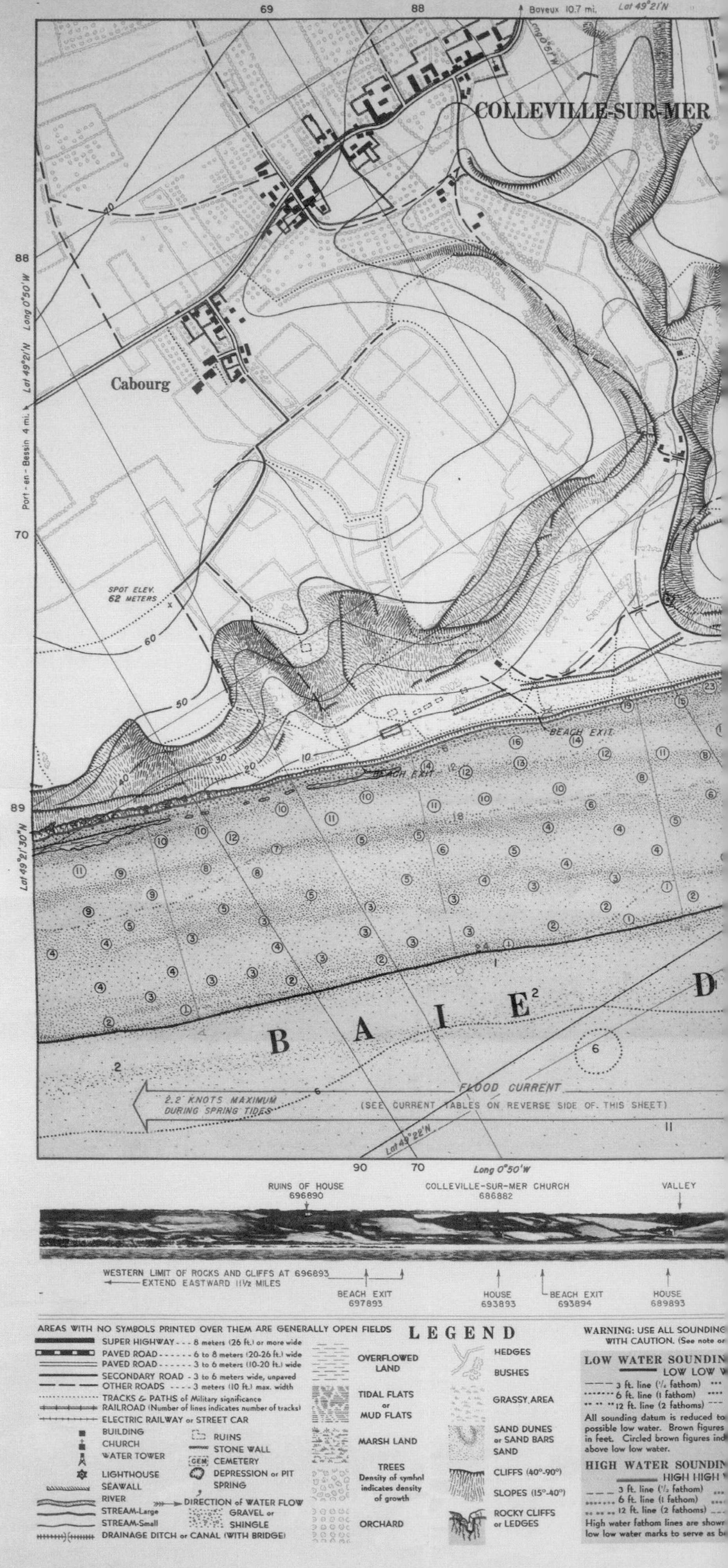

Long 0°52'W
Colleville-sur-Mer 1.7 mi.
Lat 49°21'30"N
Formigny 2.4 mi.
Long 0°53'W
St. Laurent-sur-Mer
CEM
GRID SECTION
SPOT ELEV. 54 METERS
les Moulins
RIVER RUQUET
RESERVOIR
BEACH EXIT
HIGH WATER MARK
LOW LOW WATER MARK
DEPTH OF WATER OVER SANDBARS BELOW LLW LEVEL UNCERTAIN
Lat 49°22'N
Long 0°53'W
Lat 49°22'30"N
Vierville-sur-Mer 1 mi.
This sheet joins OMAHA BEACH-WEST (Vierville-sur-Mer) at this edge — NO OVERLAP
MAGNETIC NORTH 1944
LA SEINE
EBB CURRENT
1.2 KNOTS MAXIMUM DURING NEAP TIDES
(SEE CURRENT TABLES ON REVERSE SIDE OF THIS SHEET)
2.0 KNOTS MAXIMUM DURING SPRING TIDES
Long 0°51'W
Lat 49°22'30"N
Long 0°52'W
VALLEY OF RIVER RUQUET
CONSTRUCTION ACTIVITY 669902
LES MOULINS
RUINS 684897
RUINS OF HOUSE 676900
HOUSE 676901
RUINS 671903
HOUSE 667906
Panoramic shoreline sketch above shows beach as seen from water level height, approximately 2000 yards offshore, with sea at about 16 feet above low low water
NOTE to COXSWAIN or NAVIGATOR
Building landmarks, especially near the beach, may be destroyed before any craft land. Terrain features, therefore, are much more reliable for visual navigation from panoramic shoreline sketch above. Green solid and broken lines with letter at each end on chart above refer to Beach Gradients on reverse side of this sheet. Also on reverse side are Sunlight and Moonlight Table, data on Inshore Currents and Tidal Stages.
PREPARED BY COMMANDER TASK FORCE 122, APRIL 21, 1944
OMAHA BEACH-EAST (Colleville-sur-Mer)
Contours shown are at 10 meter (approx. 33 ft.) intervals above mean sea level, which is 13 ft. above low low water.
SCALE 1:7920 (8"=1 mile; 1"=220 yds.)
Map from GSGS 4490, sheets 79 & 80 and air photo examination. Grid square equals 1 kilometer (1000 meters).
METER SCALE
1 METER = 3.3 FEET
1000 METERS
YARD SCALE
1000 YARDS
GRID REFERENCE SCALE
This scale can be used to accurately measure tenths eastward or northwards in any grid square on the map.
LONGITUDE SCALE
LATITUDE SCALE
These scales can be used to accurately measure seconds of latitude or longitude using latitude and longitude squares on the map.
TOP SECRET - BIGOT
UNTIL DEPARTURE FOR COMBAT OPERATIONS THEN THIS SHEET BECOMES RESTRICTED
M.O.D. DIRECTORATE OF MILITARY SURVEY
DATE RECEIVED
BLACK (313 EAST)

VIERVILLE-SUR-MER
Hamel au Prêtre
Gruchy
Pointe
BAIE DE LA S
FLOOD CURRENT
EBB CURRENT
LOW LOW WATER MARK
SANDBAR REPORTED ABOVE LOW LOW WATER LEVEL
RETARDS
EASTERN LIMIT OF ROCKS AT 646920, EXTEND WESTWARD 11 MILES
LEGEND
AREAS WITH NO SYMBOLS PRINTED OVER THEM ARE GENERALLY OPEN FIELDS
LOW WATER SOUNDINGS AND CONTOURS
HIGH WATER SOUNDINGS AND CONTOURS
Panoramic shoreline sketch above shows beach as seen from water level height, approximately 2000 yards offshore, with sea at about 16 feet above low low water
NOTE to COXSWAIN or NAVIGATOR
Building landmarks, especially near the beach, may be destroyed before any craft land. Terrain features, therefore, are much more reliable for visual navigation from panoramic shoreline sketch above. Green solid and broken lines with letter at each end on chart above refer to Beach Gradients on reverse side of this sheet. Also on reverse side are Sunlight and Moonlight Table, date on Inshore Currents and Tidal Stages.
PREPARED BY COMMANDER TASK FORCE 122, APRIL 21, 1944
OMAHA BEACH-WEST
Contours shown are at 10 meter (approx. 33 ft.) intervals above mean sea level, which is 13 ft. above low low water.
SCALE 1:7920
METER SCALE
YARD SCALE
GRID REFERENCE SCALE
LONGITUDE SCALE
LATITUDE SCALE
These scales can be used to accurately measure seconds of latitude or longi
TOP SECRET - BIGOT
SUPERSEDED
RECORD COPY
UNIQUE COPY

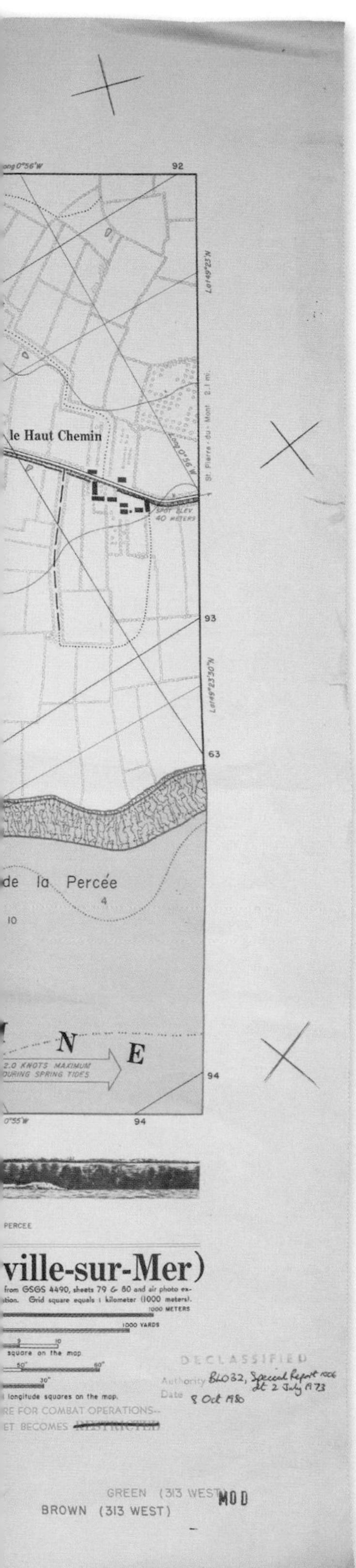

which meant that German forces were allocated with reference to threat-assessments that were at best problematic. This had been the case from early 1942: the United States' entry into the war and Germany's loss of the strategic initiative in the winter of 1941–42, more particularly on the Eastern Front, meant that defensive considerations became more pronounced. This was even more true in the winter of 1942–43. The result was a large German effort expended on the defence, indeed a larger portion of military effort than in the case of Japan. That focus lessened the resources available for attacks, and notably so because U-boats after May 1943 did not serve as a potential strategic enabler. Despite the defensive focus it remained unclear where the Allied attack was likely to come, as had been seen in 1943 when German concerns about Sardinia and Greece lessened the forces available for the defence of Sicily and mainland Italy. Moreover, this German focus did not necessarily shape the plans of the Allies as part of a defensive German strategy that was prompting a reaction from the Allies. Thus, the large German force in Norway did not lead to the Allied choice to land in Normandy. Instead, the German force simply lessened the number of troops available elsewhere.

There was also the question of how to act when the attack came. Drawing on a longstanding commitment to the manoeuvrist character of counterattacking warfare, the German preference, as shown in response to the Salerno and Anzio landings in Italy in 1943 and 1944 respectively, was to fight the Allies at or close to the waterline. This reflected concerns about the ability of the Allies to manoeuvre once they had broken forward from a beachhead, and an awareness of the vulnerability of German forces to Allied air attack. However, a forward position also lessened the flexibility and mobility of the defending forces.

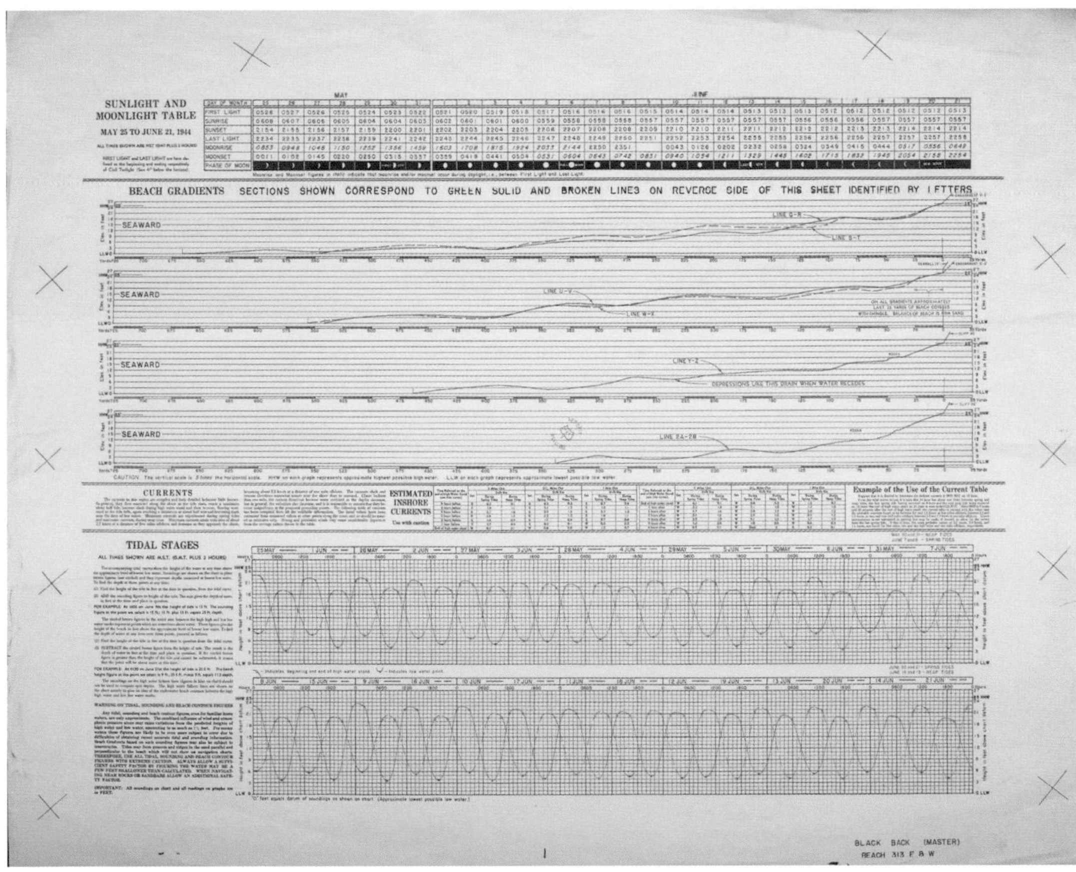

COASTAL KNOWLEDGE

Gold Beach Area, May 1944.

This map shows the Allied information, in May 1944 (at one stage the date for D-Day), about the disposition of German defences in the vicinity of Gold, Juno and Sword beaches. Sword had been added (with Utah) to the original plan in July 1943 to boost the number of invaders landing. The major priority on D-Day was to ensuring the landings on all five beaches were sufficiently strong enough to withstand anticipated counter-attacks.

One of the many essential preparatory tasks beforehand was the collection of data, notably on hydrographic and beach conditions, addressed especially by the British COPPs (Combined Operations Pilotage Parties). Comprised of navy and army personnel, COPP teams infiltrated by using submarines and folbot canoes. This work of reconnoitring the potential landing areas, helped by infrared equipment, played a major role before D-Day, just as it had played a significant role in Italy in 1943 and was to do so again in the 1945 Rhine and Elbe crossings. For example, Major Logan Scott-Bowden and Sergeant Bruce Ogden-Smith were landed at Ver-sur-Mer in order to collect samples from areas of peat marsh on what became the Normandy invasion's Gold Beach in order to establish whether the peat had a clay base treacherous to heavy vehicles if only lightly covered with sand. Due to the low-lying marsh near the coast, the Germans had assessed Gold as a flawed landing area.

The collected information was used for geological mapping that served a variety of purposes. The key element was surface geology rather than an engagement with deeper strata, and the mapping was outcome orientated, in that it had to be produced rapidly, to be readily comprehensible, to be usable in difficult conditions and to be valid for a wide range of purposes. In addition to the need for detailed charting of inshore waters and tidal conditions, the beaches had to be understandable in terms of firmness and slope profiles. Cross-beach mobility had very different meanings for vehicles and soldiers. Hydrogeological mapping was also important in order to show where boring for water would be useful.

General Sir Bernard Montgomery's 21st Army Group Planning Staff had high-grade geological advice from academics, notably Bill King and Fred Shotton. Their mapping included a preparation and presentation of material on where best to locate temporary airfields (the best place being the Calvados plateau between Bayeux and Caen) and to find sources of road metal, sand and gravel. The submarine geology of ports was also of concern. The geologists were operating without being able to acquire on-site data and to test hypotheses accordingly, but the use of material from a variety of sources, including French maps and aerial reconnaissance, enabled them to assemble and present information that was subsequently shown to have a high degree of accuracy. Geological data was also related to 'goings' needs, as with the mapping of the Caen–Falaise area.

The mapping of the coast in the landings sector captured a transition from the high limestone cliffs that stretched east from Omaha Beach (see pages 86–89), with only slight geological clefts at Port-en-Bessin and Arromanches, to stop at Le Hamel, beyond which there were wide sand or shingle beaches backed by a deep coastal plain as far as the Orne Valley. Those wide beaches provided the landing opportunities for the three eastern beaches: Gold, Juno and Sword. The easternmost beach, Sword, was afforded a slight barrier from counter-attack by the Orne River and the Caen Canal, where the crossings were to be held on D-Day by 6th (British) Airborne Division. The area was just nine miles (14 kilometres) north of Caen, which was the objective for 3rd Infantry Division sweeping through the beaches and linking up with 6th (British) Airborne Division. One of the principal strongpoints in the vicinity was Ouistreham, where a series of coastal guns and emplacements were the objectives of 4 Commando, which included two troops of Free French from 10 Commando. By the end of the day on 6 June 29,000 Allied soldiers had come ashore at Sword alone (with about 25,000 more at Gold, 21,000 at Juno, 23,000 at Utah and 34,000 at Omaha), but the Allies were still a few miles short of Caen.

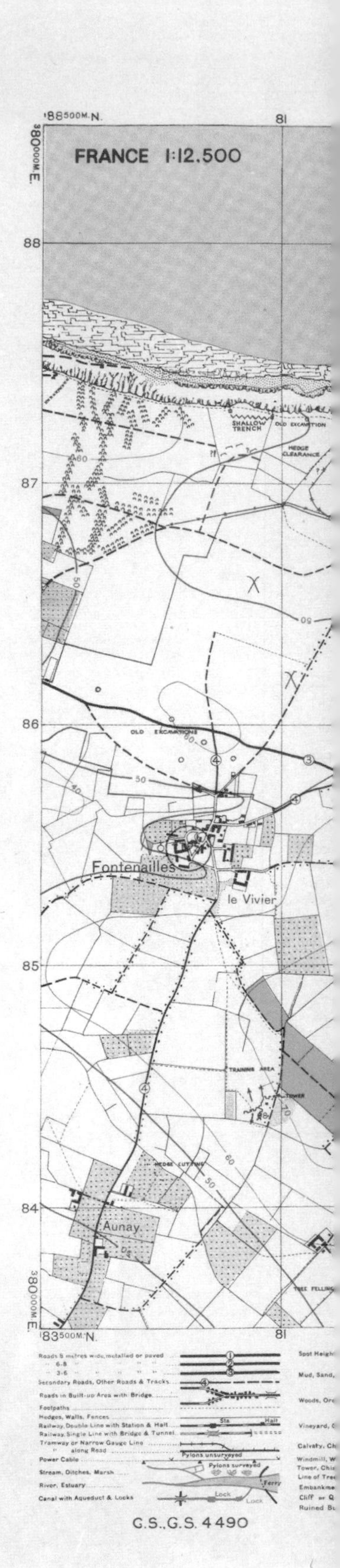

DEFENCE LEGEND-SEE BACK
.....Confirmed
DEFENCES
Information as at May 44
ARROMANCHES-LES-BAINS
Second Edition
Sheet No. 82
NOTE:
Low water as indicated on this map was plotted from air photos taken at Mean Low Water Spring Tides (approx) and NOT at Lowest Possible Low Water.
Rock formations extend below Sea Level and beyond the rock area as indicated on the map.
Approx. L.W.M.
Annual Decrease about 8'
Manvieux
la Brehollierie
Tracy-sur-mer
ARROMANCHES LES BAINS
Cabane
Asnelles sur-mer
Puits d'Herode
St. Côme-de-Fresne
Buhot
la Noe
la Rosiere
Ferme Noe
les Bergeries
les Pres Fme
RYES
Marumme
Scale 1:12,500
Approx. 5 inches to 1 Mile
INDEX TO ADJOINING SHEETS
PORT EN BESSIN
ARROMANCHES LES BAINS
VER SUR MER
THE GRID on this sheet is LAMBERT ZONE I
Compiled from Air Photographs on a control provided by existing French Triangulation
NOTES ① Road Classification is based on Michelin and other information. It has not been checked on the ground and its reliability is uncertain.
② Contours are at 10 m.V.I. They are interpolated from spot heights and hachures on the French 1:80,000 and amplified from Air Photo Examination. They should be accepted with caution.
SHEET 82.
14317.

ATTACK ON THE RIVIERA

Commander US Eighth Fleet, N-2 Section, Panoramic Beach Sketch, July 1944.

Printed by the 19th Field Survey Company, this map provides details of offshore features, the German defences and the situation on shore at 'beach 259' in Provence. The value of the information, which was compiled by the Eighth Fleet of the US Navy, was enhanced by the variety of perspectives offered. These included: a panoramic beach sketch, in turn carefully delineated between the section to scale and that in oblique perspective; the water-level silhouette in a close approach and, again, at a distance; and the overall oblique of the entire assault area, with the earlier invasion beach being but one of those marked. The defence information added was that available as of 20 July.

Operation Dragoon, the landings along the French Riviera, was mounted on 15 August 1944. It was very much an American–French operation and was pushed hard by President Roosevelt, the American Chiefs of Staff and General Dwight Eisenhower, all of whom were opposed to Britain's Mediterranean strategy, with its commitment to Italy and its interest in mounting an attack from there into Austria. Dragoon was seen as a way to gain the major port of Marseille, aid Allied logistics in France and cut off German forces in southwest France. Beach 259 at Cavalaire-sur-Mer was the most westerly of the major invasion beaches, Alpha Beach, and was attacked by the US 3rd Infantry Division, part of the US VI Corps under Major General Lucian Truscott. The French II Corps followed up the initial assault. The exploitation saw a rapid advance to Avignon, which fell on 25 August, and from this beach to Toulon, which was captured on the 26th, and Marseille, on the 28th.

The advance of Allied troops northward and their link up with those advancing from Normandy contributed greatly to the broad front of the Allied offensive, which in part reflected a 'come-as-you-go' approach, in the shape of moving forward troops from existing alignments in northern and southern France, an approach that lessened the burden on particular communication routes.

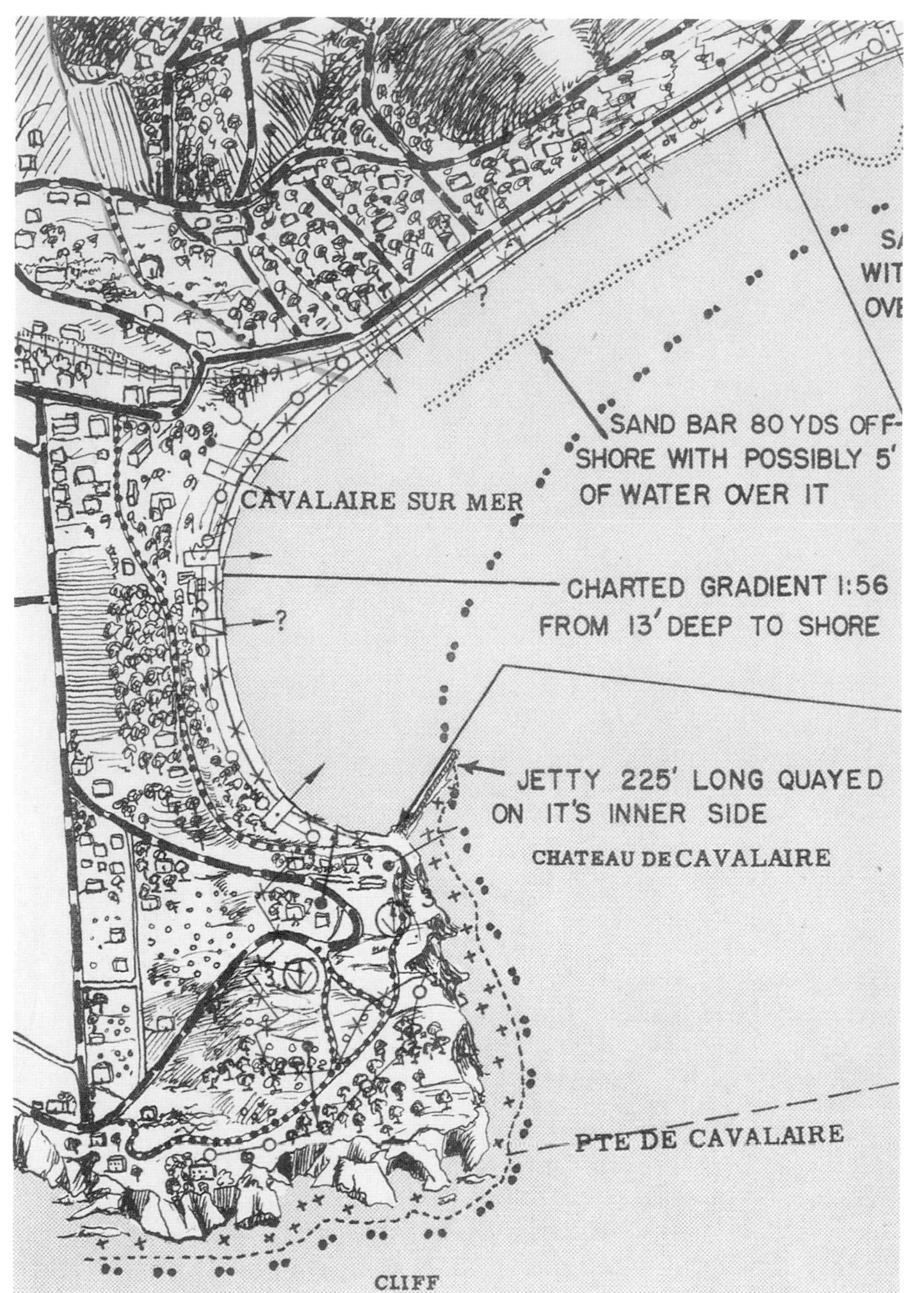

No. P — 1
BEACH 259

PANORAMIC BEACH SKETCH

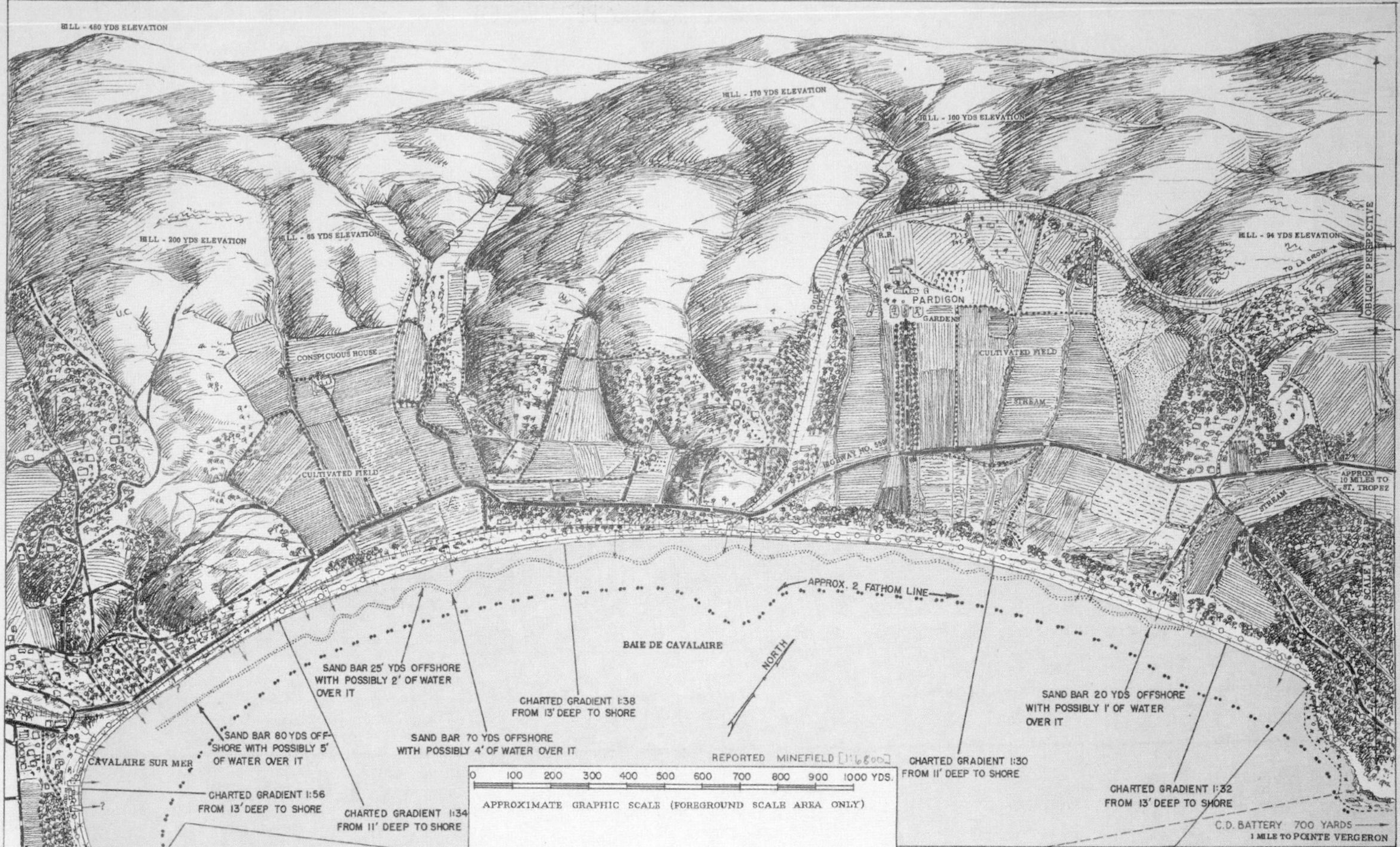

HILL - 480 YDS ELEVATION
HILL - 170 YDS ELEVATION
HILL - 160 YDS ELEVATION
HILL - 200 YDS ELEVATION
HILL - 65 YDS ELEVATION
HILL - 94 YDS ELEVATION
TO LA CROIX
PARDIGON
GARDEN
CONSPICUOUS HOUSE
CULTIVATED FIELD
CULTIVATED FIELD
STREAM
HIGHWAY NO. 559
APPROX. 10 MILES TO ST. TROPEZ
STREAM
OBLIQUE PERSPECTIVE
SCALE
APPROX. 2 FATHOM LINE
BAIE DE CAVALAIRE
NORTH
SAND BAR 25' YDS OFFSHORE WITH POSSIBLY 2' OF WATER OVER IT
CHARTED GRADIENT 1:38 FROM 13' DEEP TO SHORE
SAND BAR 20 YDS OFFSHORE WITH POSSIBLY 1' OF WATER OVER IT
SAND BAR 60 YDS OFFSHORE WITH POSSIBLY 5' OF WATER OVER IT
SAND BAR 70 YDS OFFSHORE WITH POSSIBLY 4' OF WATER OVER IT
CAVALAIRE SUR MER
REPORTED MINEFIELD [116800]
CHARTED GRADIENT 1:30 FROM 11' DEEP TO SHORE
CHARTED GRADIENT 1:56 FROM 13' DEEP TO SHORE
CHARTED GRADIENT 1:34 FROM 11' DEEP TO SHORE
CHARTED GRADIENT 1:32 FROM 13' DEEP TO SHORE
C.D. BATTERY 700 YARDS →
1 MILE TO POINTE VERGERON

0 100 200 300 400 500 600 700 800 900 1000 YDS.
APPROXIMATE GRAPHIC SCALE (FOREGROUND SCALE AREA ONLY)

JETTY 225' LONG QUAYED ON IT'S INNER SIDE
CHATEAU DE CAVALAIRE
PTE DE CAVALAIRE
CLIFF

BEACH 259

CENTER OF BEACH: LAT. 43° 10' N. LONG. 06° 32' E.
COORDINATES: U-414055 — U-434076
NATURE OF SEA BOTTOM: SAND
ANCHORAGE: 400 YARDS OFFSHORE IN 6 FATHOMS, BOTTOM OF MUD AND WEED,
LENGTH OF BEACH: 3,900 YARDS
WIDTH: 20 TO 30 YARDS
SUITABILITY FOR CRAFT: LCAs, LCVPs, LCMs, LCT (5)s ANYWHERE (LANDING WET IN PLACES); PONTOONS FOR LARGER LCTs AND LSTs IN WESTERN PORTION WHERE SAND BAR EXISTS.
BEACH No. 259

LEGEND OF ROADS	
MAIN TRAFFIC	————
SECONDARY	══ ══
OTHER ROADS	— — —
TRACKS	....

DECLASSIFIED
Authority CMRO MRLG (D/Svy 3/3/23/1 dd 17-4-73)
Date 12.1.81.

~~TOP SECRET — BIGOT~~
~~(Until departure for combat operation when this sheet becomes Restricted)~~

Defense information as of July 20 1944.

LEGEND OF SYMBOLS ON REVERSE SIDE OF THIS SHEET

PREPARED BY COMMANDER U.S. EIGHTH FLEET
N-2 SECTION

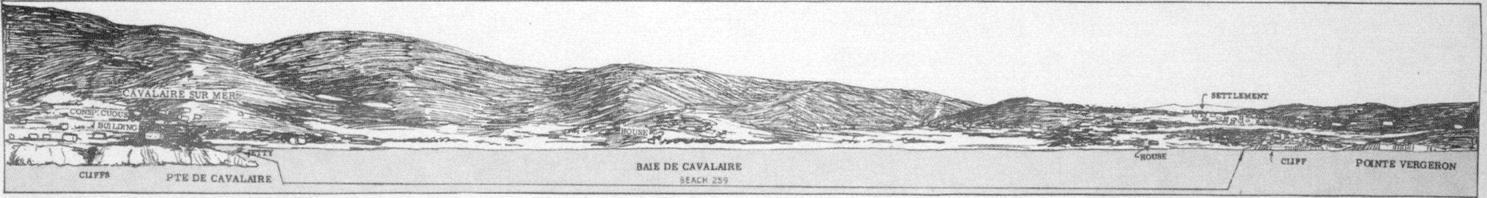

WATER LEVEL SILHOUETTE (CLOSE APPROACH)

CONSPIC HILL
CAVALAIRE SUR MER
PTE DU DATTIER
CAP CAVALAIRE
BEACH 259
BAIE DE CAVALAIRE
PTE VERGERON
BEACH 260
PTE DU BREUIL
PTE ANDATI

WATER LEVEL SILHOUETTE (DISTANT)

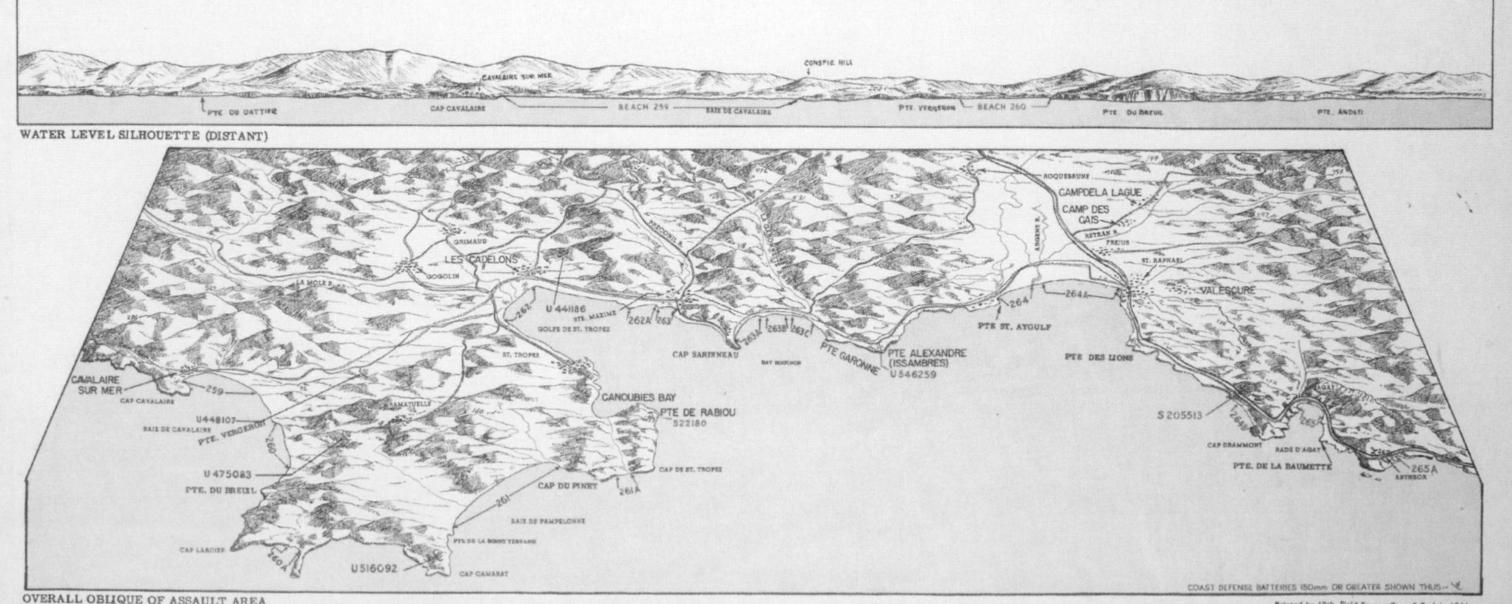

OVERALL OBLIQUE OF ASSAULT AREA

AF 4463

Printed by 19th. Field Survey Coy., R.E., July 1944

Achtung!
Der Bearbeitungsinhalt kann in Einzelheiten unvollständig sein. Alle Benutzer werden gebeten, Berichtigungen und Ergänzungen mit zugehörigen Unterlagen auf dem Dienstweg über den zuständigen Ia/Meß an Pz. AOK 2 einzusenden.
Übersichtskarte zur Geländebeurteilung von Mitteldalmatien
(Raum Zara-Knin-Šibenik)
Sonderausgabe Panzerkarte
Sd. Kdo. D/Forschungsstaffel z
Kartenstelle
eingegangen am 1. X. 44
Briefbuch Nr. 28/1 Kl I 4
Anlagen —
Ausgabe vom VIII. 194
nicht bearbeitet
KOVIGRADSKO MORE
ZARA
Knin
Šibenik
ADRIATISCHES MEER
G. MURTER
Topographische Grundlage:
DHK Südosteuropa 1 : 200000, Blätter 33/44 Zara und 34/44 Split. Stand VI. 40, Druck IV 41.
Bearbeitungsgrundlagen:
Wissenschaftliche Luftbildauswertung und Geländeerkundung durch Forschungsstaffel z. b. V.
Maßstab 1 : 200000 d. N. oder 1 cm = 2 km
Herausgegeben von Pz. AOK 2
Ia/Meß
Lageskizze
Luftbildgedeckter Raum innerhalb des bearbeiteten Gebiets
Herstellung der Folien: Reichsamt für Landesaufnahme
Druck: Armeekartenstelle (mot) 512/500/VIII. 44
Bearbeitung: Forschungsstaffel z. b. V.
Stand VI. 44
14|3a.

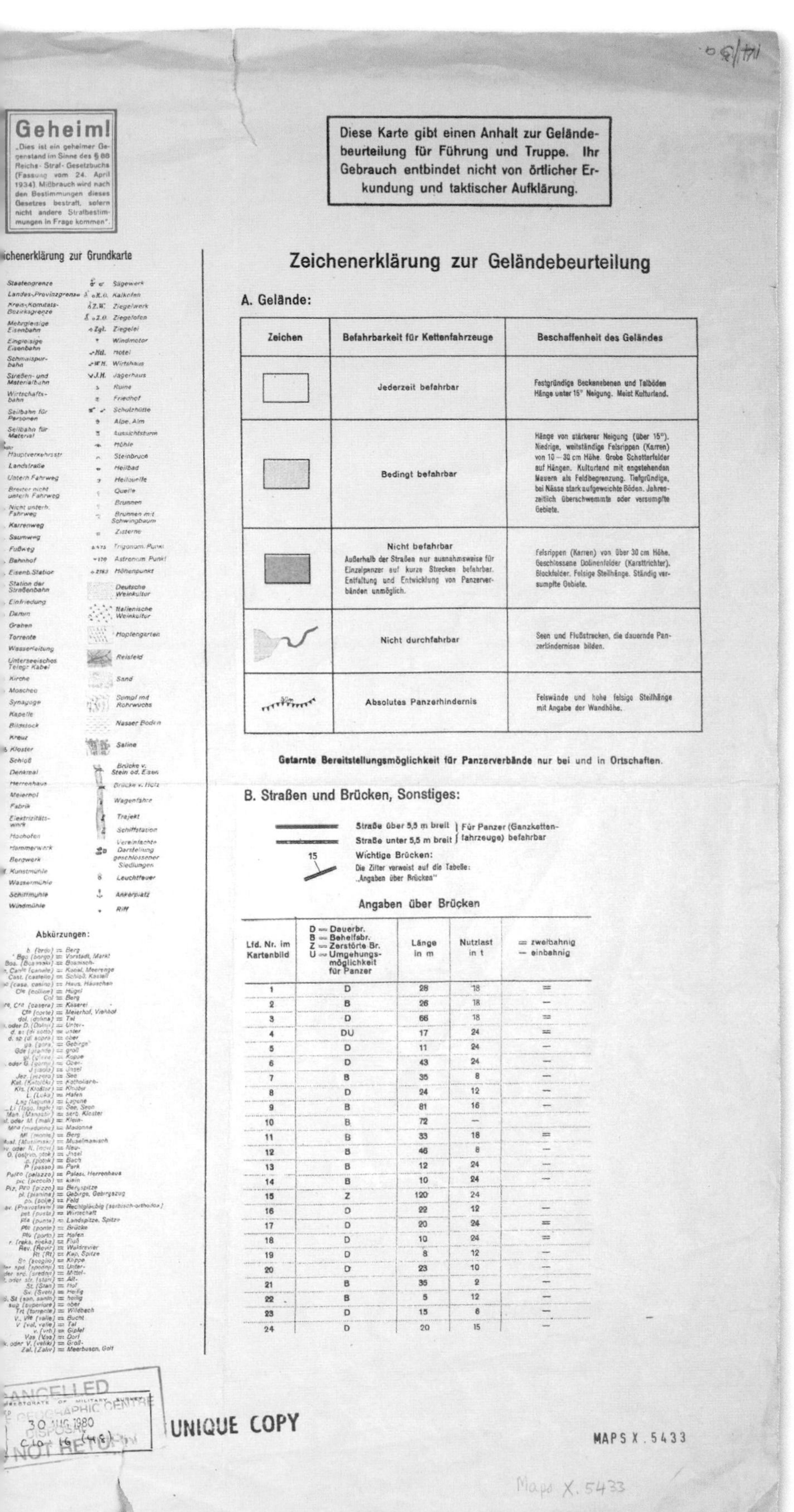

BALKAN DIFFICULTIES

German special edition tank map, overview area of central Dalmatia, August 1944.

This map of part of Dalmatia was created after the Italian occupation zone had been seized in September 1943 and it was designed to show the problems facing the use of tanks. Five types of ground are identified (three shades, two symbols), with a brief comment in the first column on 'driveability for tracked vehicles' followed by detailed comments on the condition of the terrain in the second. The terrain ranges from lighter coloured areas that are passable year round, through 'partially passable' to 'not passable'. The symbols indicate water and rocky ground presenting a 'complete barrier to tanks' (*Absolutes Panzerhindernis*), marked with lines of dark pink triangles; within the main map *nicht bearbeitet* indicates 'not edited'.

The nature of the war in Yugoslavia was greatly affected by the detailed configuration of local geography, ethnicity, politics, religion and society. Relations between sections of the Yugoslavian Resistance and the occupying power were frequently ambivalent and a complex bloody dynamic of collaboration, opposition, armed resistance and reprisals played out in the country. The same was true in Albania.

In Dalmatia, there was the added complication of British intervention from Italian bases. That made the Germans more concerned about the Dalmatian coast, which would otherwise have been of marginal importance. This threat of intervention was accentuated by the need to withdraw from the Balkans in the face of the Soviet advance, and linked concern about a British and/or Yugoslavian Resistance advance to attack German communication links through Yugoslavia. The operations of the British 2nd Commando Brigade, based in the Dalmatian Islands, were an issue for the Germans from January 1944. The situation was made more unstable for Germany by a general collapse in the Balkans, where Bulgaria declared war on 8 September. The Germans focused on trying to resist the Soviet and Bulgarian advance, which affected the situation in the rear.

Responsible for the opposition to the Yugoslav resistance movements, the 2nd Panzer Army, based at Kragujevac, in practice had no armoured divisions and very few tank battalions, which was part of a pattern in which strength was often well below the formal complement. The Germans had hit Tito's Partisan headquarters hard on 25 May, and Tito took refuge on the island of Vis under British protection, but were later driven from Belgrade when the city fell to Soviet forces and the Partisan resistance on 20 October.

ENGINEER INTELLIGENCE

US Army Corps of Engineers, mapping for the invasion of Leyte, September 1944.

The Engineer Intelligence Section of the US Army prepared these detailed maps of Leyte in September 1944 before its invasion by American forces. They draw heavily on low-oblique photography and hydrographic charts. The nature of the ground cover is explained in the legend, notably distinguishing between rainforest and the more tractable scrub or secondary growth, and marking mangrove and swamp. The possibility is highlighted that due to inadequate ground control, scaled distances might be at variance with actual distances and that inland areas were not covered by photography.

Other aspects of cartographic preparation included the map of potential airfield construction sites on Leyte prepared by the Military Geology Unit as a Strategic Engineering Study. This showed 'fair' conditions for foundation and drainage, and 'good' conditions for construction materials, clearing and water supply, as well as Japanese fields reported as under construction.

The scale and speed of American mapmaking was indicated in an article 'War Maps While You Wait' in *Popular Mechanics* in November 1943. The magazine explained: 'Data for map-making are obtained by plane, patrol and reconnaissance, brought in by radio, telephone or courier and experts quickly compile and print the maps with duplicating machines, lithograph presses, mimeograph and multigraph. Some machines turn out multi-color maps.' A sergeant of the US Army Corps Engineers was shown in the field printing a lithographic plate in an offset press: 'Fast duplication is necessary to supply all air and ground units with up-to-the-minute maps in quick time.'

The Americans landed four divisions on the east coast on 20 October, following up on 7 December with a landing in Ormoc Bay on the west coast. Leyte provided a base for airfields for the assault on the island of Mindoro, which was invaded from 31 December, and Luzon, where American forces invaded on 9 January 1945.

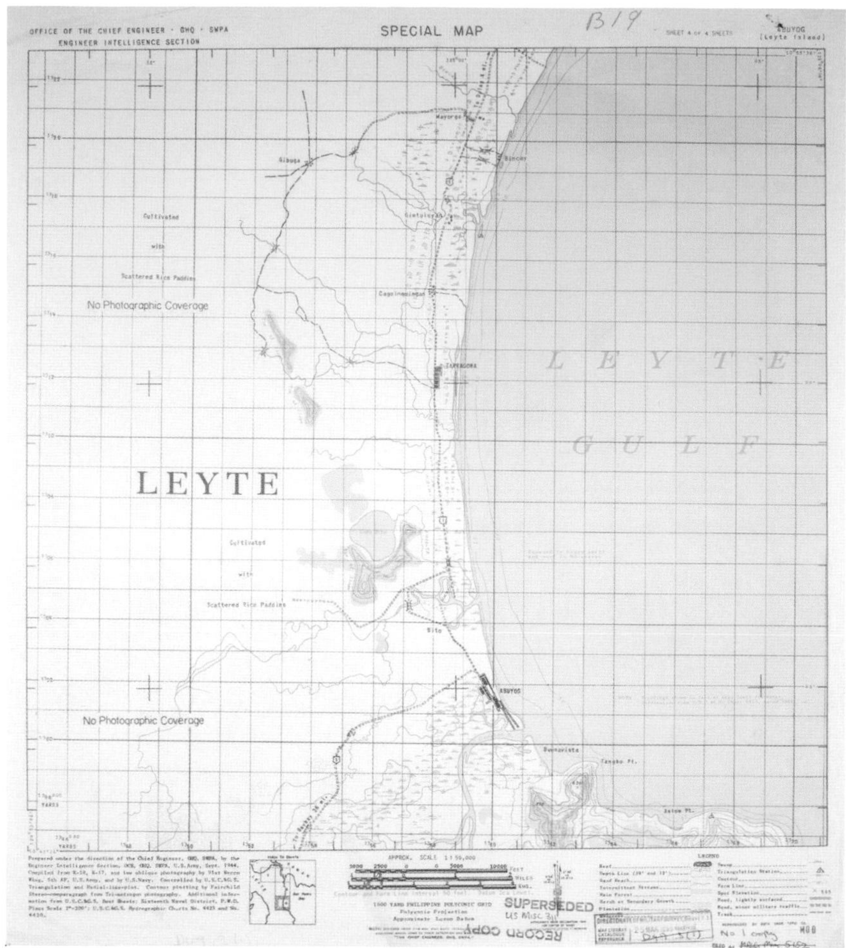

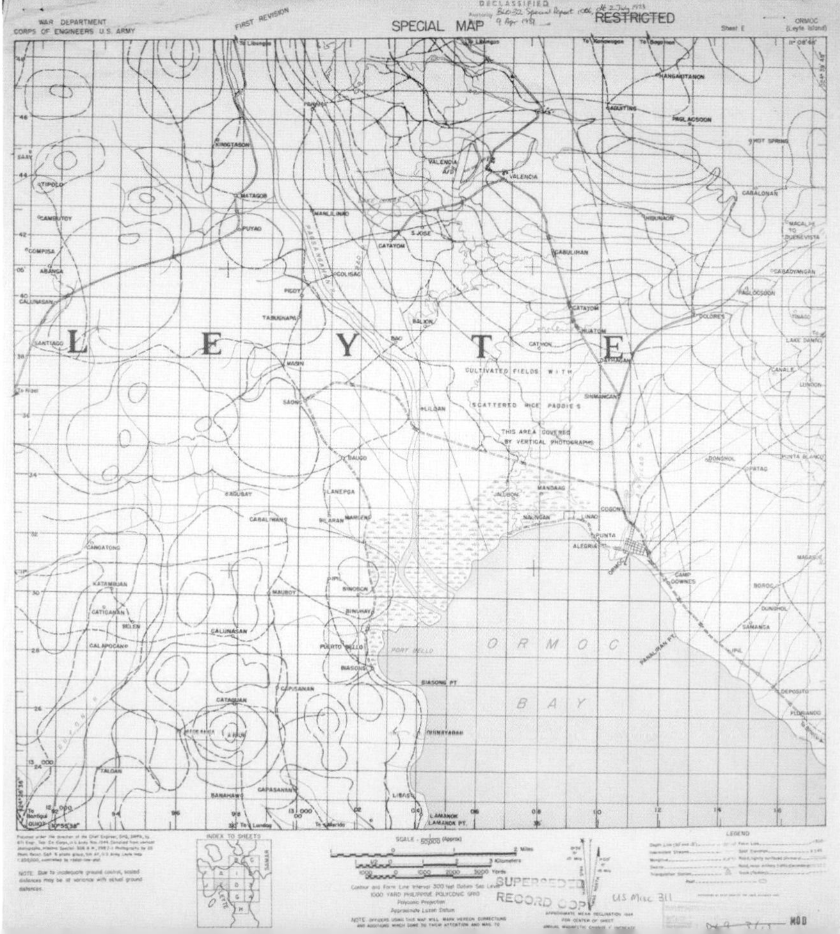

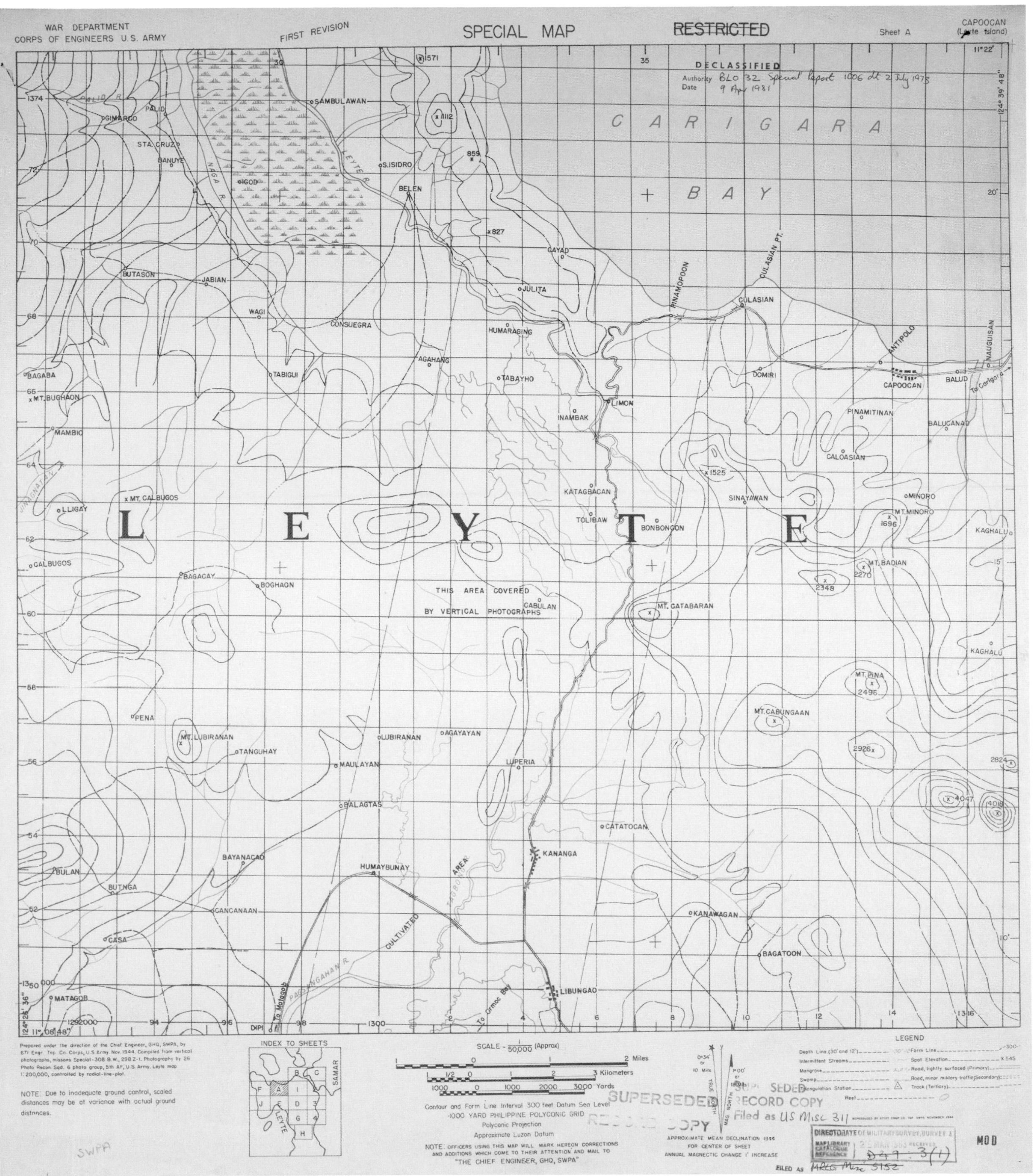
WAR DEPARTMENT
CORPS OF ENGINEERS U.S. ARMY
FIRST REVISION
SPECIAL MAP
RESTRICTED
Sheet A
CAPOOCAN
(Leyte Island)
DECLASSIFIED
CARIGARA BAY
LEYTE
THIS AREA COVERED BY VERTICAL PHOTOGRAPHS
SAMBULAWAN
S.ISIDRO
BELEN
GAYAD
JULITA
CULASIAN
CULASIAN PT.
PINAMOPOON
ANTIPOLO
CAPOOCAN
BALUD
NAUGUISAN
To Carigara
DOMIRI
LIMON
INAMBAK
TABAYHO
HUMARAGING
AGAHANG
CONSUEGRA
WAGI
JABIAN
BUTASON
TABIGUI
BAGABA
MT. BUGHAON
MAMBIC
GIMARCO
PALID
STA. CRUZ
BANUYE
IGOD
NAGA R.
LEYTE R.
PALID R.
PINAMITINAN
BALUCANAD
CALOASIAN
SINAYAWAN
MINORO
MT. MINORO
KAGHALU
MT. BADIAN
MT. CATABARAN
KATAGBACAN
TOLIBAW
BONBONGON
CABULAN
MT. CALBUGOS
LLIGAY
CALBUGOS
BAGACAY
BOGHAON
MT. PINA
MT. CABUNGAAN
PENA
MT. LUBIRANAN
TANGUHAY
LUBIRANAN
AGAYAYAN
MAULAYAN
LUPERIA
BALAGTAS
CATATOCAN
KANANGA
BAYANACAO
HUMAYBUNAY
BULAN
BUTNGA
CANCANAAN
CASA
CULTIVATED AREA
KANAWAGAN
BAGATOON
LIBUNGAO
MATAGOB
To Matagob
To Ormoc Bay
PAGSANGAHAN R.
Prepared under the direction of the Chief Engineer, GHQ, SWPA, by 671 Engr. Top. Co. Corps, U.S. Army Nov. 1944. Compiled from vertical photographs, missions Special-308 B.W., 298 Z-1. Photography by 26 Photo Recon. Sqd. 6 photo group, 5th AF, U.S. Army. Leyte map 1:200,000, controlled by radial-line-plot.
NOTE: Due to inadequate ground control, scaled distances may be at variance with actual ground distances.
INDEX TO SHEETS
SAMAR
LEYTE
SCALE - 1/50,000 (Approx)
2 Miles
3 Kilometers
3000 Yards
Contour and Form Line Interval 300 feet Datum Sea Level
1000 YARD PHILIPPINE POLYCONIC GRID
Polyconic Projection
Approximate Luzon Datum
NOTE: OFFICERS USING THIS MAP WILL MARK HEREON CORRECTIONS AND ADDITIONS WHICH COME TO THEIR ATTENTION AND MAIL TO "THE CHIEF ENGINEER, GHQ, SWPA"
SUPERSEDED RECORD COPY
Filed as US Misc 311
APPROXIMATE MEAN DECLINATION 1944 FOR CENTER OF SHEET
ANNUAL MAGNETIC CHANGE 1' INCREASE
LEGEND
MOD
SWPA

RHINE RECONNAISSANCE

Mosaic by 654th Engineer Battalion of US First Army, 'Rhine River Photomap', 15 October 1944.

Photo-reconnaissance missions flown by the United States Army Air Force (USAAF) in September 1944 were the basis for the maps of the Rhine Valley produced by the US First Army during advance planning for the push into Germany. The Allies wished to establish a bridgehead across the Rhine south of Cologne (Koln). At that stage the Allied advance from Normandy had already dramatically slowed with the US First Army tied down and taking heavy casualties from September to December in clearing the Huertgen Forest near Aachen, which only surrendered after tough streetfighting on 21 October.

The Battle of the Bulge totally reset Allied priorities that winter, but it was followed by a resumption of the Allied advance. Operations launched on 8 February 1945 drove back German forces west of the Rhine and were linked to a broad front advance on the Rhine.

The US First Army reached the Rhine opposite Cologne on 5 March, but found the bridges shown in these photographs destroyed, as the US Ninth Army had earlier found those near Düsseldorf on 2 March. However, on 7 March, a patrol of the US First Army captured the undestroyed Ludenforff railway bridge at Remagan near Bonn. A furious Hitler had those he blamed for the failure to destroy the bridge shot, and V2 rockets were unsuccessfully fired at it. The Americans were able to consolidate the bridgehead on the east shore, albeit in the face of strong resistance including counter-attacks, and added pontoon bridges to increase the flow of armour across the river.

The photo-reconnaissance missions were aided by greater Allied air dominance over Germany from March 1944 as the Americans successfully concentrated on seeking out the Luftwaffe interceptors. The latter were weakened by losses and a lack of training that reduced the skill levels of the personnel.

RHINE RIVER PHOTOMAP
ERMANY 1:9,000
SECTION "X"
SECTION "X" SHEET A ·
OSTHEIM
WESTHOVEN
WESTHOVEN
ENSEN
RHINE
Scale 1:9,000 (Approx.)
USE OF COORDINATE SCALE
ATLAS GRID
INDEX
TO SECTION "X"
UNIQUE COPY
SHEET A ·

NATIONAL LIBERATION

French Forces of the Interior (FFI), amended military map of Royan, after October 1944.

One of the major roles of the resistance movements in Europe, and one that was of strategic, operational and tactical value because it compromised German defences at every level, was providing intelligence. This was always a hazardous activity because the Germans routinely tortured and murdered those they captured. The Liberation of France in 1944 was incomplete not only because the Germans fought on in Alsace, but also because they continued to hold a number of fortified coastal positions, including Lorient, St Nazaire and La Rochelle, important submarine bases. Another position was Royan on the northern side of the Garonne estuary, where a pocket of German troops had built up.

This map was a German military map of Royan that dated to May 1943. It shows the centre of Royan and the area to the south of the town to where the Germans had retreated by October 1944, intending to get to Germany by sea. The Germans created a heavily fortified and mined zone surrounding the town. When a copy of the map fell into the hands of the FFI (Forces françaises de l'Intérieur), a unified resistance formed in February 1944, they marked on it in red the locations of the German fortifications, including the many machine gun posts, artillery positions, obstacles, lookout points and minefields. An American air raid on 5 January 1945 killed 500 civilians, wounded 1,000 and destroyed most of the town, but it did not break the German defence. Copies of the amended map were distributed to the French forces on 13 April 1945, the eve of Operation Vénérable to liberate Royan, supported by American aircraft and artillery, which took until 17 April and led to 8,000 German soldiers being taken prisoner. Over 30,000 troops were deployed, many from North Africa. The French pressed on to defeat the remaining Germans, who had retreated to the Ile d'Oléron, where they surrendered on 1 May.

There had been no large-scale resistance in France after its rapid conquest in 1940. Instead, the French Resistance, which was divided politically from the outset, took time to get going, as well as facing tensions, not least the determination of Charles de Gaulle's Free French (Française Libre) government-in-exile to direct and represent it. The German invasion of the Soviet Union in June 1941 affected the views of France's communists because prior to that, the Soviet Union had been allied to Germany. In addition, the German demands from June 1942 for forced labour (STO, *service du travail obligitaire*), to be used in Germany, contributed to support for the French Resistance, as did the awareness from the end of 1942 that the war was moving against the Axis. As a result, French Resistance membership, which in 1941–42 had been hit very hard by German repression, rose greatly from early 1943, and became more significant militarily and politically from late 1943. Groups dedicated to sending intelligence, running escape lines and producing propaganda were increasingly supplemented by those prepared to take up arms. In part this was due to the greater scale of the resistance, but the establishment of the National Council of the Resistance was also important, not least to the struggle for legitimacy against Vichy.

The forces of the Maquis – guerrilla bands of French Resistance fighters – were particularly strong in Corsica, the Massif Central, the Alps and Brittany. Everywhere, resistance activity was affected by the terrain and natural cover, which helped explain there being more in the Massif Central than in the flat and well-cultivated Loire Valley. The discouraging nature of reprisals, especially the shooting of disproportionately large numbers of civilians when German troops were killed, was also important. Memorials to those killed, including the names of streets, can be found across France. A concern for the immediate needs of family and community, including for food, discouraged civilian resistance; and there was a willingness to negotiate relations with the occupiers, which was possible as long as armed resistance was avoided. Resistance activity was also affected by the detailed configurations of local politics and society, and their relationship with the complex dynamic of collaboration and non-collaboration, which was exploited by the Germans and the Vichy paramilitary Milice, and in opposition by the French Resistance. Resisters killed few German soldiers, but there was valuable intelligence gathering, acts of sabotage, help for escapees and an important challenge to the position of Vichy.

A very different form of resistance was provided by the Free French forces, who played a major role from the outset, not least in the struggle over Vichy-run colonies, fighting in Gabon (1940) and Syria (1941). In addition, the Free French took part in operations against German forces in Libya in 1942. This capability became more significant as a result of the conquest of French North Africa in late 1942,

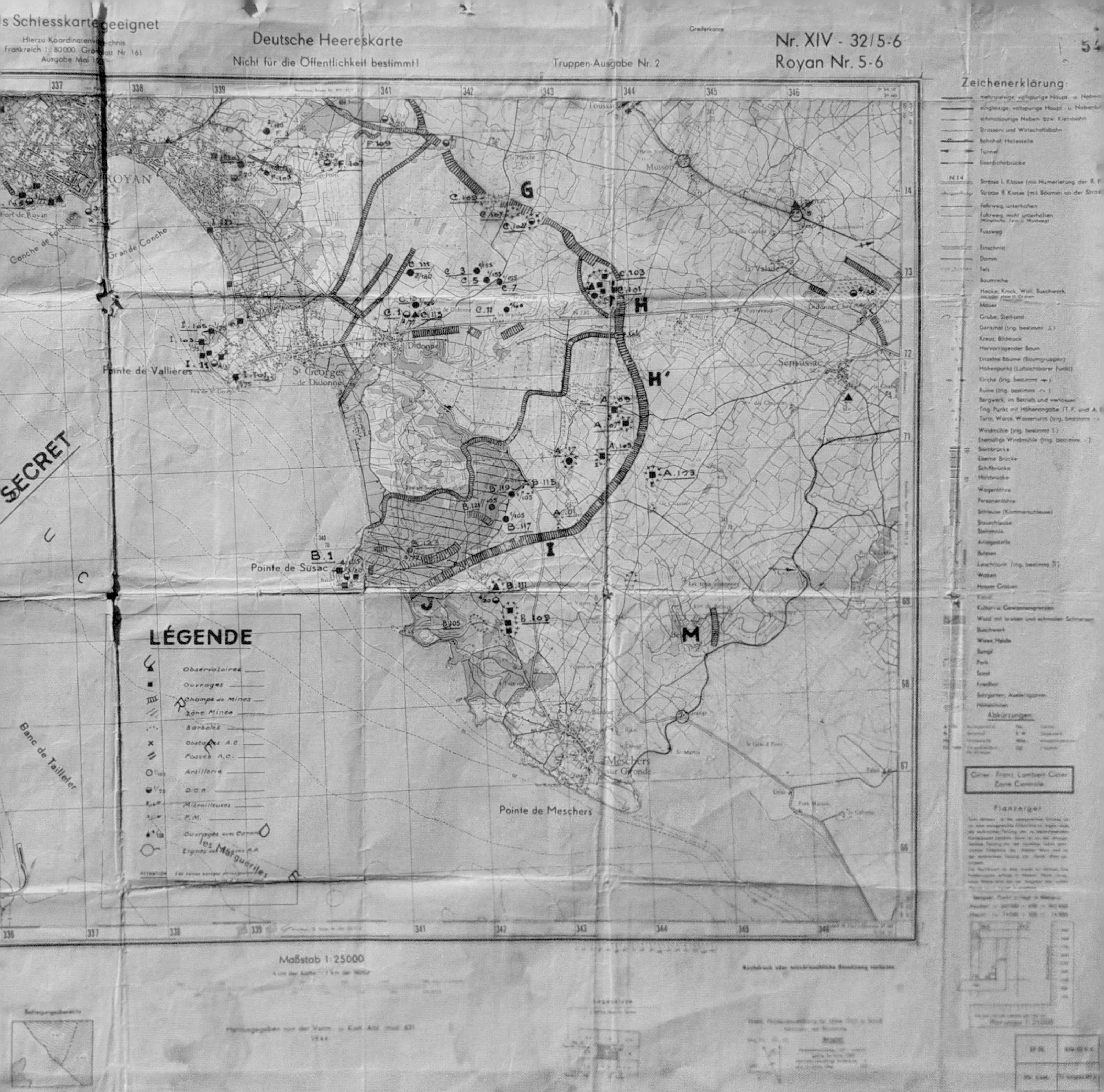

which enabled a merger with the large French Army of Africa; and American-supplied equipment also helped greatly. The French Committee of National Liberation provided a French Expeditionary Corps for the Allied conquest of Italy, as well as large forces in the reconquest of France in 1944, and notably so in southern France, especially in Operation Dragoon, the highly successful invasion of Provence, on 15 August (see page 92). These forces then advanced into eastern France, where they fought the Germans in Alsace before advancing across the Rhine. The Free French lost about 3,200 killed in 1940–42, and the French Liberation Army in 1943–45 lost 25,730 dead and 75,823 wounded, with 1945 being the most deadly year for operations. The capture of Royan was one of these operations.

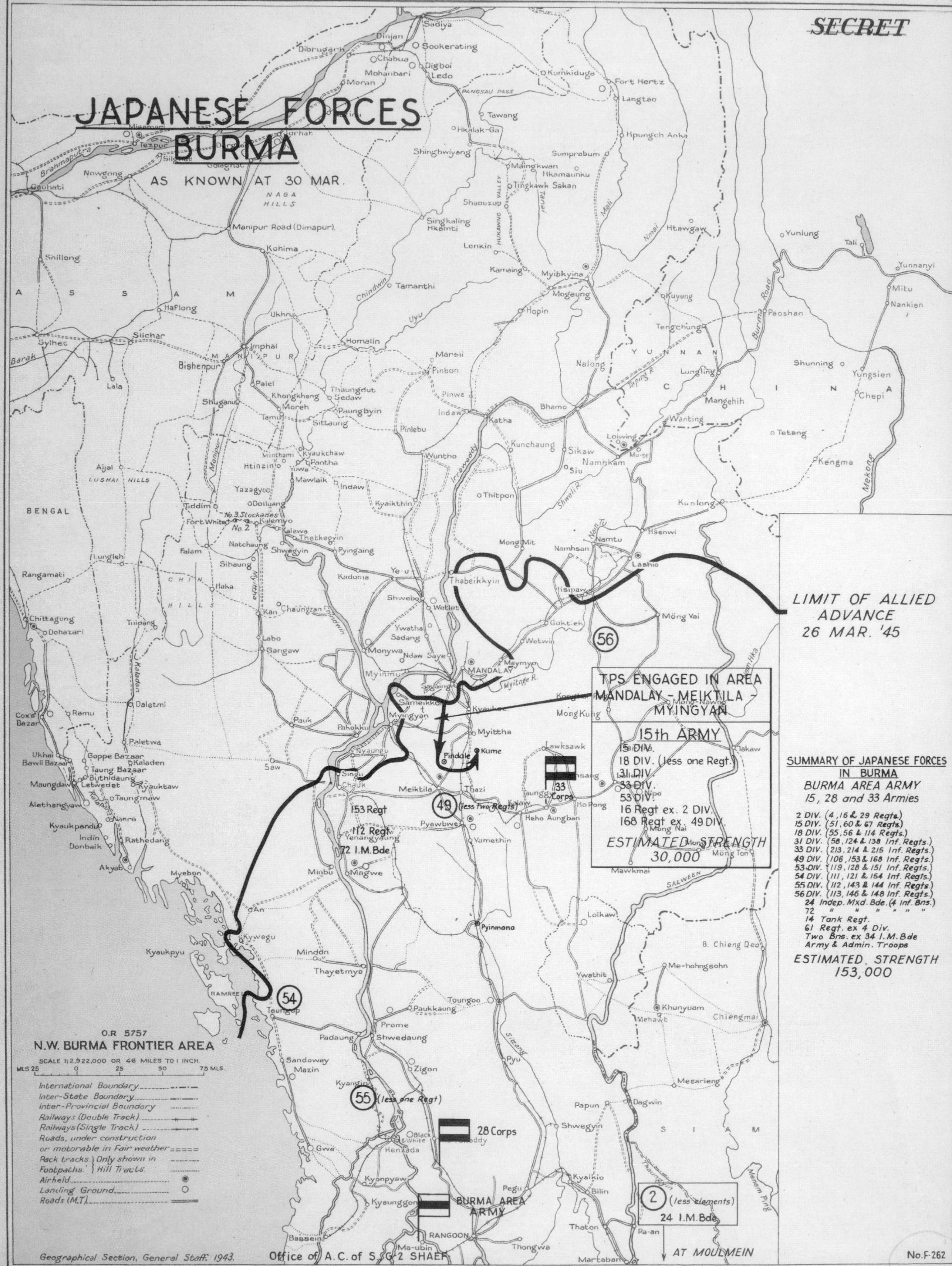

Date 8/10/81
SECRET
JAPANESE FORCES
BURMA
AS KNOWN AT 30 MAR.
LIMIT OF ALLIED ADVANCE 26 MAR. '45
TPS ENGAGED IN AREA MANDALAY - MEIKTILA - MYINGYAN
15th ARMY
15 DIV.
18 DIV. (less one Regt.)
31 DIV.
33 DIV.
53 DIV.
16 Regt ex. 2 DIV.
168 Regt ex. 49 DIV.
ESTIMATED STRENGTH 30,000
SUMMARY OF JAPANESE FORCES IN BURMA
BURMA AREA ARMY
15, 28 and 33 Armies
2 DIV. (4, 16 & 29 Regts.)
15 DIV. (51, 60 & 67 Regts.)
18 DIV. (55, 56 & 114 Regts.)
31 DIV. (58, 124 & 138 Inf. Regts.)
33 DIV. (213, 214 & 215 Inf. Regts.)
49 DIV. (106, 153 & 168 Inf. Regts.)
53 DIV. (119, 128 & 151 Inf. Regts.)
54 DIV. (111, 121 & 154 Inf. Regts.)
55 DIV. (112, 143 & 144 Inf. Regts.)
56 DIV. (113, 146 & 148 Inf. Regts.)
24 Indep. Mxd. Bde. (4 Inf. Bns.)
72 " " " " "
14 Tank Regt.
61 Regt. ex 4 Div.
Two Bns. ex 34 I.M. Bde
Army & Admin. Troops
ESTIMATED STRENGTH 153,000
56
49 (less Two Regts)
153 Regt
112 Regt
72 I.M. Bde
33 Corps
54
55 (less one Regt)
28 Corps
BURMA AREA ARMY
2 (less elements)
24 I.M. Bde
AT MOULMEIN
MANDALAY
RANGOON
ASSAM
BENGAL
MANIPUR
YUNNAN
CHINA
SIAM
O.R 5757
N.W. BURMA FRONTIER AREA
SCALE 1:2,922,000 OR 46 MILES TO 1 INCH.
MLS 25 0 25 50 75 MLS.
International Boundary
Inter-State Boundary
Inter-Provincial Boundary
Railways (Double Track)
Railways (Single Track)
Roads, under construction or motorable in Fair weather
Pack tracks. } Only shown in
Footpaths. } Hill Tracts.
Airfield
Landing Ground
Roads (M.T.)
Geographical Section, General Staff. 1943.
Office of A.C. of S. G-2 SHAEF
No. F-262
SHAEF 99 3/MAR 45/13 MRS/398 8.5/MAR 45/13 MRS/661

IRRAWADDY VALLEY VICTORY

Office of Supreme Headquarters Allied Expeditionary Force (SHAEF), map of Japanese Forces Burma, as known at 30 March 1945.

This British map is a reflection of information gathered from aerial reconnaissance, agents on the ground and signals interception. The Japanese 15th Army, under Lieutenant-General Renya Mutaguchi, had retreated in disarray after it had been defeated in the Imphal offensive, and from December 1944 the British drove successfully into Burma's Irrawaddy Valley. Fighting in the close terrain and heat of Burma was hard. Lieutenant-General Sir Henry Pownall commented in April 1944: 'It is bound to be a slow business, I fear. That jungle is so infernally thick you literally cannot see ten yards into it, and to winkle out concealed Japs, one by one (and they have to be killed to the last man) is the devil of a business.' A British Military Mission report on how best to fight the Japanese had suggested that month that 'each individual infantry soldier must be trained to be a self-reliant big game hunter imbued with a deep desire to seek out and kill his quarry'.

In practice, operational skill proved the key element. The campaigning in Burma in 1945 saw the British take the strategic, operational and tactical offensive toward Meiktila and Mandalay. General Sir William Slim, the commander of the 14th Army, proved one of the best generals of the war, with his ability to out-think the Japanese and then to implement plans in an area where terrain and climate combined to make operating very difficult. Heavy rain badly hit transport routes. Diseases such as scrub typhus were also a major problem, but these conditions also affected the Japanese who suffered, in addition, from the weakness of their supply system. Japanese logistics broke down whereas in supporting the defenders of Imphal in 1944 the RAF had mounted an extraordinary operation to replenish men and supplies.

British bridgeheads to the north of the city of Mandalay were established on 11 January 1945, and to the west and south from 13 February. The British exploited this in late February and March, advancing to seize Meiktila, the major Japanese supply base in central Burma, on 3 March and then Mandalay on 20 March. The Japanese mounted a major counter-offensive to regain Meiktila in March, cutting off the 17th Indian Division in the town, which had to be supplied by air. The 49th Division, marked on the map, was the key Japanese unit involved, commanded by Lieutenant-General Honda Masaaki, while other Japanese units were pinned down near Mandalay. On 28 March, the Japanese withdrew from near Meiktila.

Slim's success reopened the Burma Road to China and was followed by an advance from Meiktila down the Sittany Valley and then, via Pegu, south to Rangoon. Another British force advanced against the Japanese 54th Division down the Arakan coast to Gwa, which is marked on the map. Britain's 14th Army was to be referred to as a 'forgotten army', but that was not the case with the coverage at the time.

4
3
3
3
3
3

Tactical

4.

'They said, "Here are the maps"; we burned the cities.'

From *Losses* (1948) by Randall Jarrell, a USAAF celestial navigation tower operator.

While the tactical like the operational and strategic levels of warfare entailed both the use of existing maps and the creation of new ones, the issue was particularly acute at this level because the shock of combat made it suddenly necessary to have detailed and accurate maps of an area that lacked such coverage. Indeed, the range of the existing information varied greatly across the world. The Americans, after the unexpected Japanese attack in December 1941, needed to operate in the poorly mapped Pacific, much of which had been in British or Japanese hands prior to the conflict, and made extensive use of photo-reconnaissance, not least for mapping invasion beaches. Amphibious attacks required maps both of the islands and of the coastal waters. The US Navy had the old Royal Navy Admiralty charts but most of those dated from the late nineteenth or early twentieth century. While the charts were pretty good, their use posed problems, such as needing to allow for shoals created by subsequent storms. For the Solomon Islands, there were some Australian maps and charts (nautical maps) but relatively few.

Indeed, the American landing force on the island of Guadalcanal in the southwestern Pacific in August 1942 lacked adequate maps, which indicated the need for special amphibious landing maps. Moreover, the naval raid in October 1943 on Wake Island, which before the war had been an American possession, encountered the problem of inadequate charts for the surrounding waters. The issue also affected the Australians with northern Australia, New Guinea and the Solomon Islands. However, the islands, bar New Guinea and the Philippines, were small so aerial reconnaissance allowed the US Army and Marines to make up maps quickly to issue to the troops, and this was done down to the platoon level.

In contrast, the US Coast and Geodetic Survey (USCGS), a civilian agency under the Department of Commerce, surveyed the Philippines, an American colony prior to the war. Its 1933 map of the Philippines as a whole, a map with depths shown by soundings, which greatly clarified sailing routes, was reissued in 1940. The USCGS's data and charts were used for the successful American invasion of the Philippines in 1944–45. In 1944, moreover, the Army Map Service published maps of the individual islands and cities – for example, Cebu with the necessary depths of the coastal waters and location of the coral reef, both important information to amphibious operations. The USCGS also published maps for other areas of potential operations, notably Southeast China, West Java, South Borneo and North Borneo. Some of these maps were reproduced for the Aeronautical Chart and Information Service. The Royal Australian Navy Hydrographic Service surveyed New Guinea from 1939, pressing on to produce surveys for elsewhere in the South Pacific, in part using material from US Coast and Geodetic Survey mapping, notably of the Philippines.

More generally, there was a widespread use of fathometers for inshore navigation. The need for proper advance reconnaissance and beach surveys was exposed by the heavy costs incurred by the US Marines in the capture of Tarawa Atoll in the Gilbert Islands in November 1943, a landing greatly hindered by the coral reef, the depth of which had been misunderstood partly due to aerial reconnaissance. As a result, the Americans created Underwater Demolition Teams for the Pacific, pressing forward an initiative begun in late 1942. The recent development of the open-circuit scuba system allowed divers to swim in and out and actually go ashore at night to do the surveys, collect sand samples, determine the beach gradient and plot the location of reefs, obstacles and defensive arrangements such as mines, all prior to the landings.

In the final year of the war, the United States Hydrographic Office printed more than 40 million charts. Some survey ships were equipped with presses that could print 2,000 sheets an hour. That the Americans could plan where they wanted to operate and where to mount an invasion, increased the demand for maps and charts,

CHARLES LEE BURWELL, MODEL OF UTAH BEACH, 1944.

Charles Lee Burwell (1917–2016) was a Harvard graduate, fluent in French, and on 4 June 1944 he was a 27-year-old US officer with the Office of Naval Intelligence, in Plymouth, England, when he was asked to brief Generals Eisenhower and Montgomery on which day the weather would be calm enough for D-Day while the moon remained bright. Only three days offered tides low enough to see water obstacles and a moon bright enough to assist paratroopers: 5, 6 and 7 June.

For his presentation, and subsequent briefing to commanders of the amphibious assault, Burwell used this large, three-dimensional model of Utah Beach, which he had made out of rubber. The model was created at the US Navy's Special Devices Division at Camp Bradford, Virginia, and it drew on information supplied by American and British reconnaissance teams, the French Resistance and low-flying aerial reconnaissance aircraft. Burwell used bathymetric tints (colours placed between contour lines) to indicate depth, as well as capturing tide lines, the slope of the beach, buildings beyond the beach and the location of German defences, such as the steel 'hedgehog' anti-landing craft systems hidden in the water. He added numbers to correspond with traditional maps being developed for the invasion. His hedgerow-lined pastures lying inland indicated to the troops that close-quarter combat lay ahead because of the obstructed lines of sight.

Burwell also helped to plan landings in southern France, Luzon in the Philippines and on Okinawa, after which he returned to San Francisco to plan Operation Olympic: the invasion of Kyushu Island, Japan, in November 1945.

At Utah Beach the Americans benefitted greatly from the disruption to the German defence caused by having landed airborne forces behind the coastal positions. On D-Day itself Burwell was stationed on the attack transport ship USS *Bayfield*, several miles off the landing beach.

THE MILITARY GEOLOGY UNIT

A specialist unit was created by the Americans in June 1942 and led by Wilmot 'Bill' Bradley, a member of the United States Geological Survey (USGS). The Military Geology Unit produced valuable and plentiful material for operations, notably Operation Husky, the invasion of Sicily in 1943. Based on the unit's soil assessment, a diagrammatic soil map of Sicily and Calabria depicted areas favourable for invasion, rated according to the most trafficable of soils. There were also more impressively detailed maps: water-supply maps (see pages 82–83) depicted both perennial and non-perennial sources (a key distinction in Sicily) 'flowing' wells and 'deep' wells, springs, aqueducts and reservoirs; these maps were typically printed with tables giving the estimated maximum yield of sources shown on the map.

The value of this material in turn led to an increase in the unit's work in Europe. There was also coverage of Pacific operations. Thus, the unit produced Strategic Engineering Studies that included areas for potential airfields, for example on Leyte in the Philippines. The necessary data included the degree of flatness as well as conditions for foundation, drainage, construction materials, clearing and water supply. Based on existing geological maps, there was also the mapping of the type and distribution of construction materials; such maps were usually accompanied by tables summarising the suitability of the material mapped for a variety of purposes.

Mapping information was related to need. As a result, a 1945 map of southern Okinawa, included roads, the boundary of the area suitable for airfield sites and the suggested runway, to scale. The key included construction requirements (classed as minimum, moderate or maximum), the number of possible runways suggested, the length of the longest suggested runway (in terms of 6,000 feet or not more than 5,000 feet), and it reported any existing Japanese airfields. Tables that summarised ground conditions, grading and drainage problems, accessibility, and the availability of water and construction materials usually accompanied such airfield-construction maps.

There were also maps more specifically for campaigning. For example, Kikaigashima Island, northeast of Okinawa, was graded accordingly with:

> *'4 Terraces. Movement and road construction moderately difficult.*
> *4a. Movement impeded only by scarps; friable clay soil; terrace surfaces even.*
> *4b. Movement impeded by sticky clay soil; surface rolling.'*

Such information was fed into the maps produced for Intelligence summaries, such as that for Anvil, the operation later renamed Dragoon, the American–French landing in Provence on 15 August 1944 (see pages 92–93), for which the area within 200 miles (320 kilometres) of Toulon was subdivided into 17 regions described in the legend in terms of their terrain. There was more detailed consideration of the target areas for amphibious and airborne forces. Existing French maps were also overprinted with detailed information on defences, bridges, the likely capacity of harbours and the general topography:

> *'Tree and scrub covered hills with steep slopes, numerous streams, peat bogs and swamps will prevent widespread deployment of tracked and wheeled vehicles in this area.'*

as did the inherently fluid style of their operations. The British also markedly stepped up the production of hydrographic charts, notably in support of the 1944 invasion of Normandy. The danger of German bombing had led to the Admiralty's Hydrographic Office being moved from London to Bath in 1939, and then to new buildings in Taunton in 1941. The supply of, and need for, charts was more problematic for Britain in the Second World War than in the First World War due to the far greater range and speed of operations, and the resulting number of tactical possibilities, which led to increased demand that was met in part by the use of rotary offset printing machines. In the final run-up to the invasion of Normandy, the Hydrographic Office provided documents to over 6,000 vessels and authorities.

The needs-based aspect of mapping took many forms. In the Soviet Union there was a need for marine navigational-artillery charts for the Baltic Sea and Black Sea, which were designed to help warships bombard targets inland in support of ground forces, and detailed inland information and a kilometre grid were supplied accordingly. This task very much matched the general doctrines and practice for the use of Soviet warships.

Ground operations

Maps were more extensively used for ground operations than had been the case in the First World War. This was partly because of the greater mobility of units and movement of operations, not only on the Western and Eastern Fronts in Europe but also in the Mediterranean. For example, in the war in Italy in 1943–45 the comparison between the movement in the war there in 1915–18 and that in 1943–45 is instructive.

Terrain evaluation maps were important for both infantry and for vehicles operating off-road, both military and logistical. Such maps were major additions to topographical maps. German terrain evaluation maps were impressive and effective, covering a range of needs (both offensive and defensive) and regions, including the Libyan Sahara, and appeared at a number of scales. Colour was a key aspect in helping make the maps readily accessible. Aerial photography was important, although it could not replace ground information such as bridge weight limits. The information offered in German terrain evaluation maps included ground suitability, forest composition (type of tree) and density, slope gradients unsuitable for armoured vehicles and important viewpoints with their field of view. In addition, German officers and NCOs were taught map drawing so that they could produce ad hoc maps as well as additional information from the field for the cartographers.

The British established the Inter-Service Topographical Division in the autumn of 1940. The division carried out terrain analysis to supplement topographical maps, leading to the printing of information on maps in order to make them more useful for troop movements, for example with the notation 'irrigation ditches' on the maps prepared for the Salerno landing in Italy in 1943. Other remarks printed on the map included 'slopes generally unsuitable for widespread deployment' and 'limited deployment possible after reconnaissance'. Tactical needs led the British to develop what were termed 'goings' maps, which were designed to display the nature of the terrain and used colour to provide readily grasped analysis. In addition to the terrain, metalled roads were presented differently depending on their width.

American terrain appreciation maps covered a range of topics, including not only the impact of the terrain on troops and vehicles, but also additional material on slope, soil, vegetation, climate and geological features. 'Trafficability', the suitability of the terrain for cross-country movement, was the key element, and the Americans became adept at producing such material rapidly, such as in January 1945 in preparation for the successful invasion of Germany. The material was made more valuable by being accompanied by charts showing, per month, the expected number of days of 'trafficable ground'. The information available to American troops was formidable. Thus, the 'Stop Press Edition of 20 May 1944' for Isigny-sur-Mer in Normandy (see pages 142–143), a map to be used for briefing ground troops, included information on the softness of the ground, and on bridges: their overall length, the width of the stream at water level, the width of the road and the load classification of the bridge. For roadsides, the existence of banks, ditches and hedges was recorded, and on whether one or both sides of the road. The map was an update of the basic edition of May 1944, a map at a scale of 1:25,000 or 2.53 inches to the mile. The information drew heavily on aerial photographs. The contours were interpolated from spot heights on the French maps as amplified by aerial photographs. The information on magnetic declination was followed by 'compiled from Air Photographs on a control provided by existing French Triangulation'.

Reliable maps and mapping still encouraged planning to reach a specific point on a map at a certain time, planning already seen in the First World War; but that crucial planning goal was now linked to the greater mobility of troops, which meant more terrain needed to be mapped. In addition, the use of maps within the military was more widely extended. Whereas ordinary soldiers (unlike officers and senior non-commissioned officers) did not use maps extensively during the First World War, they did so in the Second World War and developed spatial awareness accordingly because maps were extensively employed at the tactical level to record both location and plans. Situation overlays were frequently used, although the means varied: from pencil or pen marked on an acetate film to a printed overlay. These overlays could reveal important aspects of the situation, as with that on Saipan for D+13, on 28 June 1944, with the fragmentary nature of the connection between the US Army units and the US Marines shown. Such maps were produced at a variety of scales, to cover units and combat at many levels. Hand-drawn maps were often part of the situation at the tactical level. Some examples have survived, such as from the US Marines at Saipan (see pages 144–145), but most have not because there was no institutional reason to keep them.

Mobility carried with it the risk of being obliged to defend an area to a greater depth than in the previous war, and this became more significant due to the Allies fighting Italy from June 1940 and Japan from December 1941. Thus, in 1940 the British mapped the border areas of Kenya in preparation for campaigning against Italian East Africa (Eritrea, Abyssinia and Italian Somaliland). In the event, the Italian invasion of Kenya was restricted to the capture of the border post at Moyale. The threat of a Japanese invasion of Australia in 1942 led to the production of large-scale maps for coastal areas, notably of Queensland, New South Wales and Victoria, and near the cities of Adelaide, Darwin and Perth. These maps were linked to the location of artillery, both to help determine and to map fields of fire from the guns – for example, to protect the naval base of Freemantle near Perth and also the sea approaches to Melbourne.

Special-purpose maps

Maps were also created for more specific purposes. For example, Britain was the first country to use 'escape and evasion maps' (see pages 120–121), notably by aircraft crew after having been shot down over enemy-held territory. They were initially printed on silk by the Edinburgh company John Bartholomew and Son as part of its business for the War Office; and then by Germany and the United States. Worn as scarfs, American escape maps included information on human settlements and ocean currents. The British produced over 1.75 million copies of about 250 separate escape and evasion maps.

Allied resource superiority and economic sophistication affected the conduct of the war at the strategic, operational and tactical levels. For example, as the Americans advanced across France in 1944 they generally did not storm villages and towns where they encountered resistance. Instead, they stopped, brought in aerial, tanks and artillery support, then heavily bombarded the site before moving in, with limited loss of American life. This approach did not lessen the contribution made by their effective infantry.

Artillery depended on a clear and accurate set of coordinates to locate their target with precision. Artillery boards were the means, with slide rules, of working out the firing data. Artillery would be sent grid references, usually a two-digit Alfa prefix (to confirm which map), and then a six-figure number of Eastings and Northings, which was computed at the gun battery command post, and the elevation and azimuth sent to each individual gun. Aircraft, in contrast, would (at the tactical level) eyeball the target using their navigator and a map.

To some of their detractors the Allies in France were overly keen in 1944 on waiting to bring up artillery, a course that could lead the Germans to disengage successfully and retreat. However, aside from self-propelled pieces like the 105mm M7, American artillery moved forward close to the line of advance, while the British had learned by hard experience at German hands in North

ALLIED SOLDIERS DOING THEIR LAUNDRY IN A CAPTURED GERMAN PILLBOX, PART OF THE ATLANTIC WALL FORTIFICATIONS.

One key element in mapping was the need to locate fortifications and to do so for strategic, operational and tactical ends, and as part of integrated defence systems. Fortifications had to be planned with reference to each other and to possible attack routes, and they had to be fitted into the terrain. The war is not generally considered in terms of fortifications, in part because of the emphasis on mobility, which for ground combat is notably placed on tanks. This emphasis, however, can lead to a serious failure to appreciate the significance of fortifications. The Soviet Red Army used artillery to blast through the Mannerheim Line in Finland in 1940, a triumph that brought their Winter War to the successful conclusion that is generally neglected due to the usual focus on earlier Finnish successes in heavily defeating Soviet attacks. In Manila Bay in the Philippines, Japanese air and artillery bombardment weakened the defences of American-held Corregidor in 1942, preparing the way for a successful amphibious assault and the rapid fall of the island. The Atlantic Wall built by the Germans to defend the French coast against Allied invasion showed in 1944 that ferroconcrete was very resistant to high explosive, but the defences were damaged or rapidly outflanked on 6 June. The Atlantic Wall neither prevented the Normandy landings nor seriously impeded them, the very difficult American experience on Omaha Beach notwithstanding. Even there, the troops still managed to get ashore and press inland in a matter of hours, and with fewer losses than anticipated.

These varied campaigns indicated that, as with other weapon systems, fortifications, while valuable, proved most effective as part of a combined arms force, while the combined dimension was also dependent on an appropriate plan. Fortifications required a flexible, supporting, counter-attack defence, because bunkers and other individual fortified positions were vulnerable and, once engaged, the occupants (as was commonly the case with fortifications) were trapped inside unless a counter-attack hit the attackers when the bunkers had held them up. All bunkers and casemates required apertures for guns, access for their crews and ventilation. With the right tactics and weapons (demolition charges, flamethrowers, grenades and tanks), most fortifications could be taken by attacking the occupants through these apertures, although it was time-consuming and tiring.

The French Maginot Line is generally held up as a failure. In practice, it provided the desired protection for much of the front and guided (diverted) the direction of the German attack in 1940 to the north of the Line through Belgium and northern France, where, however, the Germans brilliantly gained and retained the initiative in the area in which they advanced. Similarly, fortresses that eventually fell – such as Sevastopol in Crimea in 1942 and 1944, British-held Tobruk in Libya in 1942 and the German West Wall in 1945 – could delay advances and impose costs on the attackers, and thereby play a significant part in the outcome. The failure of the British defence of Singapore to do so in 1942 had serious operational and strategic consequences.

There were also ad hoc fortifications, such as those the British established near the Burma–India frontier, defending them successfully at Imphal and Kohima against Japanese attack in 1944. Fortified positions resisted the attacking efforts of not only infantry and artillery but also armour, as the Germans discovered in the Battle of Kursk in 1943 in the face of Soviet defences.

As a result of this importance, there were parallels with the mapping of opposing positions in the trench warfare of the First World War. Aerial reconnaissance again proved significant, as did the planning of artillery bombardments, for example by the British at El Alamein in Egypt in 1942 (see page 80). Greater tactical and operational mobility in the Second World War were important factors, but the similarities between the world wars can also be instructive.

Africa in 1941–42 the wisdom of methodical preparation and superior firepower when closing with the Germans. Montgomery used artillery and air support to prepare his attacks in 1944–45. His employment of the former reflected First World War doctrine and practice, as well as the defensive strength of the Germans. At the same time, Montgomery adapted to new ideas from subordinate officers and created doctrine accordingly, notably with combined arms doctrine.

In discussing American fighting style, it is necessary to allow for the variety of methods employed by the Americans in response to the many military environments in which they engaged. For example, although it was not only against the Japanese that the Americans employed close-in infantry techniques, these techniques became more significant in the Pacific in 1944–45 as the Japanese on the islands increasingly focused on resting on the defensive in well-fortified positions rather than attacking landing forces, a tactic that it was easier for the Americans to defeat. From 1944, the Americans employed 'corkscrew and blowtorch' tactics, involving satchel charges and flamethrowers, in order to kill Japanese in situ or to force them into the open to be killed by overwhelming fire. Maps were important to this process. Thus, for the successful invasion of Iwo Jima in February–March 1945 (see pages 156–157), a 'Special Air and Gunnery Target Map', produced as an annex to the operation plan, contained a special grid system to be used for pinpoint designations, and an arbitrary target square system to be used for area designations. The detailed defence symbol key indicated the extent of prior information provided by aerial reconnaissance.

In addition, bombardment plans were particularly important prior to invasions, not least to ensure the coverage of all defended positions. Alongside the extensive printed maps came more detailed manuscript maps, such as the one for the landing near Licata, Sicily, in 1943 (see page 138) that has survived in the estate of Captain Alfred Reid of the US Army's 3rd Ranger Battalion.

On Okinawa in 1945, the Japanese fortified the coral limestone hills and caves with tunnel positions, pillboxes, mortar emplacements and machine gun nests. The American suffered heavy casualties in capturing these positions in April–June, but they developed an effective co-operation of infantry, combat engineers and tanks that enabled them to seize the island. Situational awareness was very necessary in these attacks; the mapping-level was tactical rather than operational, and much of the fighting was ad hoc in character.

Resource superiority made it easier for the Americans and British to support combined arms operations; although it was also necessary to have the relevant doctrine and training. In part, this entailed knowing how best to respond to the combined arms tactics of opponents. The Germans proved unable to do so in France in 1944, where their moving armour was vulnerable to Allied close-air support. More generally, Allied air power helped to close the capability gap in combat effectiveness from which the Germans had initially benefitted in the war. This was an instance of the manner in which resource factors were very important in providing the basis for successful combined arms operations, although, again, they could not, in themselves, provide the necessary doctrine. Doctrine, which included the use of maps, had to adapt to weaponry, but also involved more factors.

Intelligence was a key element because it was necessary to locate enemy units accurately in order to mount appropriate air attacks. There were, however, contrasts between target areas and those of bomb falls, and these contrasts were demonstrated in maps – for example, the bomb fall maps covering Operation Goodwood, part of the British effort in the Battle of Normandy.

The conflict at sea

Again capturing the elements of reconnaissance and mapping, the conflict at sea – because of the greater precision that was now possible and the existence of a three-dimensional warfare environment – was increasingly about location. This situation reflected the extent to which surface ships engaged at beyond visual range, as well as the growing role of aircraft carriers and submarines. Aerial reconnaissance and radar had become of great importance. On 14 November 1942, off Guadalcanal in the Solomon Islands, the radar-controlled fire of the American battleship *Washington* wrecked the Japanese battleship *Kirishima*, which capsized on 15 November. Japanese battleships lacked radar-controlled fire. This night engagement was crucial to the American success in being able to land and fight on the island in January 1943. Until reliable, all-weather, day- and night-, reconnaissance and strike aircraft were available, which was not really until the 1950s, surface ships provided the prime means of fighting at night, although the highly successful British carrier attack on Italian warships in the harbour at Taranto in November 1940 was mounted at night. Moreover, some carrier aircraft carried radar.

On 24 May 1941, a British squadron sent to intercept the battleship *Bismarck*, the leading surface ship in the Kriegsmarine, as it tried to enter the North Atlantic in order to attack transatlantic shipping routes, was helped by radar when shadowing it off Iceland, only to suffer serious loss in the subsequent gunnery exchange, notably with the sinking of the battlecruiser *Hood*. The *Bismarck* then, initially, evaded detection, but after a successful aerial spotting it was eventually sunk on 27 May by a far larger British fleet and thanks to both aircraft and gunnery.

Submarines were less vulnerable than surface ships to blockade, detection and destruction. Moreover, they could be manufactured more rapidly and in large quantities, and required smaller crews. However, both submarines, and the warships and aircraft hunting them, faced grave difficulties given the vastness of the oceans and the problems entailed in finding and engaging targets, particularly in poor weather and low visibility. Part of the struggle involved the attempt to decipher the naval codes of opponents, in which the British and Americans proved particularly successful at the expense of Germany and Japan.

Capabilities and tactics at sea, just as with land and air warfare, have to be considered alongside resources and strategy, for the Germans had insufficient U-boats (submarines) to achieve their objectives, and those they did have lacked air support. Changing goals for both sides were significant in the struggle: for example, U-boat pressure in the Atlantic was reduced in late 1941 because submarines were moved to Norwegian and Mediterranean waters in order to attack Allied convoys to the Soviet Union and to deny the Mediterranean to British shipping. The extent to which this was the appropriate strategic response for Germany invites consideration, because although Allied aid to the Soviet Union was very important, it was not the reason for Germany's failure there in 1941.

Alongside the use of Intelligence (especially deciphered radio intercepts), air power played a key role in resisting the German submarine assault, notably by identifying the target and rapidly attacking it. Moreover, aircraft lessened the tactical capabilities of the U-boats: most U-boat attacks on shipping were made on the surface, which enabled them to use their guns, saved on torpedoes (of which they carried few) and also rendered Allied sonar less effective, but the threat from aircraft obliged U-boats to submerge, where their speed was much slower and it was harder for them to maintain visual contact with targets. On the surface, submarines operated using diesel engines that could drive the vessel three to four times faster than underwater cruising, which by necessity drew upon battery-powered electric motors, affecting cruising range and the number of days that could be spent at sea. Long-range Allied air power, both American and British, was the vital element that reduced the killing power of the German submarine fleet to the point that its effectiveness, measured by Allied tonnage lost, both in total and per submarine sunk, decreased steadily.

The British overprinted Admiralty charts in order to clarify the submarine threat and the possible response, including the range of submarine patrols between refits, the range of British destroyer escort groups and the likely limit of convoy routes. The 'Weekly Diagram of U-boat Warfare' (see pages 118–119) produced by the Admiralty's Anti-Submarine Warfare Division recorded the estimated average of submarines in an area, the anti-submarine vessels, the convoys and the shipping sunk. Coastal Command monthly reports used a chart of the waters round the British Isles to record submarine sightings, attacks on submarines, anti-submarine patrols, anti-invasion patrols and convoy escorts. Coastal Command's charts instantly showed changes in the range of German and British operations. Charts were used to record the sinking of U-boats – for example, when crossing the Bay of Biscay – and thus to gain an insight into their likely movements. Diagrams of attacks on particular convoys also aided subsequent analysis, producing assessments of the submarine threat to a convoy.

As a reminder of the key role of 'tasking' and its dependence on strategic choices, neither the RAF, which was focused on strategic bombing and theatre fighters, nor the Royal Navy, which was primarily concerned with hostile surface warships and was content to rely on convoys, hunter groups and sonar (of which there were very few sets) to limit submarine attempts, had devoted sufficient preparation to air cover against submarines. In addition, land-based aircraft operating against U-boats faced an 'Air Gap' across much of the mid-Atlantic, although Iceland's availability as an Allied base from April 1940 greatly increased the range of such air cover in the North Atlantic. Nevertheless, accurate navigation over water proved a problem for the RAF, with many missions ending as 'Convoy Not Found'.

In 1942, in contrast, there were marked improvements in Allied capability, including the increased use of shipborne radar and better sonar detection equipment, and this very local locational information powerfully supplemented that from signals Intelligence, notably Ultra, which was less specific in its locating of U-boats. In turn, enhanced weaponry, notably effective ahead-throwing, depth charge launchers and more powerful depth charges, as well as accumulated experience, gave effect to this information and capability.

There was a similar improvement with anti-submarine air resources. Again, very local target-finding was important, including ASV Mk II radar, and better searchlights. Radar sets small enough to be carried by aircraft, a key feature of applied capability, and yet capable of picking up submarine periscopes at five miles (eight kilometres), were a crucial tool. As a reminder that locational identification was part of a struggle between applied technologies, this radar lost its potency when the Germans were able to introduce detection equipment on U-boats. In turn, in March 1943 the Mk III radar, which could not be detected by these receivers, proved a crucial addition.

Allied capabilities and their implementation ensured that the number of German submarines sunk each year was much greater than in the First World War, while the percentage of Allied shipping lost was lower. This equation, important throughout, became less so when the United States entered the war, because that gave Britain access to American shipbuilding capacity. Allied success in the Battle of the Atlantic was crucial to the provision of imports to feed and fuel Britain, as well as to the build-up of military resources there in preparation for the invasion of Normandy. Thus, the Battle of the Atlantic was the background to the 'Second Front' sought by the Allies, and it underlined the strategic quandary faced by the Germans, with an intractable conflict on the Eastern Front likely to be joined by fresh commitments in France. The U-boats were proved to have had only an operational capability.

Precision targeting

Precision locational and navigational information, including on the weather, was necessary for air as well as naval warfare, and this information became more significant as the scale of each type of warfare increased – and notably so with air warfare. The identification of a target did not establish its significance, but raids were anticipated by careful briefing using target maps and photographs, for example with Luftwaffe attacks on Britain and for Allied attacks on Germany.

Nevertheless, there were serious problems with the accuracy of air raids. The British wished to destroy industrial targets in Germany, but the Butt Report on night raids in June–July 1941 showed that they were not doing so. Accuracy was difficult with night-time freefall bombing, and also, despite American bombsights, with daytime bombing, for there was no electronic navigation or target identification and the Allies did not have guided bombs.

Instead, concerned about the daytime vulnerability of their bombers, from March 1942 the British focused on night-time area bombing, helped by the Fire Hazard Maps produced for German cities, maps revised by the Allies in the light of aerial reconnaissance, as at Hanover on 14 January and 8 February 1945. Cities became Allied targets ranked on their economic importance. The capacity of bombers increased thanks to the use of four-engined aircraft. However, although an American aircraft such as the Boeing B17 Flying Fortress was a steady platform that could carry a large bomb load, precision bombing was not easy. Indeed, with bombs with restricted lethal radiuses, a tactic that involved bombardment squadrons all dropping them simultaneously could not be accurately described as precision.

Locational information, the instant mapping of a raid as it proceeded by navigators working with precise maps, was important. There were problems for the Allies with identifying the

AIRCRAFT NAVIGATION AND GUIDANCE IN THE OPERATIONS' ROOM AT NO. 10 GROUP RAF, BASED AT RAF BOX, WILTSHIRE.
Technology also affected mapping requirements. The first experiments with radar involved detecting surface ships and the ability to direct naval gunfire by radar developed. Radar was also important in finding surfaced submarines, and also in detecting periscopes. Then aircraft rapidly came into the picture. Capabilities were related to problems: radar was in part a reflection of the novel nature of the three-dimensional character of mapping that stemmed from the role of submarines at sea and from the addition of aircraft to the vertical space represented by terrain. The likely significance of air power and submarine attacks made this three-dimensional capability and therefore radar a major factor as a defensive capability. Radar offered little against a submerged submarine, where ASDIC-sonar would be useful, but the vulnerability of submarines on the surface ensured that radar played a role. As with the use of mapping for artillery, radar was a response to the need to fix position for accurate fire.

However, unlike artillery on land, which generally fired from a stationary position, aircraft and submarines posed an inherently dynamic character in location and, thus, the depiction of location. So also with naval artillery, which fired from a highly mobile platform. Thus, radar was a response to faster moving elements on the battlefield. Radar was a forerunner of what has become a key element in the depiction of the battle-space, that of the GPS (Global Positioning System). At the same time, it was necessary for radar to develop systems and practices to handle distorting features of the Earth's atmosphere.

Radio navigation systems, such as the German Knickebein system, were developed for aircraft. With Oboe, a targeting system first used in December 1941, and Gee-H, a radio navigation device introduced in 1942, the British developed accurate radio navigation systems that ensured weather, darkness and smog were less of an obstacle to bombing, which had hitherto depended on visual sighting.

Separately, Allied operations against submarines benefitted greatly at the tactical level from sonar. This technology complemented the use of signals interception to fix the general area of submarine presence.

The Operations' Room at RAF Box was established in June 1940 and moved later that year to an underground bunker in a disused quarry.

target, even in conditions of cloudless daylight and without anti-aircraft guns or enemy fighters; and accuracy remained heavily dependent on the skill of the pathfinder aircraft that preceded the bombers in order to identify targets. Nevertheless, accuracy and precision improved thanks to incremental steps, notably the role of airborne control at the target, and was significantly better in 1944 than for 1943, with the ability to mount attacks on a number of targets at the same time.

On the defensive

Mapping was also crucial for defence against air attack. This was more than simply about technical developments. Mapping is both a product of information systems and a means to their success. This was clearly shown with the Battle of Britain in 1940. Germany had radar, but it was not yet available to help the Luftwaffe. In a serious Intelligence failure, the Germans underrated the size, sophistication and strength of the British defence and, in particular, failed to understand the role of British radar and radar stations, and their place in the integrated air-defence system, which was a system of information, analysis and response. This control system, with its plotters of opposing moves and its telephone lines, was an early instance of the network-enabled capability seen in the 2000s with its plasma screens and secure data links: in each case, the targeting, sensors and shooters were linked through a network that included the decision-maker. Able command decisions and good Intelligence were linked through a system that was effective for rapid analysis and prompt response.

In turn, in reaction to the Allied air offensive, the Germans developed a complex and wide-ranging system of radar warning, with long-range, early-warning radars as well as short-range radars that guided night fighters (which also had their own radars) toward the bombers. This system caused heavy Allied losses, and led the British to map the areas and organisation of German night fighters. Radar-defence systems could be wrecked by the British use of 'window': strips of aluminium foil that appeared like bombers on the radar screens, which were, in effect, maps. In the autumn of 1943, the Germans adapted their radar to circumvent this, but, in turn, that November the British began to fly electronic counter-measures aircraft. Allied navigators used radar charts, on which Intelligence officers inked in German flak belts. The maps also marked the IP, the 'initial point' where the formation was to begin its visual bomb run.

The German ability to 'map' an Allied raid as it was proceeding was not matched by Japan, which lacked the integrated nature of the German defence system. Moreover, Japanese airfields were often without protective radar. This situation ensured that Allied aircraft were able to mount successful raids – for example, crucially, on the New Guinea airfield at Wewak in August 1943, after which Allied ground operations in New Guinea were rarely threatened by Japanese aircraft.

Air warfare tactics

The more general need to adapt tactics in order to use information more effectively was seen in the American air attacks on Japan. Initially, from November 1944, these were long-distance raids (from bases on the recently conquered island of Saipan) and unsupported by fighter cover, because fighter range was less than that of bombers. This led to American attacks from a high altitude, but that reduced their effectiveness. The raids that were launched were also hindered by: poor weather, especially cloudy conditions; strong tailwinds; difficulties with the new Boeing B29 Superfortress's reliability; and the general problems of precision bombing given the technology of the period. From February 1945, there was a switch to low-altitude night-time area-bombing of Japanese cities; and from April fighters based at the newly captured closer-in island of Iwo Jima supported these raids. The value of the American maps of Japanese cities thus depended heavily on the broader strategic situation and its consequences for the resources that could be applied.

Allied aircraft providing an increased volume of ground support also relied on information systems that included maps of the surface, targets and defences, as with the British mapping of the numerous gun emplacements and anti-aircraft batteries that hindered Allied air attempts to stop the withdrawal of German troops from Sicily in 1943 across the narrow Strait of Messina. The effectiveness of Allied ground support over Normandy in 1944 owed much to the long-term process of gaining air superiority over the Luftwaffe, but there had also been improvements in doctrine and organisation, including the use of air liaison teams with ground forces, as well as improvements in radio communication. Both attacks on specific targets and 'carpet bombing' relied on mapping. The latter was important to the American breakout from Normandy. Despite failures in coordination that led to 'friendly fire' casualties, notably in the successful breakout from the Cotentin Peninsula, the Allies had become far more skilled at integrating their forces.

Separately to maps, there was frequent use of reconnaissance to prepare models as aids to the planning of attacks and bombardments. Relief models were used by both the Allies and the Axis and helped to prepare air attacks, for example by the Japanese on Pearl Harbor (1941) and the successful attack by the British Dambusters on the German dams (1943). Models were also used to help ground forces, whether attacking defended coastlines, as with Anglo-American forces invading Sicily (1943) and Normandy (1944), or advancing overland, as with the Soviet attacks on Finnish defences in 1940 and on Berlin in 1945. That for Utah Beach, created by Charles Lee Burwell (see page 104), showed depth by bathymetric tints (colours placed between contour lines to indicate depth), as well as tide lines, the slope of the beach, buildings beyond the beach and the location of German defences.

Allies and Axis

The relationship between resources and the strategic, operational and tactical levels of war were clearly seen in the development of Allied mapping capability and its usage. Whereas Germany, Italy and Japan did not carry out systematic mapping, and generally simply overprinted, copied or enlarged existing maps, the Allies, and notably the Western Allies, carried out much new mapping. At the tactical level topographic maps were very important both for ground operations and for air offensives. New material was inputted to reflect fighting outcomes, with the British using maps based on new aerial photography in North Africa from 1941. There was also much new mapping for subsequent Allied operations in North Africa, Europe and the Pacific. In addition, Allied map preparation included the printing of maps for coalition units; thus, maps of the Monte Cassino area in Italy in 1944 were also prepared in Polish.

DUNKIRK

Royal Engineers, 'The Dunkirk Perimeter', 28 May 1940.

The German breakthrough to the Channel left an enormous pocket to the north, and this eventually became the Dunkirk position. The commander-in-chief of the British Expeditionary Force, Lord Gort, despatched Major General Thorne to Dunkirk to establish a defensive perimeter that would try to make the best use of the plentiful local water features – manmade (canals) and natural – to keep the Germans at bay. The key anchor point was Bergues in the west (with the canal from Dunkirk to Bergues and the canal from there to Furnes). The perimeter established was 25 miles long (40 kilometres) and eight miles (13 kilometres) deep. Within the zone a rear guard was established and a secondary stop line (the Dunkirk to Furnes Canal). The East Mole breakwater nearby was a crucial departure point. Plans were then put in motion to effect a timely evacuation of the many personnel corralled within.

French forces protected the perimeter's west flank. The rest of the zone was divided among three British corps, each of which was given an area in which to concentrate and a beach set aside for embarkation: Brav Dunes for 1 Corps, La Panne for 2 Corps and Malo-les-Bains for 3 Corps. An area east of Bergues was earmarked for possible flooding to create another obstacle for the pursuing Germans. The French and the British Expeditionary Force then fought ferociously to block the German advance until the evacuation could be effected. Meanwhile, a magnificent French defensive action at Lille, in which surrounded units of the First Army valiantly resisted German attacks from 28 to 31 May, played a key role in delaying the German advance, diverting German divisions and winning time for the evacuation.

From 26 May to 4 June, 213,448 British troops and 124,778 French ones were evacuated, a number far greater than had been anticipated, with the key dates being 29 May to 1 June. Unfortunately, 40,000 French troops had to be left behind. Evacuations are one of the most difficult military activities. They are a form of combined operations, always in itself problematic, but one in which the other side has set the agenda and it is necessary to evade this. Withdrawal is particularly difficult for troops who do not know what is going to happen, but can clearly hear the menacing sound of approaching foes. That was certainly the case with the Dunkirk perimeter, which was under serious attack and bombardment. The map captured the significance of inundations as a means to protect the position. A key element was the decision of the Germans on 24 May to focus instead on conquering the rest of France, a task for which they needed to replenish their military, notably the panzer divisions.

Evacuations are also strategic: you withdraw in order to fight again. This is a key element in military history on land and at sea, one that at sea is best handled by powers with a strong amphibious capability. The ability to withdraw after failures on land, for example from Gallipoli in 1915–16, was important. It became even more difficult to do so in the Second World War due to hostile air power, and this was a major factor at Dunkirk and at Crete in 1941. Nevertheless, in each case, large numbers were successfully withdrawn.

Conversely, a failure to evacuate by sea could be a serious disaster, as with Singapore and the Philippines in 1942, for the British and Americans respectively, and Tunisia in 1943 for the Germans and Italians. Each reflected the local superiority of opposing sea and, in particular, air power. As a result, large numbers of men were lost, which affected the issue of mass. The Germans proved much more successful in withdrawing troops from Sicily, Sardinia and Corsica in 1943.

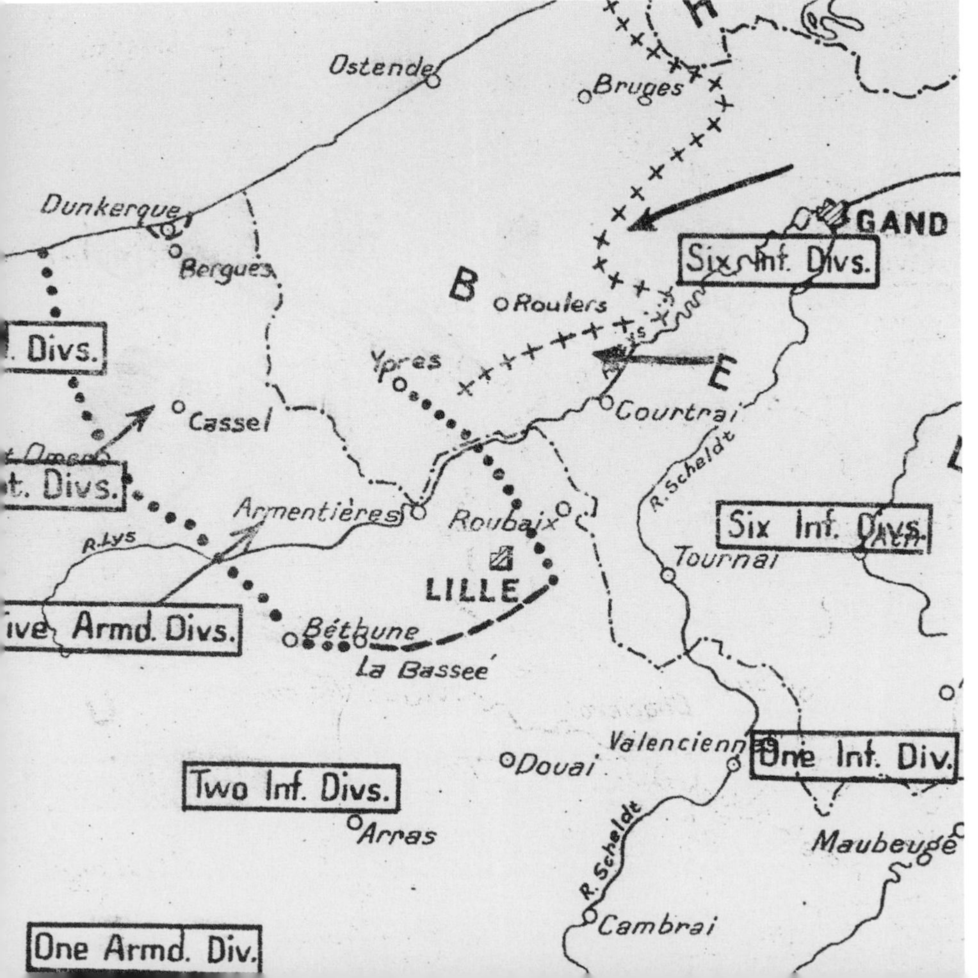

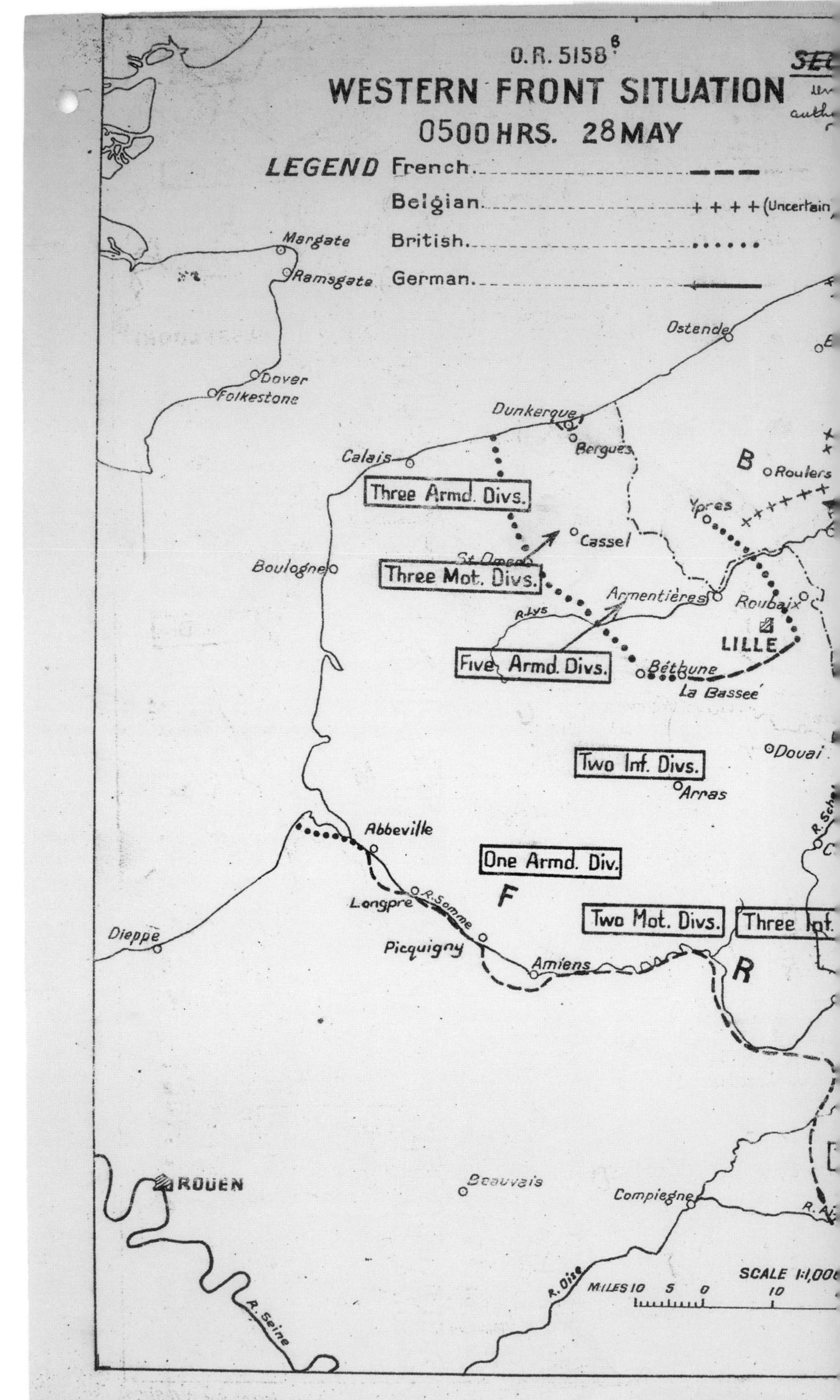
O.R. 5158
WESTERN FRONT SITUATION
0500 HRS. 28 MAY
LEGEND French.
Belgian.
British.
German.
Margate
Ramsgate
Ostende
Dover
Folkestone
Dunkerque
Bergues
Calais
B
Roulers
Three Armd. Divs.
Ypres
Cassel
St Omer
Boulogne
Three Mot. Divs.
Armentières
Roubaix
R. Lys
LILLE
Five Armd. Divs.
Béthune
La Bassee
Douai
Two Inf. Divs.
Arras
Abbeville
One Armd. Div.
F
Longpre
R. Somme
Two Mot. Divs.
Three Inf.
Dieppe
Picquigny
Amiens
R
ROUEN
Beauvais
Compiegne
R. Oise
R. Seine
SCALE 1:1,00
MILES 10 5 0 10

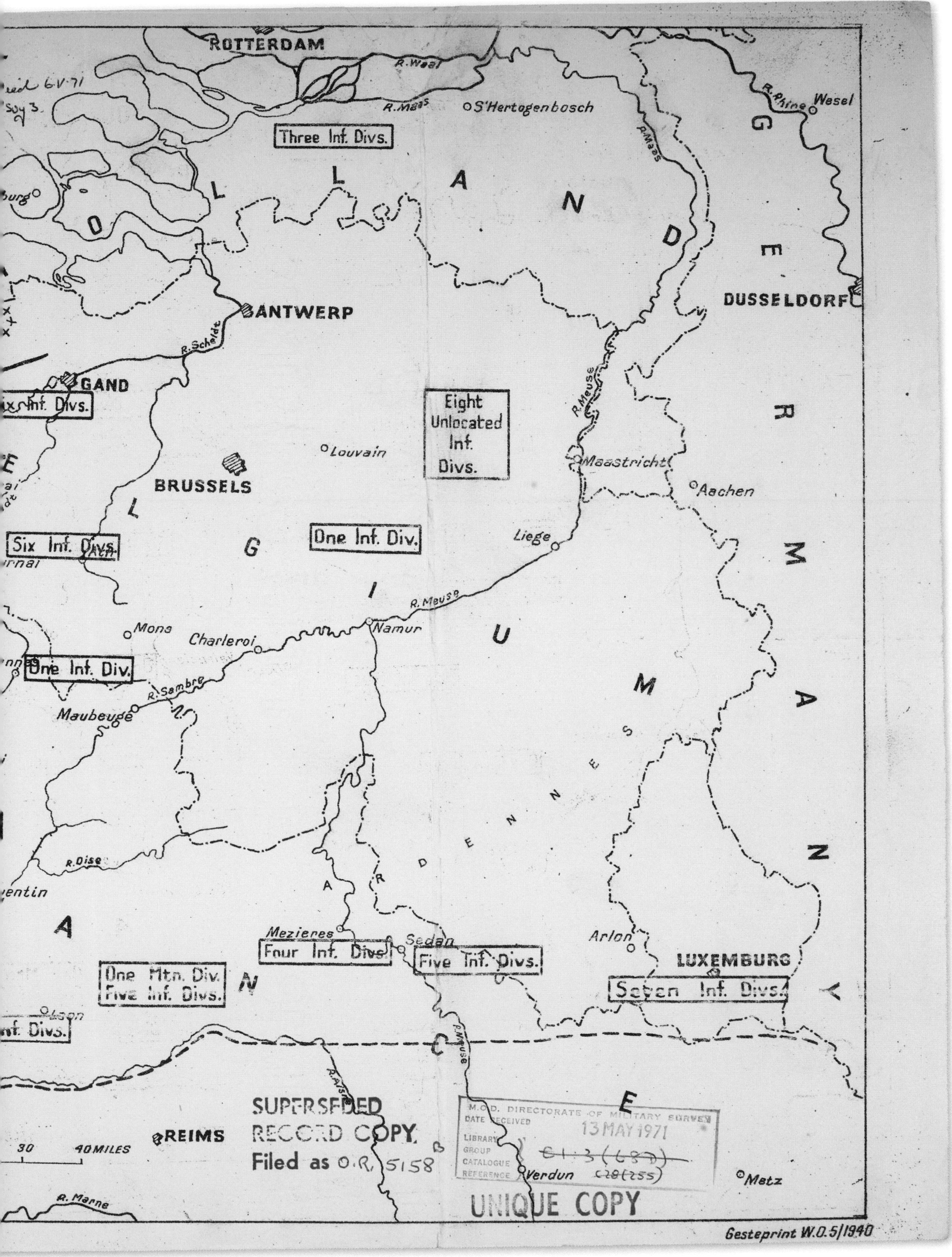
ROTTERDAM
R. Waal
R. Maas
S'Hertogenbosch
Three Inf. Divs.
R. Rhine
Wesel
R. Maas
ANTWERP
R. Scheldt
DUSSELDORF
GAND
Eight Unlocated Inf. Divs.
R. Meuse
Louvain
Maastricht
BRUSSELS
Aachen
One Inf. Div.
Liege
Six Inf. Divs.
R. Meuse
Mons
Charleroi
Namur
One Inf. Div.
R. Sambre
Maubeuge
R. Oise
A R D E N N E S
Mezieres
Sedan
Arlon
Four Inf. Divs.
Five Inf. Divs.
LUXEMBURG
One Mtn. Div.
Five Inf. Divs.
Seven Inf. Divs.
Laon
R. Meuse
R. Aisne
REIMS
30
40 MILES
Verdun
Metz
R. Marne
G E R M A N Y
H O L L A N D
B E L G I U M
F R A N C E
Gesteprint W.O.5/1940

THE BLITZ

Luftwaffe target map for London, September 1940.

Target acquisition is always a key process in bombing, but the detailed maps provided can be highly misleading as far as implementation is concerned. This map indicates the more general process by which the Germans used British Ordnance Survey maps that they had overprinted in order to identify prime targets to bomb, which in this case included the War Office '(1311)' and the Admiralty '(1312)', as well as the area of neutral embassies ('*Neutrale Botlschaften u.s.w.*') that should be avoided. A kilometre scale is overprinted at the base of the map, and 'Lft Kdo.2' identifies the map as for Luftflotte 2, which was formed from Luftwaffengruppenkommando 2 and was involved in the air assault on Britain. However, as happened later with Allied planning, the identification of targets was often inadequate. The BBC (British Broadcasting Corporation) was more significant than its absence here suggests. Furthermore, the Heinkel He 111, the Luftwaffe's fast medium bomber, had a variety of specifications, depending on the variant used, but the key elements were that it was using free-fall bombs and was flying at over 200mph (320kmh; maximum speed 270mph/435kmh). The first compromised the accuracy of bombing, while the latter ensured that time-over-target was restricted. Moreover, although the aircraft had a low-level capability, the presence of barrage balloons, ground anti-aircraft fire and intercepting aircraft encouraged the Luftwaffe to fly closer to the service ceiling of 21,300 feet (6,500 metres), which affected accuracy. As a result, it was not possible to achieve the accuracy in targeting suggested by such maps.

From 7 September 1940, the Luftwaffe bombed London heavily and repeatedly. The German word *blitzkrieg* ('lightning war') was shortened by the British to give a name to this terrifying new form of conflict: the Blitz. That day, 348 German bombers attacked. The inadequately prepared defences were taken by surprise and failed to respond adequately. There were only 92 anti-aircraft guns ready for action, and their fire-control system failed. The British fighter squadrons proved unable to stop the powerful air assault. Some 430 people were killed in the East End, where the docks on the River Thames were hit hard. This was what became a German strategy to starve Britain into surrender, in part by destroying the docks through which food was imported. Liverpool, Plymouth and Southampton were also to be heavily attacked for that reason, as well as to deny the Royal Navy bases.

Another 412 people were killed in a night raid on London on 8 September. It was only on 11 September that anti-aircraft fire was sufficiently active to force the Germans to bomb from a greater height. Barrage balloons also forced bombers to fly higher, thus reducing bombing accuracy. In comparison with later figures, these casualties were modest, but they were far higher than those from individual bombing raids during the First World War.

From 7 September until mid-November 1940, the Germans bombed London every night bar one. There were also large-scale daylight attacks, as well as hit-and-run attacks. The attack on the moonlit night of 15–16 October proved particularly serious, with 400 bombers active, of which only one was shot down by the RAF's 41 fighters. The railway stations were hard hit, while Battersea Power Station was struck, as was BBC headquarters. By mid-November, when the attacks became less focused on London, over 13,000 tons of high explosive bombs had been dropped on the city, as well as nearly one million incendiaries.

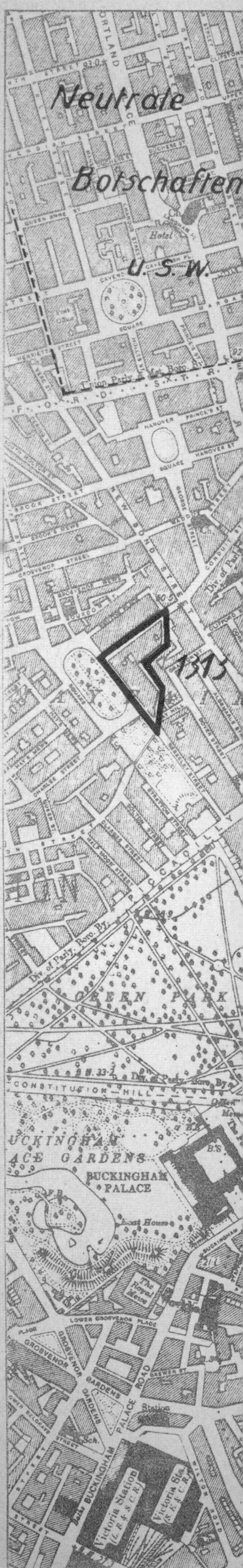

0' 30" N
7' 15" W Kartenmitte

London

Mayfair Square
Neues Luftfahrtministerium
Techn. Amt

1: 100 000 Engl. Bl. ~~34~~ 29
1: 63 360 Engl. Bl. 107 u. 115

WESTERN APPROACHES

British Anti-Submarine Warfare Division, 'Weekly Diagram of U-Boat Warfare', 30 September–27 October 1940.

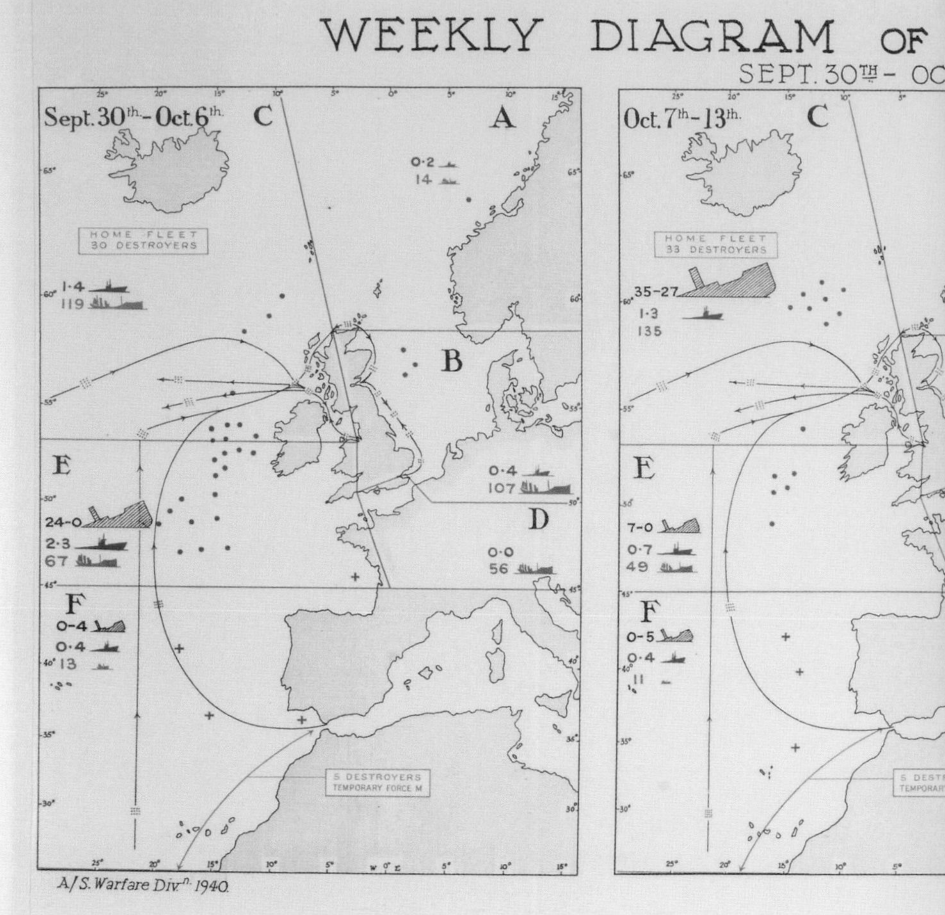

The First World War had demonstrated the impact that U-boats could have on Britain's ability to wage war, but in September 1939 Germany had just 57, only half of which had the range to enable operations in the Atlantic. There was a lull in U-boat activity during early 1940, but at the end of May the Germans announced that U-boat warfare was recommencing (a warning to neutrals). The amount of Allied shipping sunk by U-boats rose in the summer of 1940 when the tactic was adopted of surfaced attacks at night. The U-boats began to make pack attacks and by October 1940 the Admiralty feared the U-boats were winning.

The weekly diagrams mapped the challenge facing British anti-submarine warfare but they also showed how that challenge varied depending on the opportunities apparently facing the U-boats. The maps did not differentiate between the U-boats en route to or from cruising areas and those on station. Whereas the beginning of the Arctic convoys (of supplies to the Soviet Union after the Germans attacked in June 1941) was to dramatically increase activity in zone A, in late 1940 there was nearly none because Britain was not trading with occupied Norway or with the Soviet Union, which was then an ally of Germany. Zone B was essentially U-boats in transit to the Atlantic and attacks on the British inshore east coast convoy route, which was reasonably well protected. Zones E and F were both ones of transit and also, with C, areas of attacks on Atlantic convoys.

This map for September–October is a snapshot of the growing threat to the Western Approaches: shipping losses in October 1940 were the worst of the war to date, especially around the still quite full moon on 18–19 October. The cluster of red dots northwest of Scotland in the map for 14–20 October is largely accounted for by wolfpacks lying in wait and encountering two convoys. Convoy SC7, en route from Canada to Liverpool with 30 ships, was attacked repeatedly on 16–19 October; 20 ships were sunk and six damaged. On 19–20 October, Convoy HX79 was also en route to Liverpool with war materials, losing 12 sunk and two damaged of its 49 ships.

The British struggle against German submarines became more serious once the Germans established bases in Norway and France, although the occupation of Danish-ruled Iceland in May 1940 helped Britain. There was less anxiety about Greenland, which backed the Free Denmark movement. The Germans benefitted from Italy's entry into the war in June, which increased pressure on the Royal Navy. The Germans also had operational, from winter 1940–41, the Type VIIc submarine; its range (ultimately) of 12,600 nautical miles (23,350 kilometres) meant it was able to operate in the North Atlantic, although it was very limited compared to the Type XXI that could travel fast underwater, but that only came into active service at the end of the war. The fall of France enabled the Germans to use the French Atlantic ports as U-boat bases, from where they could reach out into the western parts of the North and Central Atlantic. British concern about the U-boat threat had resulted in planning, in early 1941, for landings in the Azores and the Canary Islands to pre-empt possible German moves. In the event they were not carried out.

As dangerous as the U-boat attacks were, the Germans lost U-boats more speedily than new ones

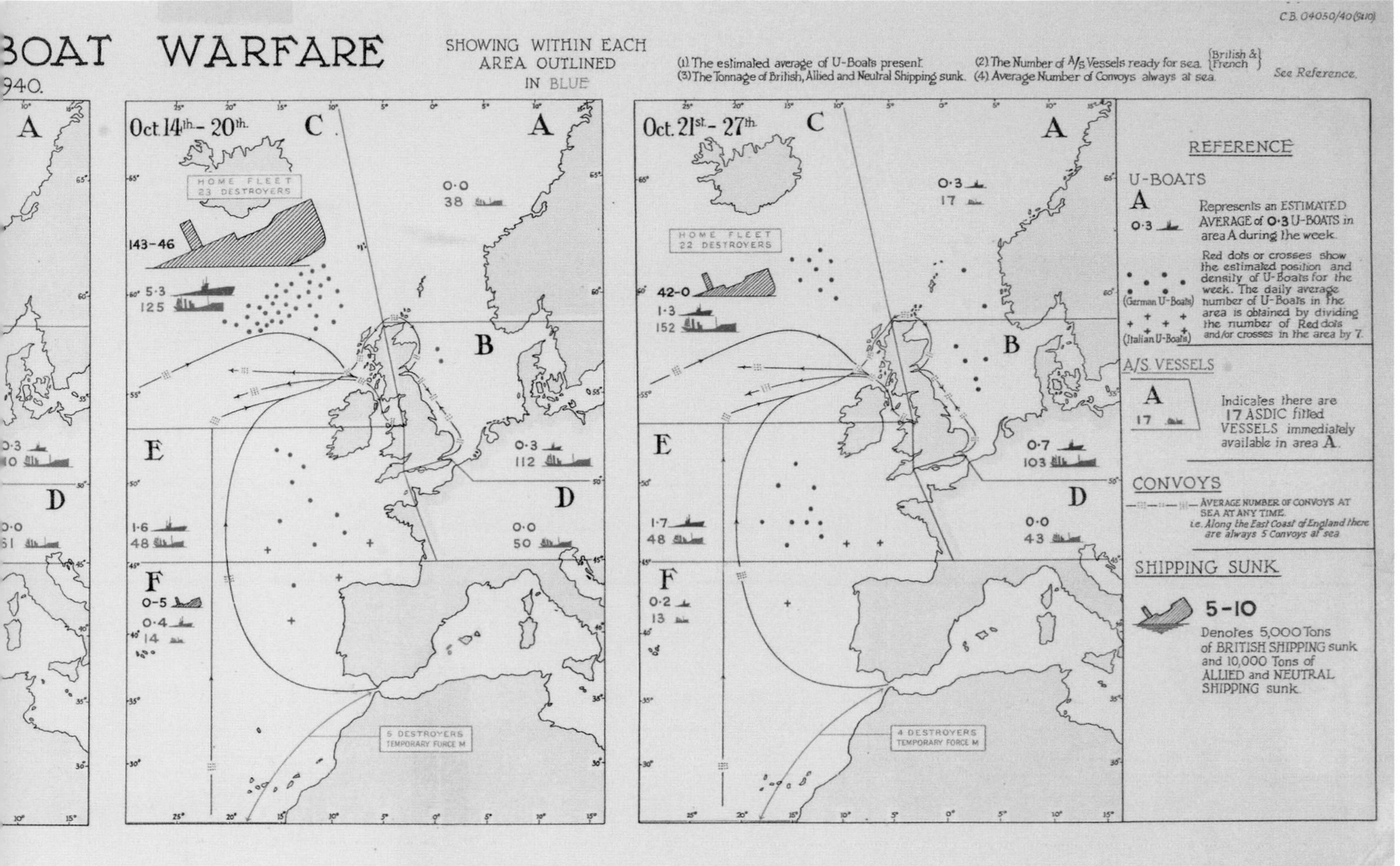

were being brought into service. The British used ASDIC (sonar), but there were insufficient sets, while the sonar was less effective when U-boat 'kills' of shipping were made by attack on the surface. In contrast, aircraft forced U-boats to submerge, where their speed was much slower and it was harder to maintain visual contact with targets. On the surface, submarines operated using diesel engines that could drive the vessel three or four times faster than underwater cruising, which was powered, by necessity, from battery-powered electric motors. This affected cruising range and the number of days that could be spent at sea. The U-boat assault, which was supported by Italian submarines, notably based in Bordeaux, was linked to the air attacks on British ports, although there was no specific coordination.

Despite the hopes placed by their advocates, and the enhancement offered by 'wolfpack' tactics from the winter of 1940–41 and by the submarine snorkel from 1943, the U-boats were proved to have only an operational capability, repeating the situation in the First World War. Aside from blockading Britain into surrender, the U-boats were also part of a defensive strategy because only once the U-boats had been contained as a serious operational challenge (and thus eliminated as a real strategic threat) could an invasion of occupied Europe be planned; a major maritime invasion was not practical otherwise. In 1943 the operational capability of the U-boats was eclipsed by improvements in anti-submarine warfare and large-scale and rapid American shipbuilding. Allied merchant tonnage sunk by U-boats fell from 5,800,000 in 1942 to 2,300,000 in 1943. Most convoys crossed the Atlantic without suffering any or many losses. Allied air attacks on the well-protected submarine bases in France, notably Brest, Lorient and Saint-Nazaire, did not end the threat, but increased the burden of the Atlantic war on both sides. By 1944 the merchant tonnage lost (600,000) was the same as that of 1939, and well below every other year of the war so far. Moreover, the Allies built over 13 million tons of shipping that year.

ESCAPE MAPS

War Office and John Waddington Ltd., Leeds, Schaffhausen escape map provided to British aircrew, c.1940.

This silk map shows a section of the German-Swiss border (at 1:100,000 scale) and is known as the Schaffhausen escape map. The map is also often identified with Lieutenant Airey Neave, who escaped from Colditz in January 1942, after which he crossed from Singen in Germany to Ramsen in Switzerland. Upon his return he was recruited into the British Directorate of Military Intelligence Section 9 (MI9), which was responsible for supporting resistance networks in Europe and also entrusted with helping Allied servicemen, airmen in particular, to return to Britain.

The creation of the Schaffhausen map relied on aerial photography as well as Intelligence both from escapees and from Britain's Intelligence operation in neutral Switzerland. Information was obtained from local smugglers. The instructions printed at the top of the map are explicit: 'Escapes into Switzerland have the greatest chance of success if attempted across the frontier of the Canton Schaffhausen. The region around LAKE CONSTANCE is to be avoided.' Lake Constance (the Bodensee) was to be avoided because it was heavily fortified with boats on the lake, searchlights, shore patrols and barbed wire. The frontier was harder for the Germans to police near Schaffhausen, a Swiss enclave to the north of the Rhine. MI9 had selected this area precisely because it offered the safest crossing point for escapers to reach neutral Switzerland. MI9 also helped to foster in British servicemen a culture of escape-mindedness, to defy the enemy and seek to return home, boost morale and continue the war effort.

The German civilian population was generally very hostile to Allied aircrew, many of whom were shot down, which made the provision of anything to help effect an escape even more necessary. This map and its embedded information provided plentiful detail but the hazards were many. The text advises which areas must be avoided (like German posts) and detailed descriptions of the characteristic landmarks. Mountain paths that follow the crest of hills are marked with a red dotted line.

Escape maps were the brainchild of the polymath Christopher Clayton Hutton, who was Technical Officer to the Escape Department at the War Office. He realised that maps needed to be thin enough to be able to be secreted, without revealing noise, in a boot or a jacket lining; durable and waterproof enough to survive action in the field; and to contain sufficient detail to be of vital use to someone in an unfamiliar location. He settled eventually on silk printed with a mixture of ink and pectin. However, he anticipated that silk supplies might be prioritised for parachutes and after he became aware that Japan's lightweight *fugo* balloon bombs used mulberry leaf paper, he obtained some and tested it with sensational results. The makers of the Monopoly board game set up a secret printing space at their factory in Leeds to produce MI9's maps. By 1945 the equivalent of three British Army divisions of escapees and evaders had managed to get back to Britain, many of them using a copy of one of more than 240 individual map designs created for the escape and evasion programme.

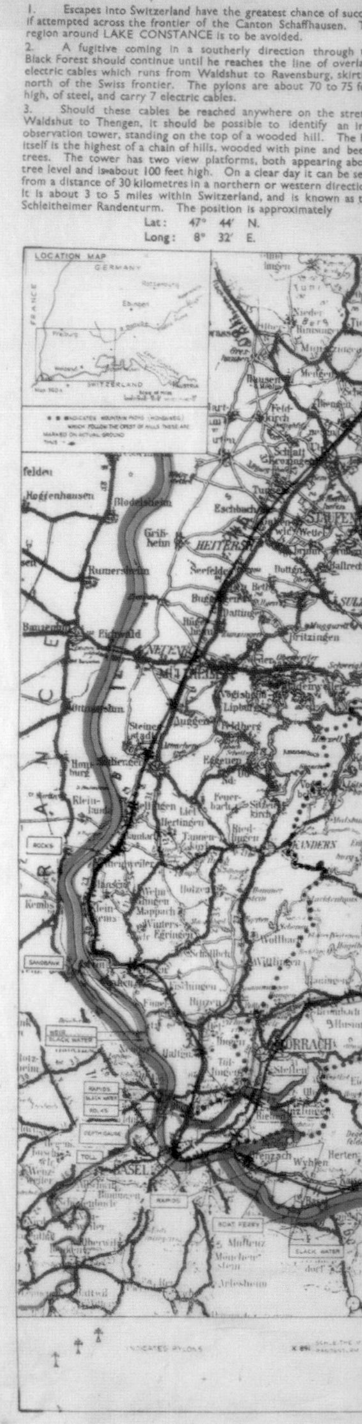

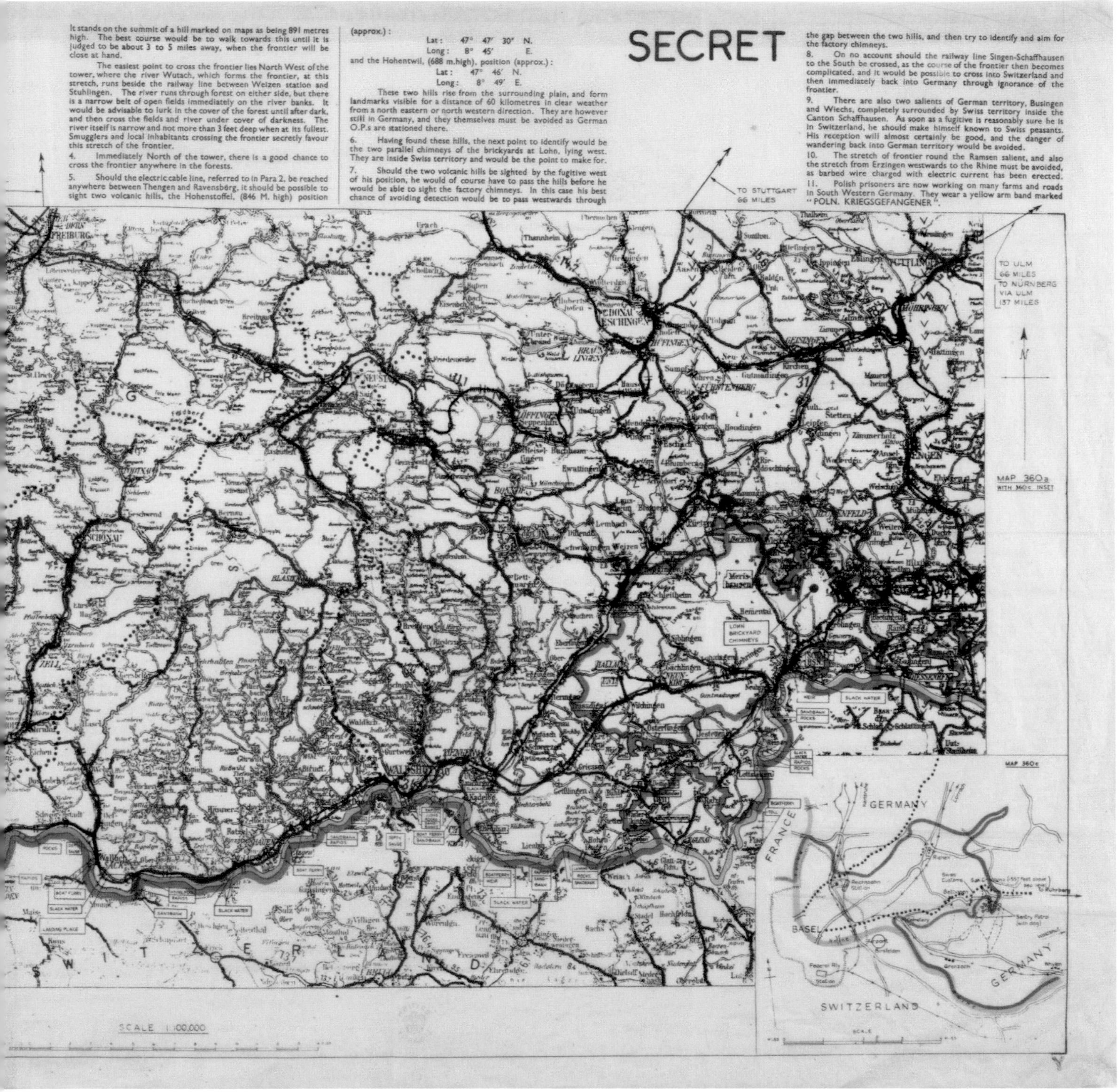

SECRET

It stands on the summit of a hill marked on maps as being 891 metres high. The best course would be to walk towards this until it is judged to be about 3 to 5 miles away, when the frontier will be close at hand.

The easiest point to cross the frontier lies North West of the tower, where the river Wutach, which forms the frontier, at this stretch, runs beside the railway line between Weizen station and Stuhlingen. The river runs through forest on either side, but there is a narrow belt of open fields immediately on the river banks. It would be advisable to lurk in the cover of the forest until after dark, and then cross the fields and river under cover of darkness. The river itself is narrow and not more than 3 feet deep when at its fullest. Smugglers and local inhabitants crossing the frontier secretly favour this stretch of the frontier.

4. Immediately North of the tower, there is a good chance to cross the frontier anywhere in the forests.

5. Should the electric cable line, referred to in Para 2, be reached anywhere between Thengen and Ravensbürg, it should be possible to sight two volcanic hills, the Hohenstoffel, (846 M. high) position (approx.):

Lat: 47° 47′ 30″ N.
Long: 8° 45′ E.

and the Hohentwil, (688 m.high), position (approx.):

Lat: 47° 46′ N.
Long: 8° 49′ E.

These two hills rise from the surrounding plain, and form landmarks visible for a distance of 60 kilometres in clear weather from a north eastern or north western direction. They are however still in Germany, and they themselves must be avoided as German O.P.s are stationed there.

6. Having found these hills, the next point to identify would be the two parallel chimneys of the brickyards at Lohn, lying west. They are inside Swiss territory and would be the point to make for.

7. Should the two volcanic hills be sighted by the fugitive west of his position, he would of course have to pass the hills before he would be able to sight the factory chimneys. In this case his best chance of avoiding detection would be to pass westwards through the gap between the two hills, and then try to identify and aim for the factory chimneys.

8. On no account should the railway line Singen-Schaffhausen to the South be crossed, as the course of the frontier then becomes complicated, and it would be possible to cross into Switzerland and then immediately back into Germany through ignorance of the frontier.

9. There are also two salients of German territory, Busingen and Wiechs, completely surrounded by Swiss territory inside the Canton Schaffhausen. As soon as a fugitive is reasonably sure he is in Switzerland, he should make himself known to Swiss peasants. His reception will almost certainly be good, and the danger of wandering back into German territory would be avoided.

10. The stretch of frontier round the Ramsen salient, and also the stretch from Erzingen westwards to the Rhine must be avoided, as barbed wire charged with electric current has been erected.

11. Polish prisoners are now working on many farms and roads in South Western Germany. They wear a yellow arm band marked "POLN. KRIEGSGEFANGENER".

TO STUTTGART
66 MILES

TO ULM
66 MILES
TO NÜRNBERG
VIA ULM
137 MILES

MAP 360a
WITH 360c INSET

LOHN
BRICKYARD
CHIMNEYS

MAP 360c

GERMANY
FRANCE
BASEL
SWITZERLAND

SCALE 1:100,000

LUBLIN GHETTO

Announcing the formation of a ghetto in Lublin, 24 March 1941.

The city of Lublin in Poland was seized on 18 September 1939 and allocated to the General Government of Poland region established by the Germans. Lublin was one of the most important cultural centres for Eastern European Jewry. Jews had thrived there for 500 years, and in 1939 accounted for around 43,000 of the city's 120,000 population, supporting many synagogues and rabbinical schools.

The Nazis moved German settlers into the city and persecution of the city's Jews began. In March 1941 a proclamation, with a map (right), was issued by Lublin's Nazi governor Ernst Zörner. It announced 'with immediate effect' the formation of a ghetto for 'all Jews resident in Lublin' within the Podzamcze district, the poorest part of the historical Old Town beneath the castle. The notice states that the ghetto boundaries are deliniated by the following streets: 'From the corner of Kowalska, down Kowalska and Krawiecka and along the block of houses marked in the plan, crossing the open ground of Sienna to Kalinowszczyzna, up to the corner of Franciszkańska; along Franciszkańska until it reaches Unicka Street and the corner of Lubartowska; down Lubartowska to the corner of Kowalska.' One of the ghetto gates stood at the corner of Lubartowska and Kowalska.

In October 1941 Lublin became the location for the headquarters of Operation Reinhard, the SS plan for the systematic slaughter of 2,284,000 Jews in the General Government area. Since November 1939 SS General Odilo Globocnik had been the brutal head of police in Lublin district. By early 1942 he had established deportation teams and the building of killing centres, such as Majdanek, Sobibór and Bełżec. Most of the Jews from Lublin were murdered in the Bełżec extermination camp, and Lublin contained the SS clothing warehouses where the clothing, shoes, jewellery and gold of victims was sorted, some of it for use in Germany.

At Bełżec, the Germans began gassing on 17 March 1942, and could kill 1,500 people there daily. Jews from Lublin were followed by those from Lwów (Lvov) and then Kraków (Cracow), and over 600,000 Jews were slaughtered there, as were 1,500 Poles who tried to assist Jews. In November 1942, Jan Karski (real name Jan Kozielewski), a Polish agent who had entered and left Bełżec disguised as a guard, arrived in London with a full account of the killing. In response, on 17 December, the Allied governments issued a declaration attacking 'this bestial policy of cold-blooded extermination'. By the time the Soviets captured Lublin on 23 July 1944, the ghetto had been liquidated. It was in Lublin, on 31 December 1944, that the Lublin Committee (formed, in agreement with the Soviet government, in the city that July) was transformed into the (pro-Moscow) Provisional Government of the Polish Republic.

BEKANN

betr. die Bildung eines

in d

Aus zwingenden öffentlichen

ein geschlossener jüdischer Wohnbezirk

ordne ich an:

1. Das Ghetto der Stadt Lu

Kowalska über Kowalska, Krawie

serblock entlang, über das freie Feld

dend zur Kalinowszczyzna bis zu

kańska, Franciszkańska über U

Ecke Lubartowska, Lubartowska

Kowalska. Öffentliche Gebäude, so

den und Formationen belegte Häuser u

ben hiervon unberührt.

Lublin, den 24. März 1941.

Der Chef d

gez.

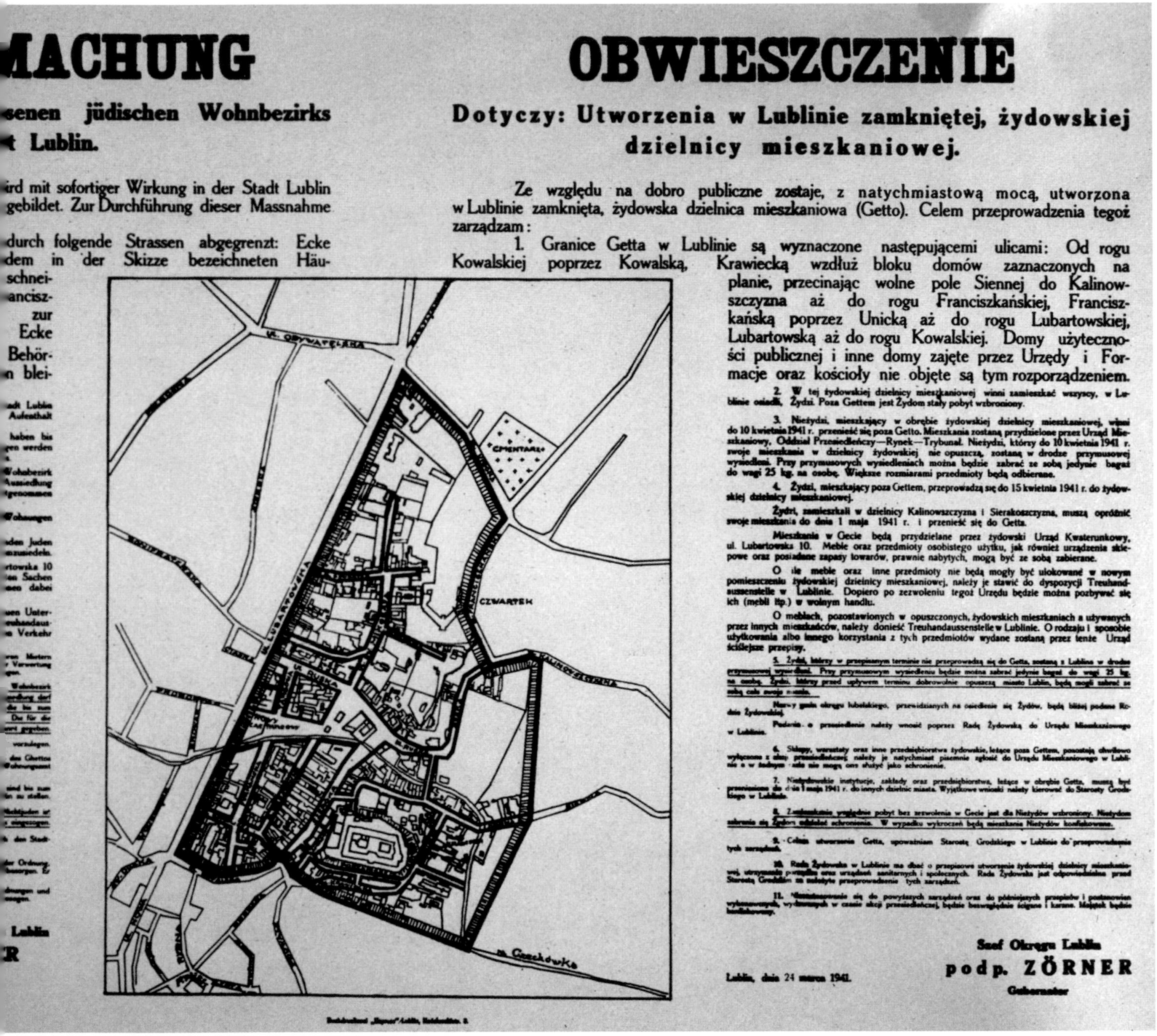

[...]MACHUNG

[...]senen jüdischen Wohnbezirks
[...]t Lublin.

[...]ird mit sofortiger Wirkung in der Stadt Lublin
[...]gebildet. Zur Durchführung dieser Massnahme

[...]durch folgende Strassen abgegrenzt: Ecke
[...]dem in der Skizze bezeichneten Häu-
[...]schnei-
[...]ancisz-
[...]zur
[...]Ecke
[...]Behör-
[...]n blei-

[...]adt Lublin
[...]Aufenthalt

[...]haben bis
[...]gen werden

[...]Wohnbezirk
[...]Aussiedlung
[...]genommen

[...]Wohnungen

[...]den Juden
[...]uzusiedeln.
[...]rtowska 10
[...]en Sachen
[...]een dabei

[...]uen Unter-
[...]handaus-
[...]Verkehr

[...]Lublin
[...]R

OBWIESZCZENIE

Dotyczy: Utworzenia w Lublinie zamkniętej, żydowskiej dzielnicy mieszkaniowej.

Ze względu na dobro publiczne zostaje, z natychmiastową mocą, utworzona w Lublinie zamknięta, żydowska dzielnica mieszkaniowa (Getto). Celem przeprowadzenia tegoż zarządzam:

1. Granice Getta w Lublinie są wyznaczone następującemi ulicami: Od rogu Kowalskiej poprzez Kowalską, Krawiecką wzdłuż bloku domów zaznaczonych na planie, przecinając wolne pole Siennej do Kalinowszczyzna aż do rogu Franciszkańskiej, Franciszkańską poprzez Unicką aż do rogu Lubartowskiej, Lubartowską aż do rogu Kowalskiej. Domy użyteczności publicznej i inne domy zajęte przez Urzędy i Formacje oraz kościoły nie objęte są tym rozporządzeniem.

2. W tej żydowskiej dzielnicy mieszkaniowej winni zamieszkać wszyscy, w Lublinie osiadli, Żydzi. Poza Gettem jest Żydom stały pobyt wzbroniony.

3. Nieżydzi, mieszkający w obrębie żydowskiej dzielnicy mieszkaniowej, winni do 10 kwietnia 1941 r. przenieść się poza Getto. Mieszkania zostaną przydzielone przez Urząd Mieszkaniowy, Oddział Przesiedleńczy—Rynek—Trybunał. Nieżydzi, którzy do 10 kwietnia 1941 r. swoje mieszkania w dzielnicy żydowskiej nie opuszczą, zostaną w drodze przymusowej wysiedleni. Przy przymusowych wysiedleniach można będzie zabrać ze sobą jedynie bagaż do wagi 25 kg. na osobę. Większe rozmiarami przedmioty będą odbierane.

4. Żydzi, mieszkający poza Gettem, przeprowadzą się do 15 kwietnia 1941 r. do żydowskiej dzielnicy mieszkaniowej.

Żydzi, zamieszkali w dzielnicy Kalinowszczyzna i Sierakoszczyzna, muszą opróżnić swoje mieszkania do dnia 1 maja 1941 r. i przenieść się do Getta.

Mieszkania w Gecie będą przydzielane przez żydowski Urząd Kwaterunkowy, ul. Lubartowska 10. Meble oraz przedmioty osobistego użytku, jak również urządzenia sklepowe oraz posiadane zapasy towarów, prawnie nabytych, mogą być ze sobą zabierane.

O ile meble oraz inne przedmioty nie będą mogły być ulokowane w nowym pomieszczeniu żydowskiej dzielnicy mieszkaniowej, należy je stawić do dyspozycji Treuhandaussenstelle w Lublinie. Dopiero po zezwoleniu tegoż Urzędu będzie można pozbywać się ich (mebli itp.) w wolnym handlu.

O meblach, pozostawionych w opuszczonych, żydowskich mieszkaniach a używanych przez innych mieszkańców, należy donieść Treuhandaussenstelle w Lublinie. O rodzaju i sposobie użytkowania albo innego korzystania z tych przedmiotów wydane zostaną przez tenże Urząd ściślejsze przepisy.

5. Żydzi, którzy w przepisanym terminie nie przeprowadzą się do Getta, zostaną z Lublina w drodze przymusowej wysiedleni. Przy przymusowym wysiedleniu będzie można zabrać jedynie bagaż do wagi 25 kg. na osobę. Żydzi, którzy przed upływem terminu dobrowolnie opuszczą miasto Lublin, będą mogli zabrać ze sobą całe swoje mienie.

Nazwy gmin okręgu lubelskiego, przewidzianych na osiedlenie się Żydów, będą bliżej podane Radzie Żydowskiej.

Podania o przesiedlenie należy wnosić poprzez Radę Żydowską do Urzędu Mieszkaniowego w Lublinie.

6. Sklepy, warsztaty oraz inne przedsiębiorstwa żydowskie, leżące poza Gettem, pozostają chwilowo wyłączone z akcji przesiedleńczej; należy je natychmiast pisemnie zgłosić do Urzędu Mieszkaniowego w Lublinie a w żadnym razie nie mogą one służyć jako schronienie.

7. Nieżydowskie instytucje, zakłady oraz przedsiębiorstwa, leżące w obrębie Getta, muszą być przeniesione do dnia 1 maja 1941 r. do innych dzielnic miasta. Wyjątkowe wnioski należy kierować do Starosty Grodzkiego w Lublinie.

8. Zamieszkanie względnie pobyt bez zezwolenia w Gecie jest dla Nieżydów wzbroniony. Nieżydom zabrania się Żydom udzielać schronienia. W wypadku wykroczeń będą mieszkania Nieżydów konfiskowane.

9. Celem utworzenia Getta, upoważniam Starostę Grodzkiego w Lublinie do przeprowadzenia tych zarządzeń.

10. Rada Żydowska w Lublinie ma dbać o przepisowe utworzenie żydowskiej dzielnicy mieszkaniowej, utrzymanie porządku oraz urządzeń sanitarnych i społecznych. Rada Żydowska jest odpowiedzialna przed Starostą Grodzkim za należyte przeprowadzenie tych zarządzeń.

11. Niezastosowanie się do powyższych zarządzeń oraz do późniejszych przepisów i postanowień wykonawczych, wydawanych w czasie akcji przesiedleńczej, będzie bezwzględnie ścigane i karane. Majątek będzie konfiskowany.

Lublin, dnia 24 marca 1941.

Szef Okręgu Lublin
podp. ZÖRNER
Gubernator

Bearbeitet vom OKH – GenStdH – Abt. Fremde Heere West (III)
unter Verwendung von Unterlagen der Stabia des Gen d L beim Ob d H

Befestigungsskizze Tobruk 1:50000

Geheim!

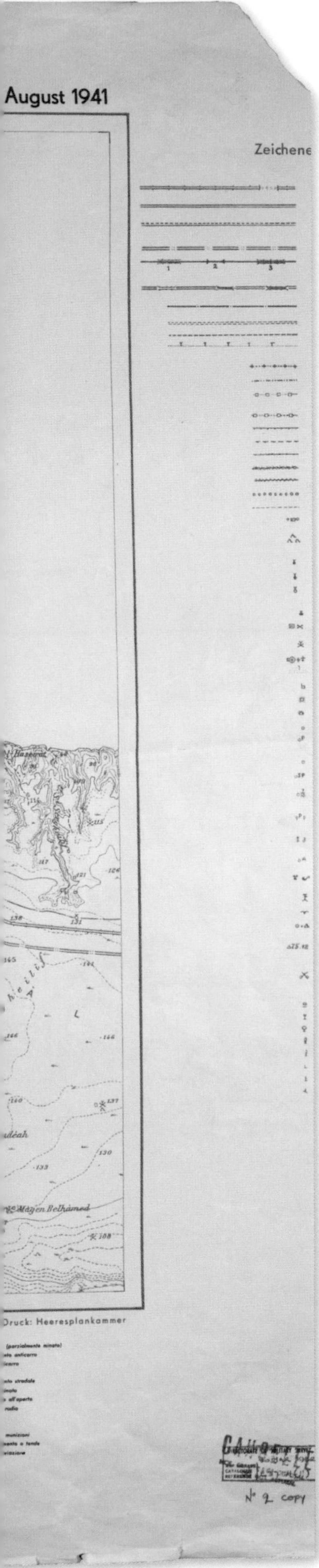

FORTRESS TOBRUK

German map of Tobruk, August 1941.

The most significant fortified position in Libya during the North African campaign turned out to be Tobruk, which provided a port and airbases close to the Egyptian frontier. Tobruk's location had made it a key base for the Italians during their unsuccessful invasion of Egypt, launched on 13 September 1940. The Italian force was totally defeated by much smaller British forces at Sidi Barrani in December, and after advancing into Libya the pursuing British easily took Tobruk on 22 January 1940. The Allied Western Desert Force pressed on but failed to anticipate the potential for a counter-attack. That attack, launched by the newly formed Afrika Korps under Irwin Rommel, was successful from late March 1941 and led to the first siege of Tobruk. This map derives from that siege and shows the extent of German knowledge of the defences, which was a product of aerial reconnaissance and the fact that the position had earlier been developed by Germany's ally Italy.

In the 1930s the Italians had built an outer defensive perimeter (known to the besieged Allies as the Red Line) with concrete trenches, anti-tank ditches, a bomb-proof bunker and dozens of strongpoints, surrounded by barbed wire and landmines, connected by a supply road. Tobruk's defences constituted a rough semi-circle across the desert from the coast eight miles (13 kilometres) east of the harbour, which lay at the heart of the fortress, to the coast nine miles (14 kilometres) west of it. Within the Red Line lay a secondary perimeter, the Blue Line, intended to contain any enemy breakthrough. A final line, the Green Line, protected the harbour. The entire area lay in a fairly flat plateau, offering an attacker little cover. The Allies defended it with regular, aggressive, combat patrolling, especially at night, which was aimed at making the besiegers the besieged and keeping the Germans at bay.

As long as the enclave of Tobruk remained in Allied hands it was a thorn in the side of the Afrika Korps with the potential to threaten its vulnerable logistical links, which had to be brought overland from Tripoli across more than 900 miles (1,500 kilometres) of desert. The successful retention of Tobruk hindered Rommel's advance east and showed the significance of defensive positions when ably defended. The besieged Allies also benefitted from the maintenance of maritime supply lines (the 'Tobruk ferry' provided by the Royal Navy and Royal Australian Navy) and from the extent to which Rommel had to divide his forces in order to battle against British forces advancing from Egypt. The last led in fact to the relief of Tobruk in December 1941.

However, in 1942 Rommel advanced anew and this time with greater success. The British were defeated at Gazala and nearby Tobruk was captured on 20/21 June. Rommel had proved an adroit commander, able to outthink his opponents and better able to direct the battle, although the margin of advantage was very narrow and luck played an important role. The Axis army was superior in fighting quality and more skilled in tank warfare, not least at coordination between armour and artillery cover, while Rommel also benefitted from control of the air. The Allies lost 33,000 prisoners in Tobruk. Ultimately, Allied overreliance on defensive boxes able to provide all-round defence had proved harmful because they enabled Rommel to direct the flow of the battle and, once heavily attacked, could not be held. The defeat triggered a motion of no-confidence in Churchill in Parliament, and, although he won a very clear majority, the political need for continued success was made clear.

PEARL HARBOR

Commander Mitsuo Fuchida, hand-drawn estimated damage map of Pearl Harbor, 8 December 1941.

This aerial photograph (right), looking southwest on 30 October 1941, of Pearl Harbor, the base of the US Pacific Fleet on the island of Oahu in the Hawaiian archipelago, makes for an interesting comparison with the hand-drawn map (opposite) by Commander Mitsuo Fuchida of the Imperial Japanese Navy Air Service, with his estimate, headed 'Top Secret', of the damage inflicted against surface ships following the air attack on 8 December 1941.

The American ships are named and their types referred to, while, the amount and type of munitions used during the attack is reported in the legend (opposite, top). Red line strokes indicate the severity of the damage in terms of minor, moderate, serious and sunk. Red arrows indicate the direction and location of where torpedoes struck, and cross marks indicate the impacts of bombs. These hits were thought to signify severe damage (but sinking a vessel at anchorage in harbour means a ship essentially settles in shallow water and is easy to salvage, and the skill of subsequent American repair efforts was underestimated).

To emphasise the raid's apparent success, Fuchida's map was photographed, blown up and shown by him – in his capacity of leader of the first wave of attackers who sent out the now famous message 'Tora! Tora! Tora!' – to Emperor Hirohito on 26 December.

In 1927, as part of his graduation exercises at Japan's Naval War College, Takagi Sokichi had planned an attack by two Japanese carriers on Pearl Harbor, although in the evaluation he was held to have suffered heavy losses. The Japanese were encouraged by the attack mounted by two waves of 21 British carrier-launched torpedo aircraft attack on Italian warships in the naval base of Taranto on 11/12 November 1940, especially the technique of shallow running the torpedoes, which badly damaged three battleships and caused the Italian fleet to seek safer harbours.

In reality, the target-prioritisation scheme of the Japanese raid was poor, the attack routes were conflicted and the torpedo attack lacked simultaneity. Moreover, there was no third-wave attack on the fuel and other harbour installations. Over 350 Japanese aircraft from six carriers destroyed two US battleships and damaged six more, damaged six cruisers and destroyers, and also destroyed or damaged 347 aircraft, but the attack was less successful strategically. Moreover, some of the damaged ships were to be returned to service. Had the oil farms (stores) been destroyed, the US Pacific Fleet would probably have had to fall back to the base at San Diego, which would have gravely inhibited American operations in the Pacific. The damage to American battleships, several of which were of limited value because of their slow speed, forced a shift in naval planning toward an emphasis on the carriers which, despite Japanese expectations, were not in Pearl Harbor when it was attacked.

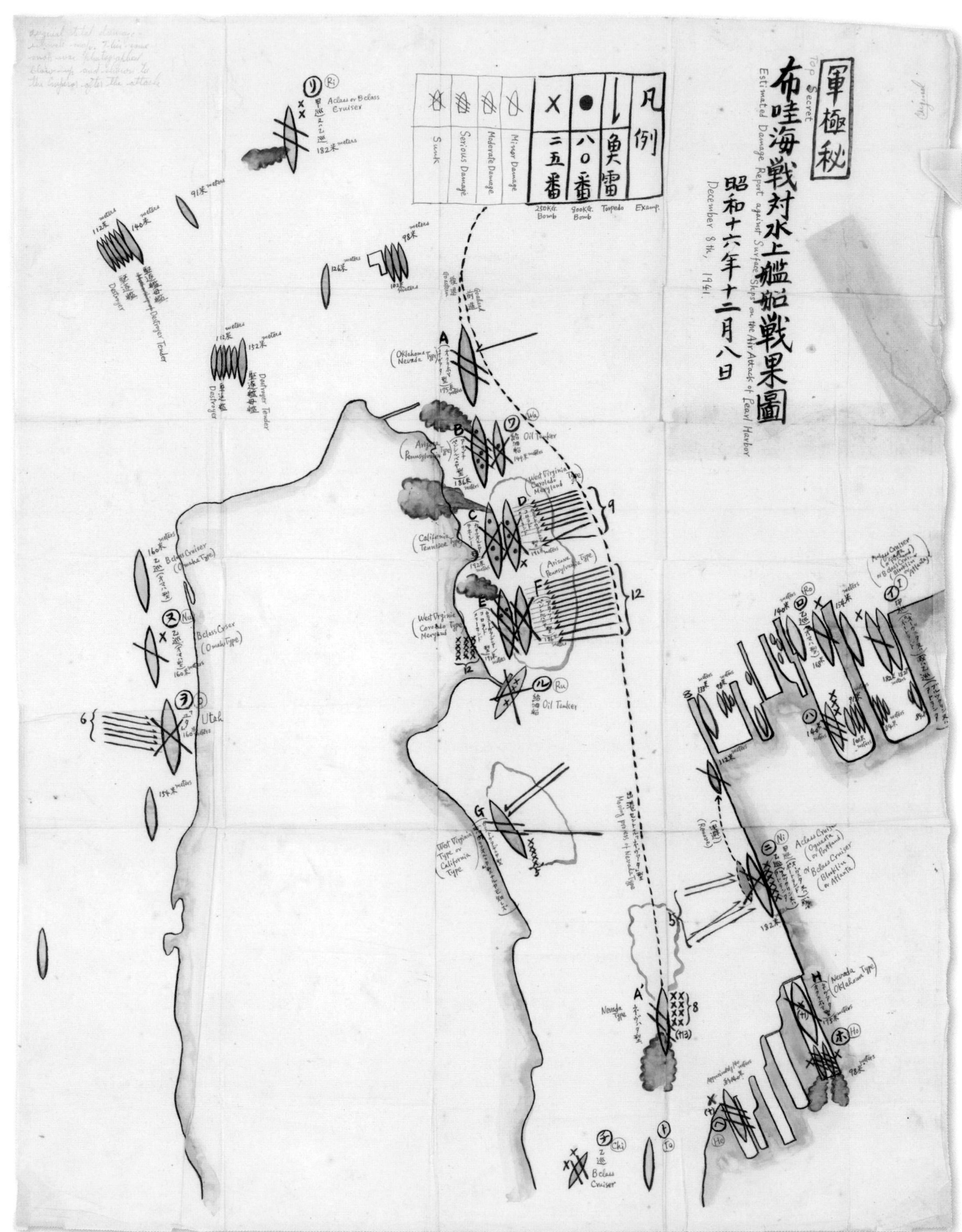
軍極秘
Top Secret
布哇海戦対水上艦船戦果圖
Estimated Damage Report against Surface Ships on the Air Attack of Pearl Harbor
昭和十六年十二月八日
December 8th, 1941
凡例
Examp.
魚雷
Torpedo
八〇番
800KG. Bomb
二五番
250KG. Bomb
Minor Damage
Moderate Damage
Serious Damage
Sunk
A class or B class Cruiser
182米
91米
112米
140米
98米
126米
Destroyer
Destroyer Tender
132米
Oklahoma Nevada Type
Oil Tanker
Arizona Pennsylvania
West Virginia Colorado Maryland Type
California Tennessee Type
Arizona Pennsylvania Type
West Virginia Colorado Type Maryland
9
12
160米
B class Cruiser (Omaha Type)
Utah
6
Oil Tanker
West Virginia Type or California Type
Moving process of Nevada Type
A class Cruiser (Augusta or Portland)
B class Cruiser (Blacklin or Atlanta)
5
Nevada Type
8
Nevada Oklahoma Type
B class Cruiser

MASTERY IN THE MEDITERRANEAN

Survey Directorate of GHQ Middle East, 'The Battle of Matapan', 1942.

This illustration of the naval night action Battle of Cape Matapan in March 1941 was drawn by Private Chemis based on a sketch by Lieutenant-Commander G.M. Stitt, but only reproduced in 1942 by the Survey Directorate of GHQ Middle East. The sketch captures the role of movement in formation but also the complexity of needing to cover shellfire, torpedo attack and air attack.

The large, and mostly modern, Italian navy (Regia Marina) was essentially deployed in the Mediterranean and not in a position to co-operate with the German surface fleet in the Atlantic, other than in the important indirect sense of keeping British warships busy in the Mediterranean, which was a major reason why Hitler welcomed Italy's entry into the war.

The Royal Navy benefitted from knowing that Germany's Operation Sealion had been cancelled, which enabled it to focus its naval forces in the Atlantic and Mediterranean. The British had superiority in battleships and carriers in the Mediterranean, although the lack of Italian carriers was countered by the role of land-based air power, a situation that was accentuated when German aircraft were moved to Sicily in early 1942. However, there were major problems in coordination between the Italians and the Germans, not least lack of provision of weaponry to Italy. Both its impressive navy and its small air force were handicapped by limited fuel.

Crucially, the Italian supply routes to North Africa remained open and most tonnage and personnel reached their destinations. In contrast, the dangers of the Mediterranean meant that the British were obliged to use the longer Cape Route to the Indian Ocean, which hit their shipping availability and the tonnage of goods moved, while the Royal Navy's commitment in the Mediterranean affected the availability of British warships elsewhere.

Like the successful air attack on Italian warships in Taranto on 11/12 November 1940, the outcome of events at Cape Matapan was important in affecting the naval balance in the Mediterranean as well as in improving British morale.

In response to the Taranto attack, the Italians withdrew units northward, and thus lessened the vulnerability of British maritime routes and naval forces in the Mediterranean, notably by increasing the problems of concentrating Italian naval forces and maintaining secrecy. In addition to Italy's admirals and senior commanders having understood the limitations of their ships and industrial base, they were also averse to taking risks because they believed the war would be won or lost by Germany and that the navy would be Italy's most important military asset in a post-war world.

On 28/29 March 1941, off Cape Matapan, southern Greece, the Italian navy's defeat at the hands of the Royal Navy ended meaningful Italian fleet operations. Thanks to torpedo bombers from the carrier *Formidable*, battleship firepower (15-inch gunfire wrecked the Italian cruisers) and ships' radar, the British, who had broken the Italian codes, sank three of the best Italian cruisers – *Fiume*, *Pola* and *Zara* – and damaged the modern battleship *Vitorio Veneto*.

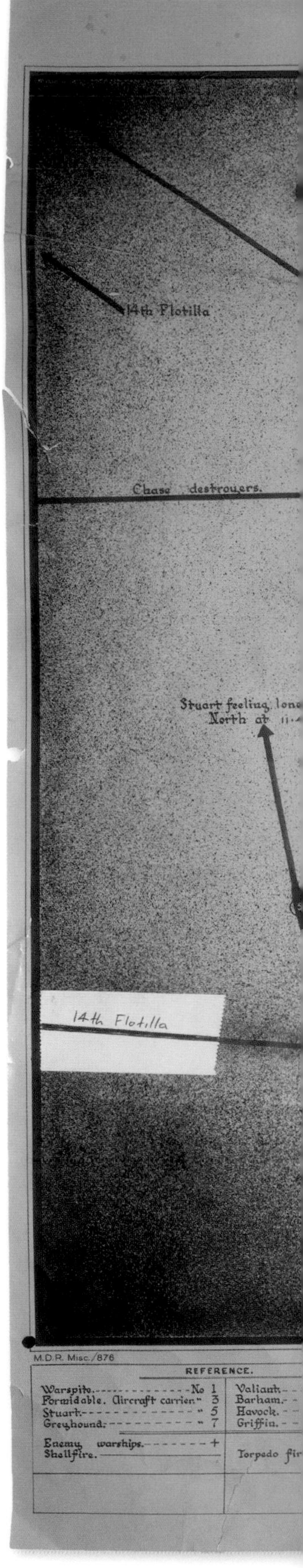

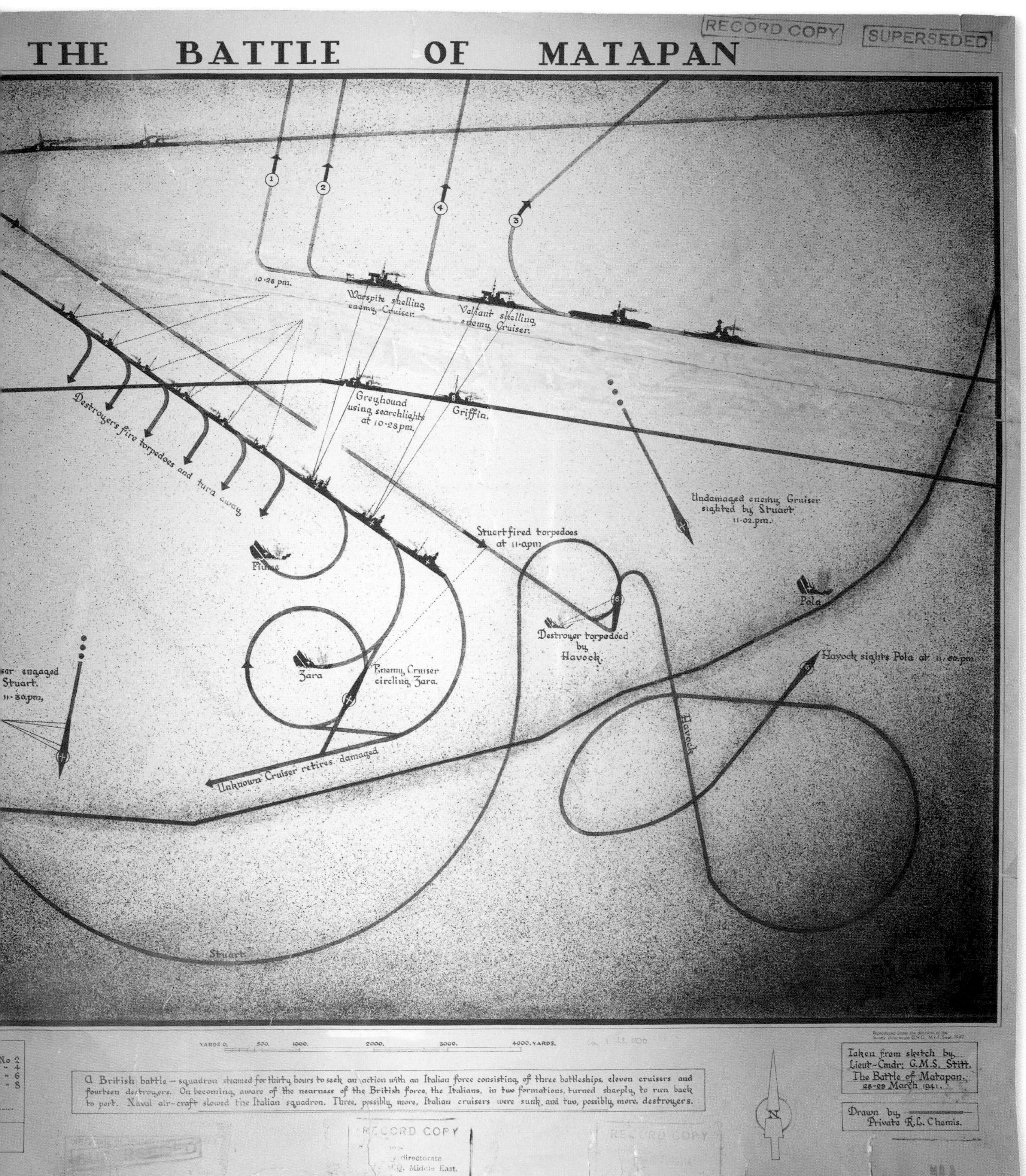
THE BATTLE OF MATAPAN
RECORD COPY
SUPERSEDED
10-28 pm.
Warspite shelling enemy Cruiser.
Valiant shelling enemy Cruiser.
Greyhound using searchlights at 10-28 pm.
Griffin.
Destroyers fire torpedoes and turn away
Undamaged enemy Cruiser sighted by Stuart 11-02 pm.
Stuart fired torpedoes at 11-apm
Fiume
Pola
Destroyer torpedoed by Havock.
Havock sights Pola at 11.50 pm
Zara
Enemy Cruiser circling Zara.
Stuart. 11-30 pm.
Unknown Cruiser retires damaged
Havock
Stuart
YARDS 0. 500. 1000. 2000. 3000. 4000. YARDS.
A British battle-squadron steamed for thirty hours to seek an action with an Italian force consisting of three battleships, eleven cruisers and fourteen destroyers. On becoming aware of the nearness of the British force, the Italians, in two formations, turned sharply to run back to port. Naval air-craft slowed the Italian squadron. Three, possibly more, Italian cruisers were sunk, and two, possibly more, destroyers.
Taken from sketch by Lieut-Cmdr: G.M.S. Stitt. The Battle of Matapan. 28-29 March 1941.
Drawn by Private R.L. Chemis.
N
RECORD COPY
Directorate G.H.Q. Middle East.
RECORD COPY

OPERATION JUBILEE

Planning maps for the Dieppe raid, including objectives, 1942.

Mounted on 19 August 1942 to test German defences against a cross-Channel attack, including a beach assault, the Dieppe raid suffered from a lack of prior bombardment from the air and sea, in part because the RAF, which had not as yet deployed effective tactical support aircraft, was committed to the bombing campaign against Germany. In addition, it was hoped the raid would benefit from the element of surprise, but unfortunately that was lost. There was a large air battle, in which the Germans were able to deploy considerable numbers of aircraft and, helped by anti-aircraft fire, lost only 48 to the RAF's 106. The attacking force was mostly from the Canadian 2nd Division, because the Canadian government wanted their troops in action, but not in the Mediterranean. Some 4,963 Canadian troops, 1,075 British and 50 US Rangers took part, supported by warships; but not, due to fears about their safety, by the heavy guns of the battleships used in the Normandy landing of 1944. Aerial reconnaissance provided information on the defences, but it was not comprehensive, and the Intelligence analysis of the task, both the physical environment and the defences, was inadequate. Serious problems were posed by the lack of sufficient covering fire, lateness in landing some units and the strength of the German response. The majority of the Canadians landed were killed, wounded or captured.

The poorly conceived plan was a prelude to a similar failure in a British airborne assault at Arnhem in 1944. The necessary capacity to implement a dubious strategic idea was absent, but so, even more, was the sense to say no to a rash course of action undertaken without due cause and proper preparations. The only objective achieved in the raid was the capture of the coastal battery at Varengeville, to the west of Dieppe (see page 133, top).

It is always helpful to consider attacks in a comparative context. Operation Jubilee was the most unsuccessful amphibious attack in 1942, contrasting with the repeated Japanese successes as well as the

PLAN OF TOWN, SHOWING OBJECTIVES

MAP I

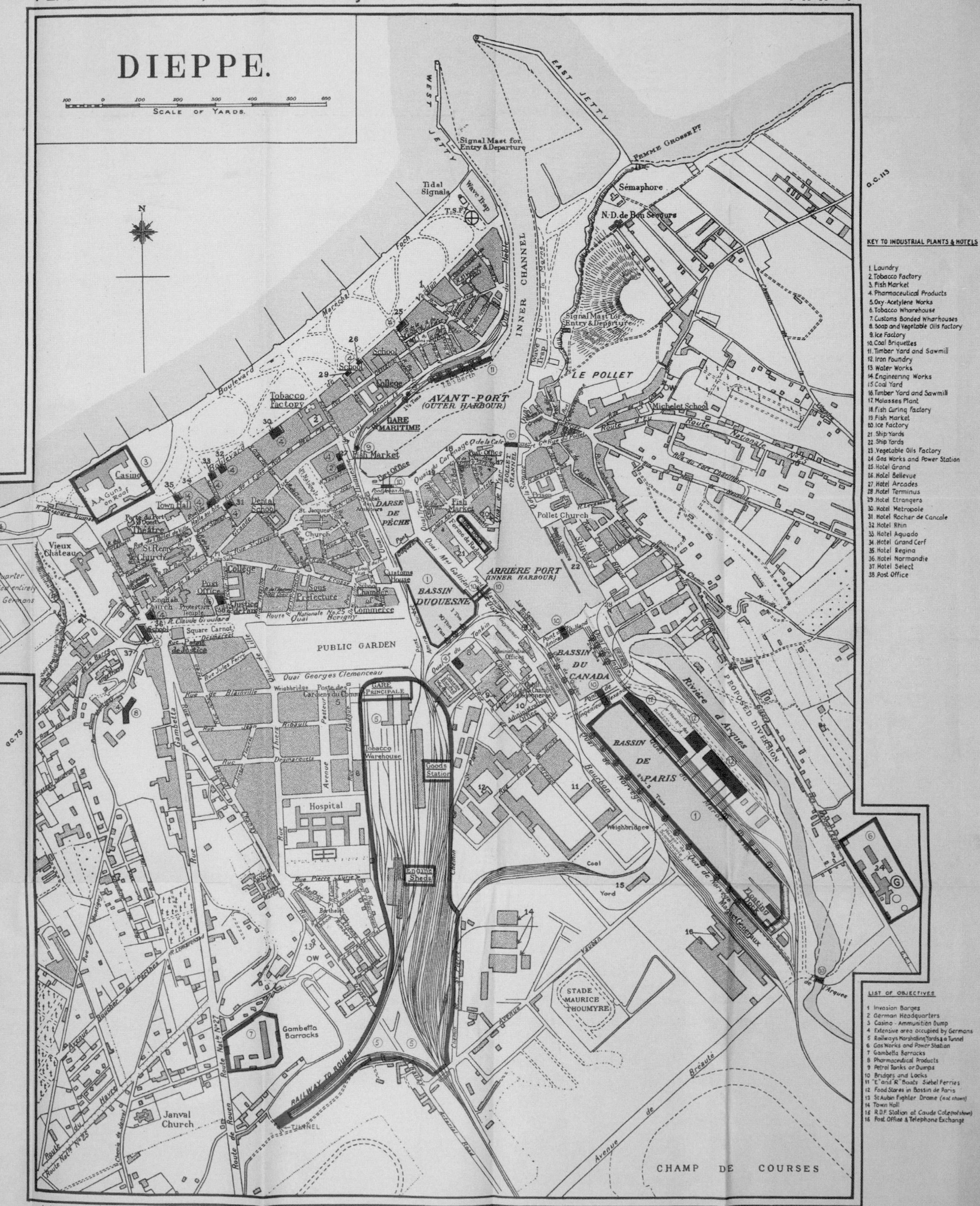

DIEPPE.
100 0 100 200 300 400 500 600
SCALE OF YARDS.
N
WEST JETTY
EAST JETTY
Signal Mast for Entry & Departure
Tidal Signals
Wave Trap
T.S.F.
INNER CHANNEL
Sémaphore
N.D. de Bon Secours
FEMME GROSSE Pᵀ
G.C. 113
Boulevard Maréchal Foch
Signal Mast for Entry & Departure
LE POLLET
Michelet School
Route Nationale Nº 25
School
College
Tobacco Factory
AVANT-PORT (OUTER HARBOUR)
GARE MARITIME
Fish Market
Post Office
Casino
A.A. Guns on Roof
Town Hall
Dental School
Théâtre
St. Remy Church
Vieux Château
This Quarter Occupied entirely by Germans
DARSE DE PÊCHE
St. Jacques
Church
Fish Market
Pollet Church
POLLET CHANNEL
ARRIÈRE PORT (INNER HARBOUR)
College
Post Office
Sous Préfecture
Chamber of Commerce
Customs House
BASSIN DUQUESNE
English Church
Protestant Temple
Justice de Paix
Square Carnot
Palais de Justice
School
PUBLIC GARDEN
Quai Georges Clemenceau
BASSIN DU CANADA
Administrative Offices
GARE PRINCIPALE
Tobacco Warehouse
Goods Station
BASSIN DE PARIS
Rivière d'Arques
PROPOSED DIVERSION
Hospital
Engine Sheds
Weighbridge
Coal Yard
Gambetta Barracks
RAILWAY TO ROUEN
TUNNEL
Janval Church
STADE MAURICE THOUMYRE
CHAMP DE COURSES
G.C. 75
G.C. 154
KEY TO INDUSTRIAL PLANTS & HOTELS
1. Laundry
2. Tobacco Factory
3. Fish Market
4. Pharmaceutical Products
5. Oxy-Acetylene Works
6. Tobacco Wharehouse
7. Customs Bonded Wharhouses
8. Soap and Vegetable Oils Factory
9. Ice Factory
10. Coal Briquettes
11. Timber Yard and Sawmill
12. Iron Foundry
13. Water Works
14. Engineering Works
15. Coal Yard
16. Timber Yard and Sawmill
17. Molasses Plant
18. Fish Curing Factory
19. Fish Market
20. Ice Factory
21. Ship Yards
22. Ship Yards
23. Vegetable Oils Factory
24. Gas Works and Power Station
25. Hotel Grand
26. Hotel Bellevue
27. Hotel Arcades
28. Hotel Terminus
29. Hotel Etrangers
30. Hotel Metropole
31. Hotel Rocher de Cancale
32. Hotel Rhin
33. Hotel Aquado
34. Hotel Grand Cerf
35. Hotel Regina
36. Hotel Normandie
37. Hotel Select
38. Post Office
LIST OF OBJECTIVES
1 Invasion Barges
2 German Headquarters
3 Casino - Ammunition Dump
4 Extensive area occupied by Germans
5 Railways Marshalling Yards & a Tunnel
6 Gas Works and Power Station
7 Gambetta Barracks
8 Pharmaceutical Products
9 Petrol Tanks or Dumps
10 Bridges and Locks
11 "E" and "R" Boats Siebel Ferries
12 Food Stores in Bassin de Paris
13 St Aubin Fighter Drome (not shown)
14 Town Hall
15 R.D.F. Station at Caude Cote (not shown)
16 Post Office & Telephone Exchange
Inter Service Topographical Department

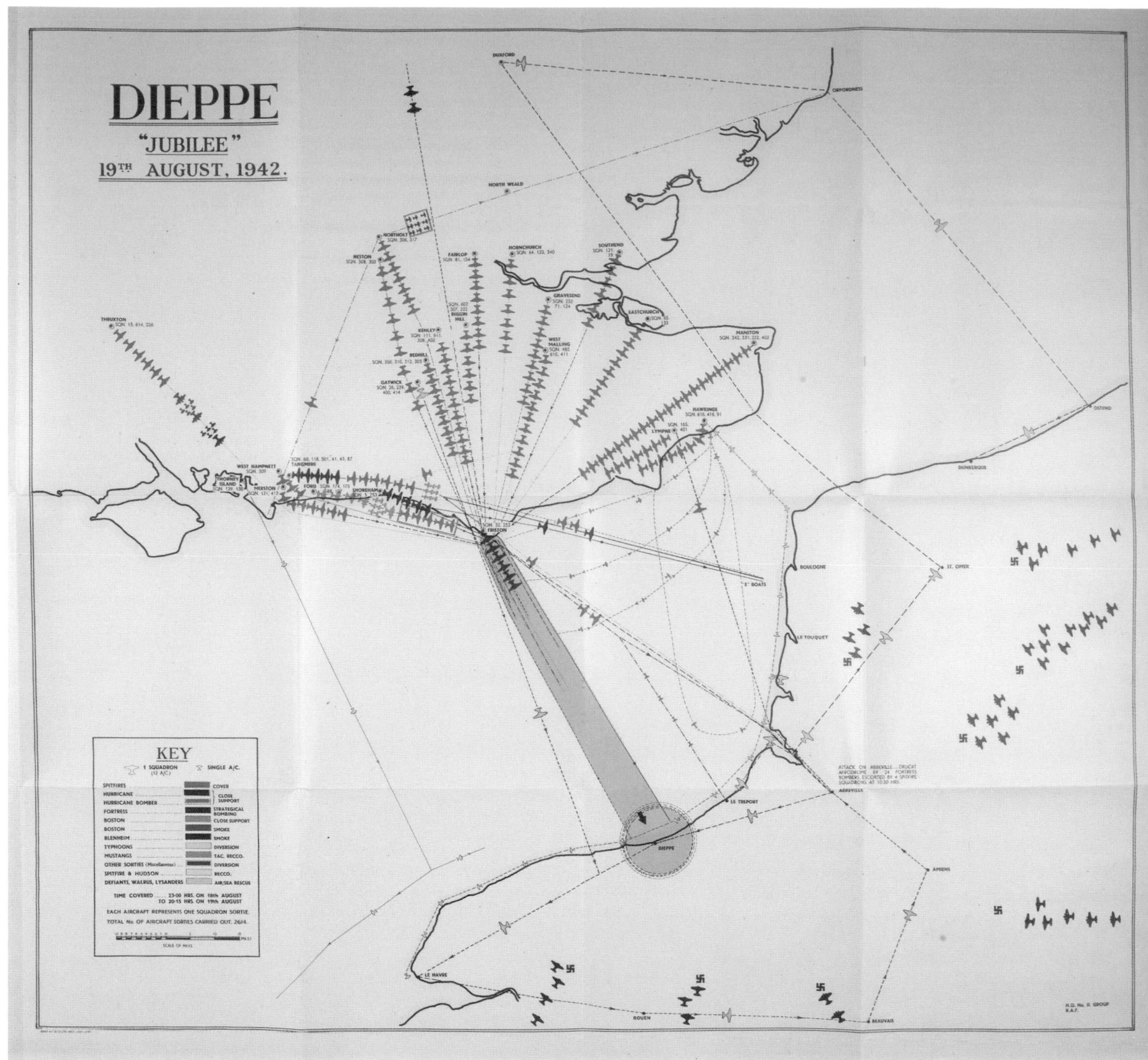

British ability to conquer Madagascar from Vichy France and the Anglo-American (mostly American) invasion of Morocco and Algeria in Operation Torch.

In large part, there was the question of the density of defence forces and their ability as a result to withstand the element of surprise achieved by the attackers. Moreover, the former requires success in translating that surprise into taking and using the initiative. Air command was also significant in that it should restrict the response of the defenders and provide the attackers with a form of mobile artillery. Success in Madagascar owed a lot to surprise, sea control and situational awareness based on sound intelligence that had fed through into an effective planning process. The landing there was handled by experienced troops who had rehearsed, drawing explicitly on the example of Japanese amphibious assaults in 1941–42 – a landing on a broad front. In Morocco, as at Dieppe, there was inadequate planning and a poor tempo of landing and exploitation, but the Americans prevailed because of the relatively light opposition.

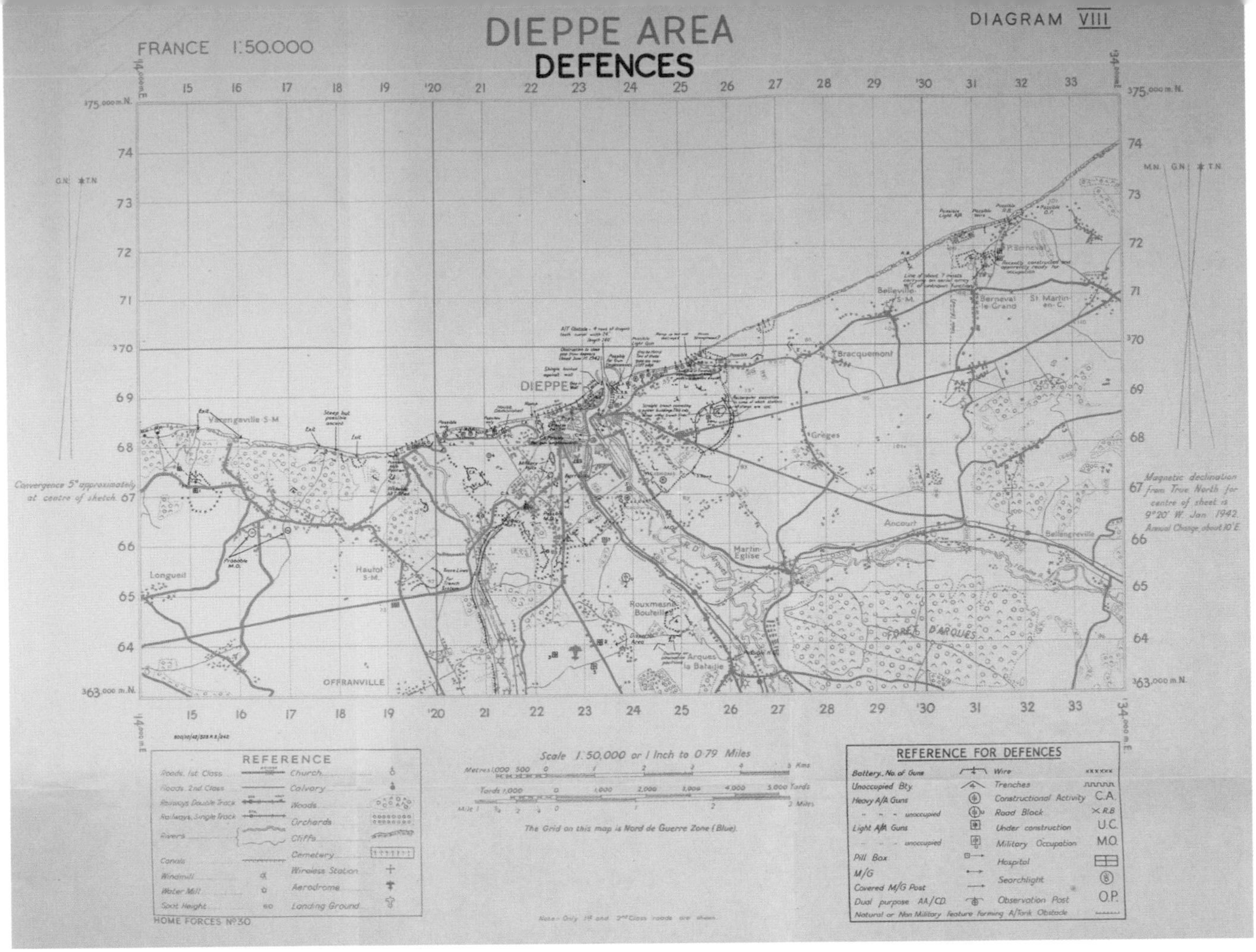

DIAGRAM VIII

FRANCE 1:50,000

DIEPPE AREA
DEFENCES

Convergence 5° approximately at centre of sketch

Magnetic declination from True North for centre of sheet is 9°20' W. Jan 1942. Annual Change about 10'E.

Varengeville S-M, DIEPPE, Bracquemont, Belleville S-M, Berneval le Grand, St Martin-en-C., Greges, Ancourt, Bellengreville, Martin-Eglise, Rouxmesnil Bouteilles, Arques la Bataille, Forêt d'Arques, Hautot S-M, Longueil, Offranville

HOME FORCES Nº30

REFERENCE

Roads, 1st Class		Church	
Roads, 2nd Class		Calvary	
Railways Double Track		Woods	
Railways, Single Track		Orchards	
Rivers		Cliffs	
		Cemetery	
Canals		Wireless Station	
Windmill		Aerodrome	
Water Mill		Landing Ground	
Spot Height	60		

Scale 1:50,000 or 1 Inch to 0.79 Miles

The Grid on this map is Nord de Guerre Zone (Blue).

Note – Only 1st and 2nd Class roads are shown.

REFERENCE FOR DEFENCES

Battery, No. of Guns		Wire	xxxxxx
Unoccupied Bty		Trenches	
Heavy A/A Guns		Constructional Activity	C.A.
" " unoccupied		Road Block	× R.B
Light A/A Guns		Under construction	U.C.
" " unoccupied		Military Occupation	M.O.
Pill Box		Hospital	
M/G		Searchlight	
Covered M/G Post			
Dual purpose AA/CD		Observation Post	O.P.
Natural or Non Military feature forming A/Tank Obstacle			

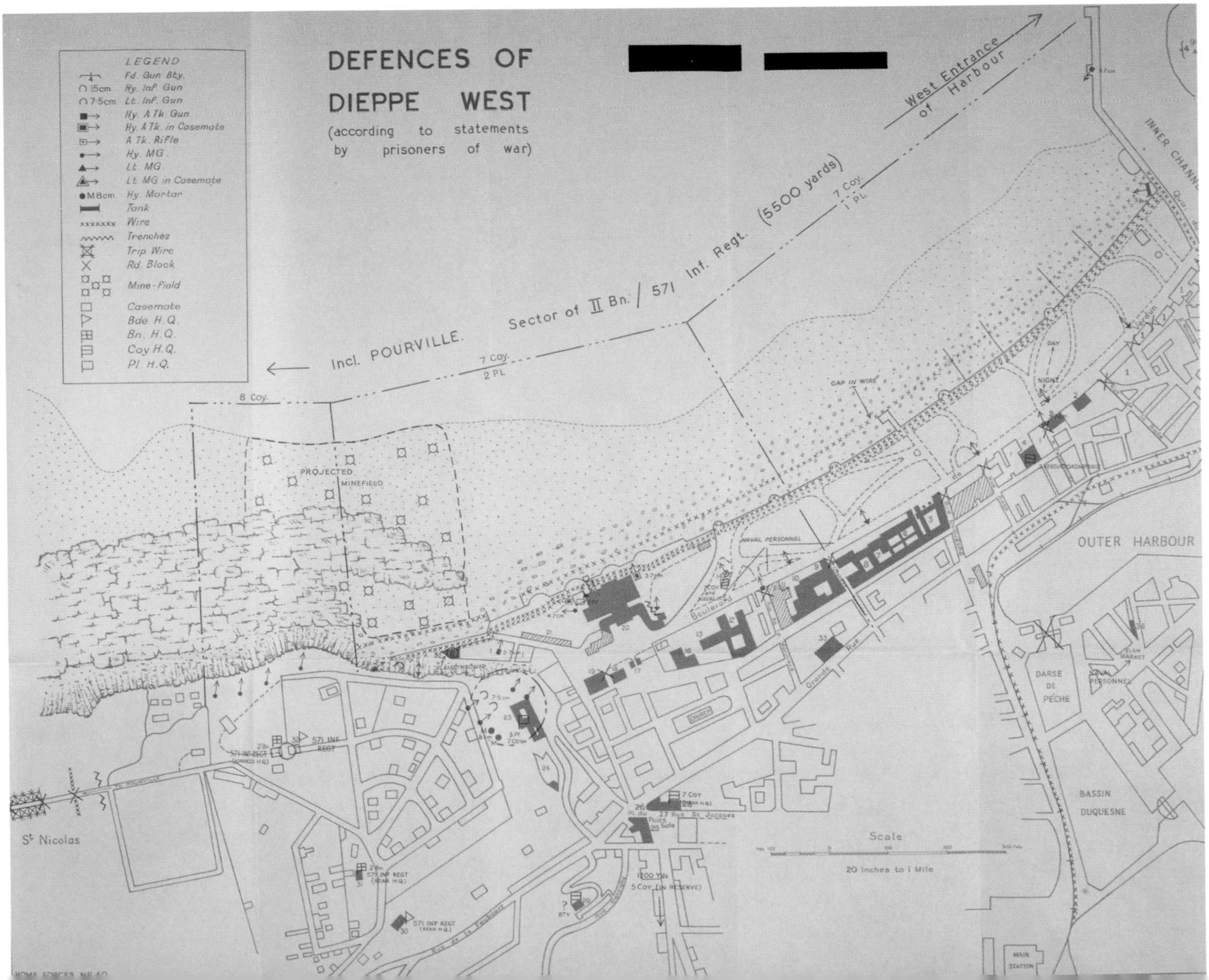

DEFENCES OF DIEPPE WEST

(according to statements by prisoners of war)

LEGEND

	Fd. Gun Bty.
15cm	Hy. Inf. Gun
7·5cm.	Lt. Inf. Gun
	Hy. A Tk. Gun
	Hy. A Tk. in Casemate
	A Tk. Rifle
	Hy. MG.
	Lt. MG.
	Lt. MG in Casemate
M 8cm.	Hy. Mortar
	Tank
xxxxxxx	Wire
	Trenches
	Trip Wire
	Rd. Block
	Mine-Field
	Casemate
	Bde. H.Q.
	Bn. H.Q.
	Coy H.Q.
	Pl. H.Q.

← Incl. POURVILLE. Sector of II Bn./571 Inf. Regt. (5500 yards)

West Entrance of Harbour

7 Coy. 2 PL; 8 Coy.; 7 Coy. 1 PL

PROJECTED MINEFIELD

GAP IN WIRE; NAVAL PERSONNEL; DAY; NIGHT

INNER CHANNEL; OUTER HARBOUR; DARSE DE PÊCHE; BASSIN DUQUESNE

571 INF REGT; 571 INF REGT (REAR H.Q.); 7 Coy; 5 Coy (IN RESERVE)

St. Nicolas

Scale

20 Inches to 1 Mile

MAIN STATION

HOME FORCES Nº 40

THE ROCK OF GIBRALTAR

Geographical Section of the General Staff, Gibraltar, November 1942.

This plan was produced by the Royal Engineers in 1942 for the Map Room of the Geographical Section of the General Staff. It focuses on the approach lines into Gibraltar for land and sea planes, and a sea plane alighting area. The commanding position of the rocky promontory of Gibraltar made it crucial for British (both the Royal Navy and Merchant Marine) entry into the western Mediterranean. The key context was strategic, because Spain's entry into the war could have led to a land attack on Gibraltar, or at least to an air attack if the Germans had gained air bases in southern Spain. It would then have been harder to protect than Malta.

For centuries Gibraltar's defences had been much improved by tunnelling, and during the Second World War around 25 miles (40 kilometres) of tunnels were dug out of the rock to create workshops, quarters, supply dumps and a fully equipped hospital. The excavated rock was used to build new runways from which the RAF could operate. In November 1942, General Dwight D. Eisenhower established his Allied Command Headquarters inside the rock for Operation Torch. After the North African campaign and the surrender of Italy in 1943, Gibraltar's role shifted from a forward operating base to a rear-area supply position. The harbour continued to operate dry docks and supply depots for the convoy routes through the Mediterranean.

Spain's leader General Francisco Franco was a keen supporter of Hitler's cause, a point greatly played down after the war. To Franco, Hitler was bound to win, which provided Spain with an opportunity to gain control of Gibraltar, Morocco and even Portugal, a traditional ally of Britain. However, Franco's demands for territorial gains from French North Africa, particularly Morocco, were seen as excessive by Hitler, and as likely to weaken Vichy France, which, with its empire and fleet, was regarded as more important to Germany politically and militarily. Franco's requests for food, raw materials, manufactured goods and armaments were also unacceptable.

To Hitler, Spain was largely inconsequential, a source of minor advantages that were not worth major effort, not least due to the clear weakness of the Spanish economy following its civil war. Indeed, due to the Allies' sea blockade, Spain was dependent on the Allies for fuel and food. The multiple problems posed by the alliance with Mussolini discouraged Hitler from adding Franco as an ally, and the latter was believed to offer little bar distraction from the goal of war with the Soviet Union. In any case, Franco claimed that Spain was exhausted, and that war with Britain, or giving German forces transit permission to attack Gibraltar, would lead to attacks on Spanish possessions overseas. Spain was a self-declared 'non-belligerent'.

Nevertheless, Franco actively collaborated – for example, he provided bases for German reconnaissance aircraft and Italian human torpedo units, facilitated German espionage and propaganda, and refuelled U-boats. Spain also provided not only raw materials, but manpower: around 47,000 Spaniards volunteered to serve in the 'Blue Division', which fought on the Eastern Front against the Soviet Union in what was presented as a crusade against communism. The División Española de Voluntarios was known as the 'Blue Division' because many of the early volunteers were Falange (Fascist) militia who wore blue shirts. This was not a regular force but it was justified as a response to Soviet intervention in Spain's civil war, although Franco did not declare war on the Soviet Union. The division fought bravely and well but by late 1942 it was short of volunteers.

The Spanish regime itself was divided. There were tensions within the government, within the army, within the Falange and among the right-wing as a whole. In 1942, in the Basilica of Begoña episode in Bilbao, Falangists and Carlists clashed, with grenades thrown, and Falange leaders were punished. Franco's brother-in-law, Ramón Serrano Suñer, the leader of the Falange, was a strongly anti-Allied Foreign Secretary (1940–42), who was replaced in 1942, thus moving Spain away from being so close to the Axis.

In turn, when the war started to go very badly for Germany, and notably after Sicily was invaded and Mussolini overthrown in July 1943, the Franco regime became more accommodating to the Allies. The Allied conquest of Vichy Morocco and Algeria in November 1942, and then the defeat of Axis forces in Tunisia, had already made Spain appear vulnerable, even though the German occupation of Vichy France in November 1942 increased the German presence on Spain's northern border. In 1943, Portugal also moved toward the Allies, allowing them to establish air bases in the Azores. As the war continued to look less good for the Axis Powers, the Franco regime tried to make itself look less Fascist. For example, Franco became more sympathetic to Jewish refugees.

The 'Blue Division' was run down in size before being withdrawn in 1943; 1,500 men remained as the Spanish Legion, but in February 1944 it was ordered

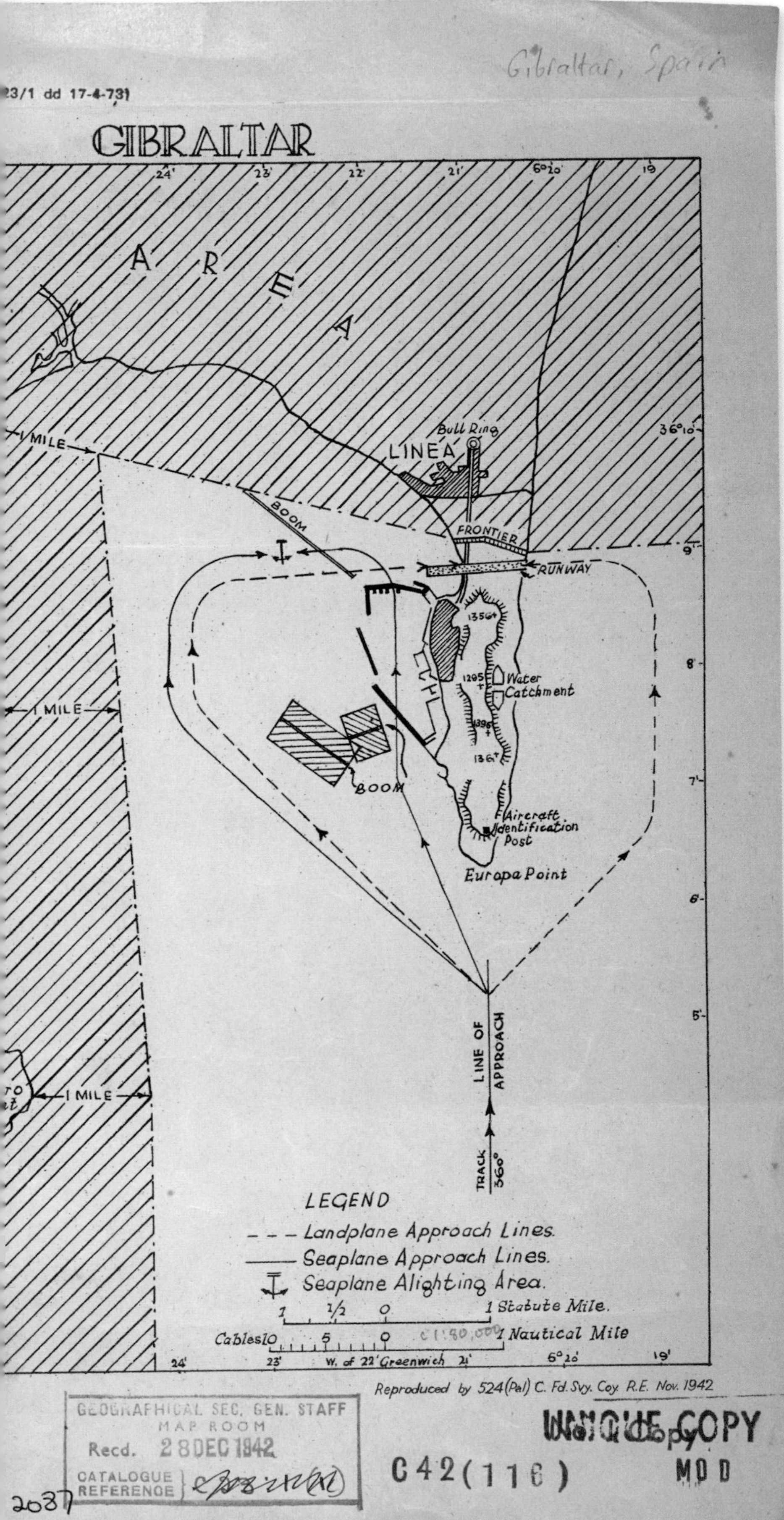

home. Under pressure from a threat by the United States to cut off oil supplies, Franco agreed in May 1944 to hand over all interned Italian ships, to expel all German agents and largely to cut off the supply of tungsten to Germany. In October 1944, Spain recognised the government of General de Gaulle in France. However, diplomatic relations with Germany were not cut until April 1945, while Intelligence aid to Germany was provided up to the end of the war. Moreover, German and Vichy French figures were sheltered at the war's end, and some were helped in their flight to South America.

As with Thailand, another pro-Axis 'neutral', the process of wartime change prepared the way for the post-war myth of neutrality, a myth that facilitated Franco's military co-operation with the USA during the Cold War. Spain's rewriting of the Second World War since 1975 has been linked to political trends and is subordinate to dissension over the civil war.

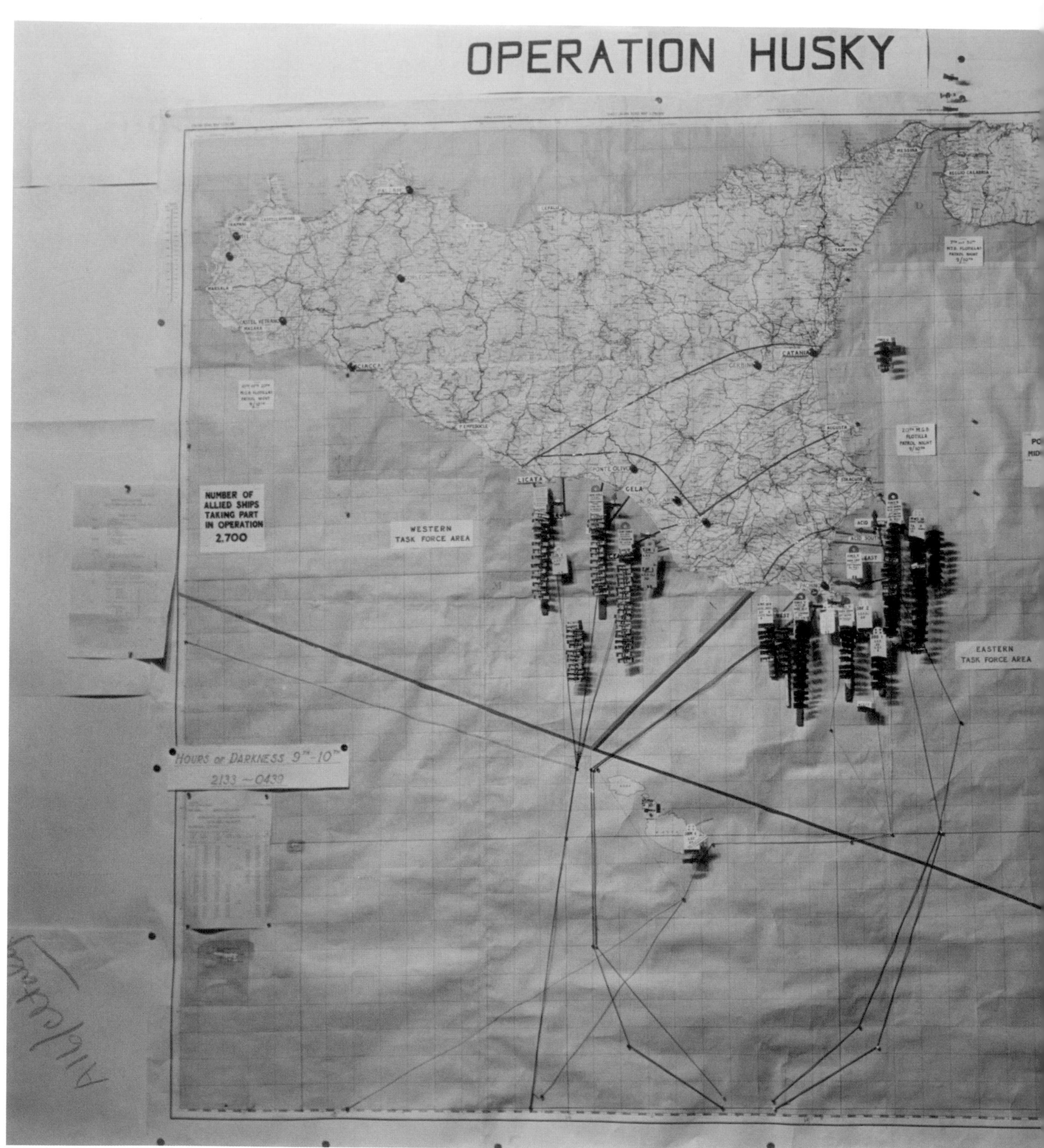
OPERATION HUSKY
NUMBER OF ALLIED SHIPS TAKING PART IN OPERATION 2.700
WESTERN TASK FORCE AREA
EASTERN TASK FORCE AREA
HOURS OF DARKNESS 9TH-10TH 2133 – 0439
20TH M.G.B. FLOTILLA PATROL NIGHT 9/10TH
LICATA
GELA
CATANIA

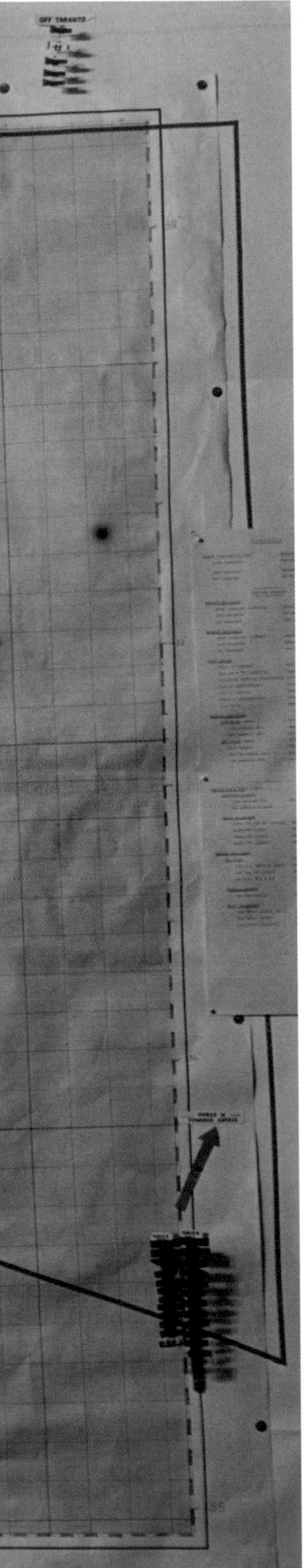

THE INVASION OF SICILY

Battle plan of Operation Husky, 1943.

A photograph of part of the planning stage for the deployment of the shipping taking part in Operation Husky, the Allied invasion of Sicily. Amphibious power and air support allowed the Allies to seize the initiative and helped ensure a wide-ranging attack on Sicily, the largest island in the Mediterranean. Husky, launched before dawn on 10 July, initially involved 180,000 Allied troops, alongside 750 warships, 2,500 transport ships and 400 landing craft. Only Operation Overlord in Normandy in 1944 was a larger amphibious action in Europe. The landings were more sophisticated than those mounted in Operation Torch in northwest Africa in November 1942. Operation Husky used appropriate ship-to-shore techniques, notably amphibious pontoon causeways and trained beach parties. The British landed in southeast Sicily and the Americans farther west between Cape Scaramia and Licata.

Once landed, the Allies faced problems from the defenders of the island. General Alfredo Guzzoni commanded the Axis forces' defence on Sicily. Including navy and air force personnel, Guzzoni had 315,000 Italians and 40,000 Germans. His troops were primarily composed of Coastal Divisions – five of the nine – and he had only about 148 tanks, many of which were obsolete. After considering options, he concentrated his force in the southeastern sector, which was the Allied territorial objective. The SIM – Italian Military Intelligence Service – was sure the Allies were going to land in Sicily, after precise information had been received from its agents in Lisbon, while the Germans believed the Allies would land in Sardinia or Greece. Kesselring, the German commander, disregarded the Italian intelligence reports and deployed the German tanks of 14th Panzer Corps in central and western Sicily. On 7 July, Guzzoni was warned that a landing was expected within two or three days, but despite the information from Lisbon and the Allied seizure of the offshore island of Pantellaria on 11 June after heavy bombing, Kesselring kept his troops in the interior and did not defend the coast. In contrast, Italian troops who were deployed there were able, on 10 July, to blow up the ports of Gela and Licata to deny them to the Allies.

The Italian forces resisted but were overwhelmed. For example, when the Eighth Army landed, that part of the coast was defended by the 206th Coastal Division, which was scattered along a line more than 80 miles (130 kilometres) long. The British first wave alone outnumbered the Italian defenders three to one. The Italians, who had no tanks, were annihilated. However, when the British moved on to attack the Italian-defended city of Catania, it took 23 days of hard fighting to fall. The US Seventh Army landed south of Gela. Italian coastal units resisted as best they could, but suffered heavy casualties: the 429th Coastal Battalion lost 45 per cent of its men. The Livorno Division counter-attacked, supported by 50 obsolete Italian tanks and by aircraft bombing the beachhead, but the attack failed with the loss of 50 per cent of the troops. There were 176 German tanks in Sicily, part of the three German divisions there, but they were too far away to offer initial support. When the panzer units finally arrived, the attack was resumed with initial success, but the American troops, supported by naval firepower, fought bravely and enlarged their beachhead. The Allies benefitted from support provided by the Sicilian Mafia, which had hungered for revenge against the Fascist authorities since the 1920s and helped Allied Intelligence to build up detailed local information. Some Italian officers seeking to organise opposition to the Allies were shot, while some American units received guidance from local Mafiosi.

Subsequently, with Montgomery held before Catania, Lieutenant General George S. Patton commanding US Seventh Army, moved north to Palermo, creating and exploiting opportunities for outflanking Axis opposition. The heavily outnumbered Assietta Infantry Division was unable to stop him and, having taken Palermo, Patton moved eastward on Messina after fighting off counter-attacks at Troina.

The war had taken on an attritional character for Germany and Japan, and for Italy the conflict had been forced onto home terrain. The availability of massive resources enabled the Allies to maintain their attack.

LICATA ASSAULT

Manuscript planning map for the American landing near Licata, Sicily, on 10 July 1943.

This is the map used by Captain Alfred J. Reid of the US Army's 3rd Ranger Battalion. Reid's unit was one of two new battalions formed in North Africa in spring 1943 by Lieutenant Colonel William O. Darby, in preparation for the Allied invasion of Europe through the Italian peninsula, beginning with Operation Husky in Sicily. Captain Reed (1925–87) survived the war.

Darby's three Ranger battalions (the veterans of the 1st, plus the newly formed 3rd and 4th) were instructed to support Lieutenant General George S. Patton's Western Task Force in its assault on Palermo and they began a three-week training programme for Husky at Nemours in Algeria. The core training was long marches, cliff climbing, weapons training and amphibious operations. Each battalion was organised with a headquarters company and six line companies, each with around 65 men.

The three Rangers units were to act as the spearhead for the American landings, attached to Major General Terry de la Mesa Allen's 1st Infantry Division. Reid's 3rd Battalion, under Major Herman W. Dammer, was to land at the westernmost objective: the town of Licata. The other battalions were formed into Force X to capture the key port of Gela, around 20 miles (32 kilometres) to the east.

Licata was just to the west of the River Salso (marked clearly on Reid's map) and had an artificial harbour, with two moles and a breakwater, capable of handling hundreds of tons of cargo a day. Monte Economo overlooked the town and was the location for a signal station and a fortress-villa, Castel San Angelo, with a gun battery and a searchlight.

The landing plan assigned five beaches on either side of Licata divided into four sectors and attack groups. The 3rd Battalion was in the Green, or Molla, Attack Group, which was responsible for two of the five beaches. The first Rangers went ashore just before 3 a.m. and found little opposition. The Rangers quickly seized the high ground around the landing beaches prior to the infantry assault waves coming in a short while afterwards.

Once the infantry had moved through the beach the Rangers targeted and seized Castel San Angelo, where they remained until Licata had fallen to the infantry. In the days that followed the American forces drove westward with the Rangers screening ahead of the advance. The battalion played a key role in the capture of Porto Empedocle on 16 July, taking 700 prisoners, before the fall of Agrigento on 17 July.

The capture of Palermo on 22 July meant that American attention switched to Messina. The 3rd Battalion had by then been assigned to Major General Lucian K. Truscott's 3rd US Infantry Division and it undertook the last Ranger action in Sicily on 11/12 August, near Brolo. All three Ranger battalions subsequently regrouped at Corleone, before leaving to train up for Operation Avalanche at Salerno.

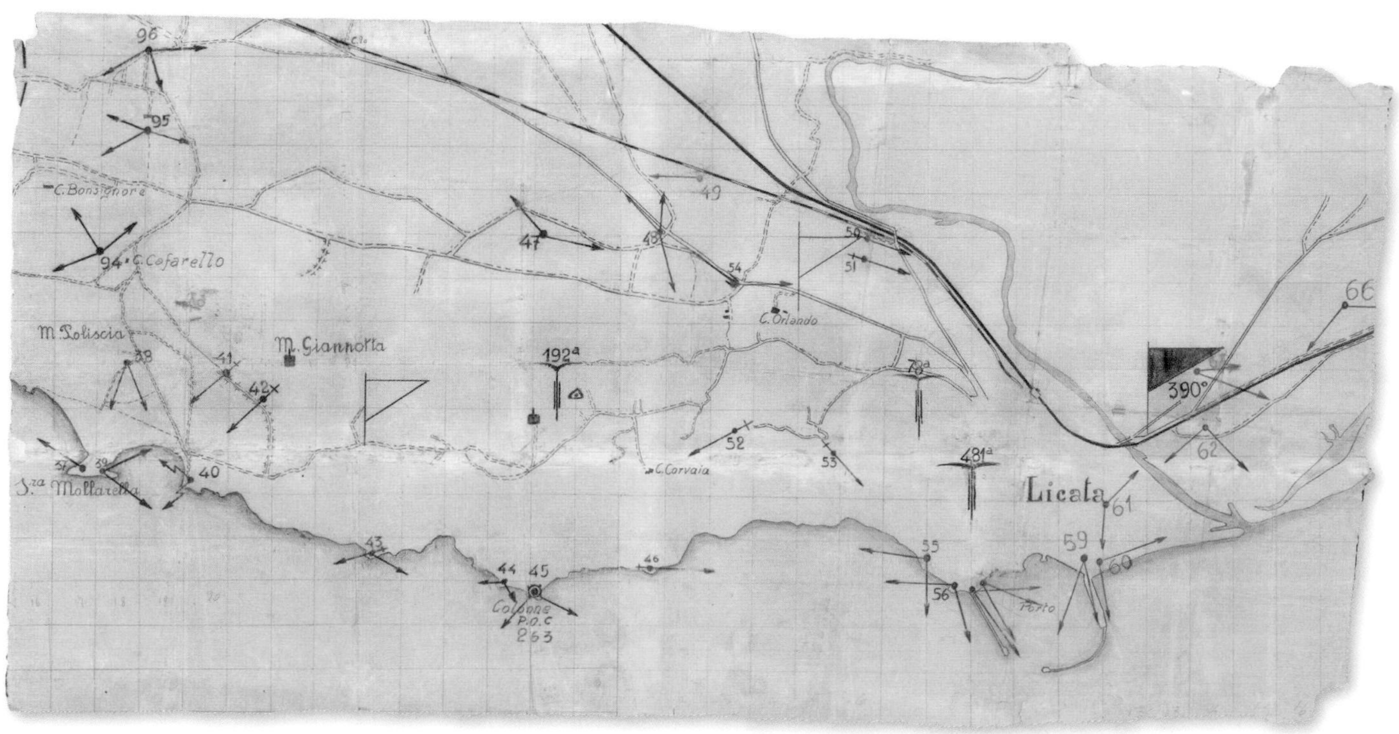

BREAKOUT AND ADVANCE

War Office, Geographical Section of the General Staff, 'goings' maps of Erkelenz and Mezidon, 1944.

Although both of these maps were produced by the British, that of Erkelenz in Germany (below) was copied from a German topographical map, while that of Mezidon in France (overleaf) was compiled from air photographs on a control, or standardised base, provided by existing French triangulation.

Erkelenz in the Rhineland, southwest of Mönchengladbach, was captured on 26 February 1945 as part of Operation Grenade in which the US Ninth Army, from 23 February onwards, had advanced to the Rhine. The Americans reached the river opposite Düsseldorf on 2 March. This was an advance in which there was much hard fighting through prepared German defences, notably those of the Siegfried Line, which proved vulnerable to concentrated and mobile firepower. Tank-infantry teams, supported by artillery fire and demolitions, had a major impact on individual positions. Aside from tanks, the Allied used self-propelled tank destroyers and motorised gun-carriages. After heavy bombardment, the

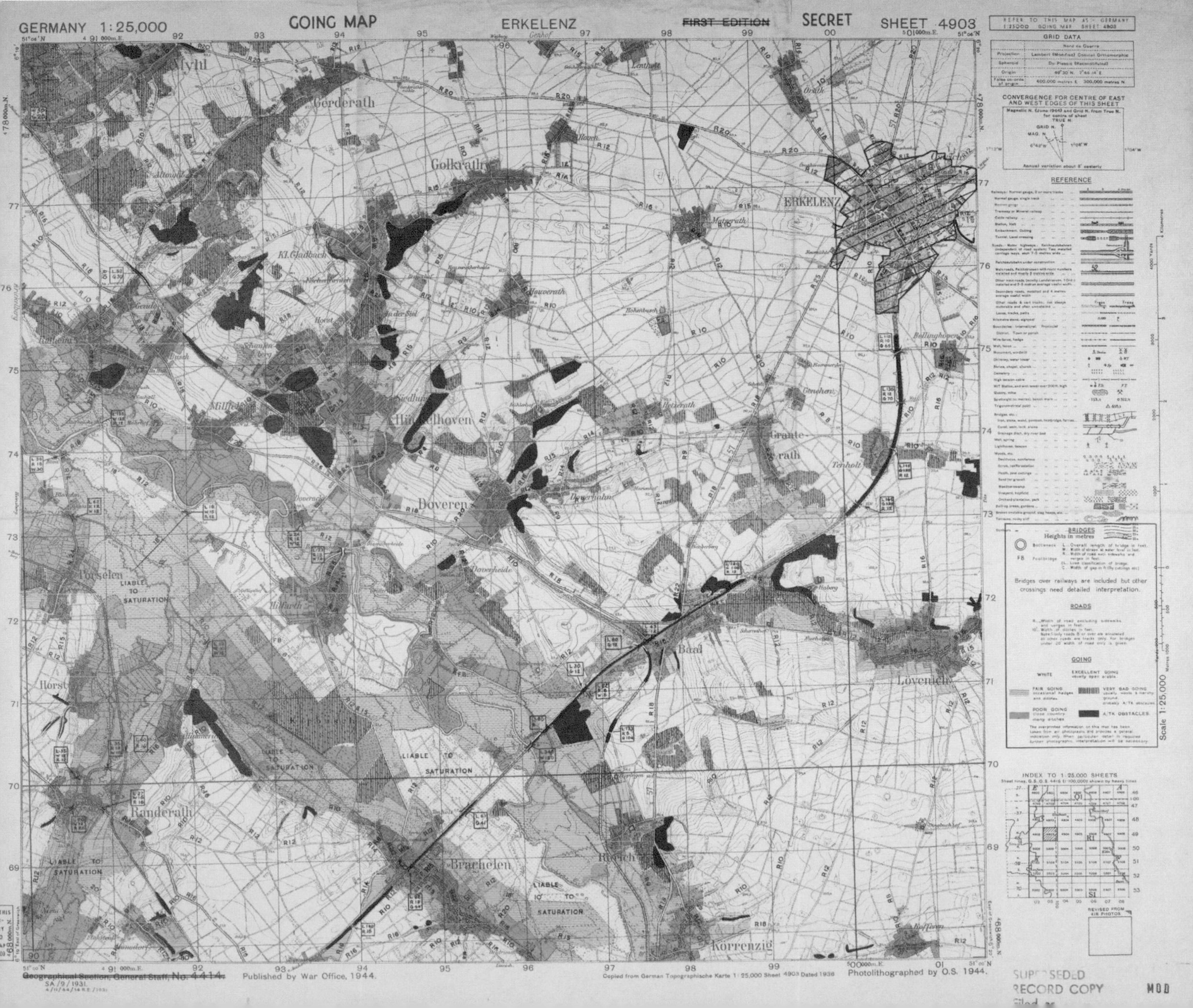

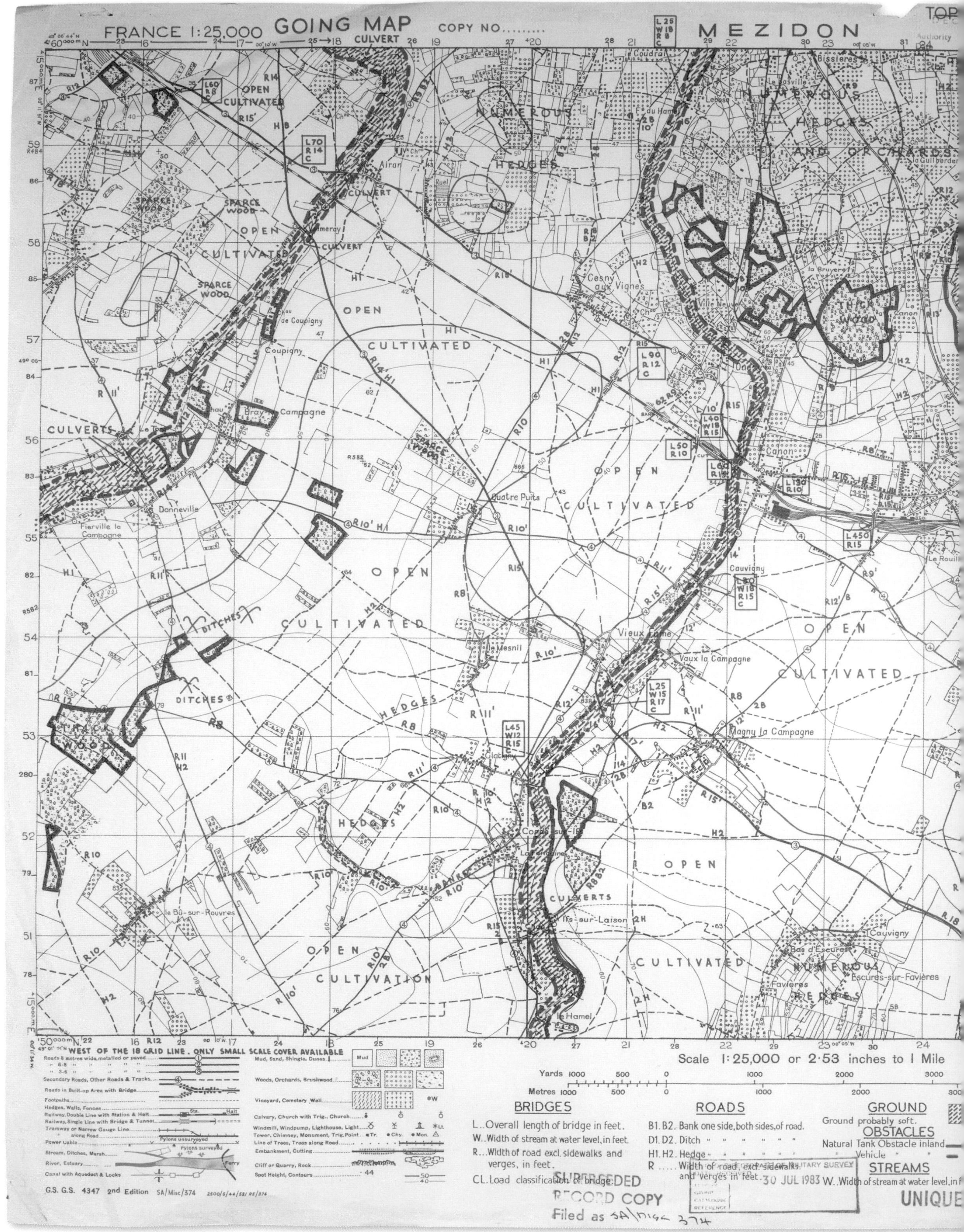

FRANCE 1:25,000
GOING MAP
COPY NO.........
MEZIDON
CULVERT
OPEN CULTIVATED
NUMEROUS HEDGES
NUMEROUS HEDGES AND ORCHARDS
Airan
Ruel
SPARCE WOOD
OPEN CULTIVATED
SPARCE WOOD
Cesny aux Vignes
Ville Neuve
THICK WOOD
la Bruyere
Chau de Coupigny
Coupigny
OPEN CULTIVATED
CULVERTS
Bray Campagne
SPARCE WOOD
Canon
Donneville
Fierville la Campagne
Quatre Puits
OPEN CULTIVATED
Cauvigny
OPEN CULTIVATED
DITCHES
Le Mesnil
Vieux Fame
Vaux la Campagne
OPEN CULTIVATED
DITCHES
HEDGES
THICK WOOD
Clabigny
Magny la Campagne
HEDGES
le Bû-sur-Rouvres
CULVERTS
Ifs-sur-Laison
OPEN
OPEN CULTIVATION
CULTIVATED
Bas d'Escures
NUMEROUS HEDGES
Favieres
Escures-sur-Favières
le Hamel
WEST OF THE 18 GRID LINE. ONLY SMALL SCALE COVER AVAILABLE
Scale 1:25,000 or 2·53 inches to 1 Mile
Yards 1000 500 0 1000 2000 3000
Metres 1000 500 0 1000 2000 3000
BRIDGES
L..Overall length of bridge in feet.
W..Width of stream at water level, in feet.
R...Width of road excl. sidewalks and verges, in feet.
CL. Load classification of bridge.
ROADS
B1. B2. Bank one side, both sides, of road.
D1. D2. Ditch " " " " "
H1. H2. Hedge " " " " "
R ..... Width of road, excl. sidewalks and verges in feet.
GROUND
Ground probably soft.
OBSTACLES
Natural Tank Obstacle inland
Vehicle " "
STREAMS
W.. Width of stream at water level, in f
G.S.G.S. 4347 2nd Edition SA/Misc/374
SUPERSEDED RECORD COPY
Filed as SA/Misc 374
30 JUL 1983
UNIQUE

defences were overcome in operations that proved much like the fighting of the First World War. At the same time, American tanks not only supported infantry advances but were also involved in tank conflict with German tanks. Moreover, there were divisional-level attacks, as with the US 14th Armored Division breaking through the Siegfried Line and advancing to the Rhine.

The map provided overprinted information that was taken from aerial photographs. There is colour-coded goings information, as well as relevant data on roads and bridges. White signifies 'excellent going, usually open arable'; orange with horizontal red lines is 'fair going, occasional hedges and ditches', plain orange is 'poor going, close country, many ditches'; red vertical lines is 'very bad going, usually woods & marshy ground, probably A/TK obstacles', and plain red is 'A/TK obstacles'.

Mezidon was attacked during the Battle of Normandy breakout in Operation Totalize of 8–9 August 1944, when First Canadian Army tried to close the 'Falaise Pocket' and trap German Army Group B. The openness of the land to tank advances was the crucial element. The 'softness' of the ground was significant, as were natural tank obstacles and the width of streams. Alongside a detailed legend of symbols (encompassing mud, sand, shingle, dune, woods, orchards, brushwood, trees along a road, culverts and much more), there were written descriptions on the map that helped in a rapid assessment: 'numerous hedges', 'cultivated', 'open' and 'liable to flood'. As the legend made clear, it was not possible from the scale of the photographs to make accurate measurements of minor roads and tracks, and therefore they must be considered a guide rather than a statement of fact.

ISIGNY-SUR-MER

British War Office, map of the defences of Isigny-sur-Mer, 'stop press edition' of 20 May 1944.

Created in preparation for D-Day, this map of part of the invasion zone presents information as at 19 May 1944 providing amendments and additions since the basic edition of May, which itself was printed on a map produced in April. The map was to be used for briefing troops. The key elements shown are defences, ground firmness and bridges. The role played by aerial intelligence in gathering the information is indicated by the note 'Heavy flak fire reported by aircrews in this area' west of Isigny-sur-Mer. Again, for the contours: 'They are interpolated from spot heights and hachures on the French 1:80,000 and amplified from Air Photo Examination. They should be accepted with caution.' A prominent note at the top states about the colour-coding in the defence legend ('see back'): 'Blue......Confirmed, Purple...NOT Confirmed.'

The section depicted was important to any immediate exploitation of the landings because it was the area in which the US 82nd and 101st Airborne divisions had landed, to the southwest of Omaha Beach and to the south of Utah Beach. This importance helps to explain the stress on bridges, roads and on whether the ground was drying out. Isigny-sur-Mer was one of the mission objectives of the US V Corps, but it took some time for the Americans to occupy the area and it was not until 8 June, having helped the US Rangers to clear Pointe du Hoc, that the town fell to troops of the 29th Infantry Division advancing from Omaha Beach, and Carentan on 11 June to units advancing from both beaches. The Germans failed to organise a defence in this area comparable to that seen at Salerno in 1943 and Anzio in 1944. Moreover, the key German units were located further east.

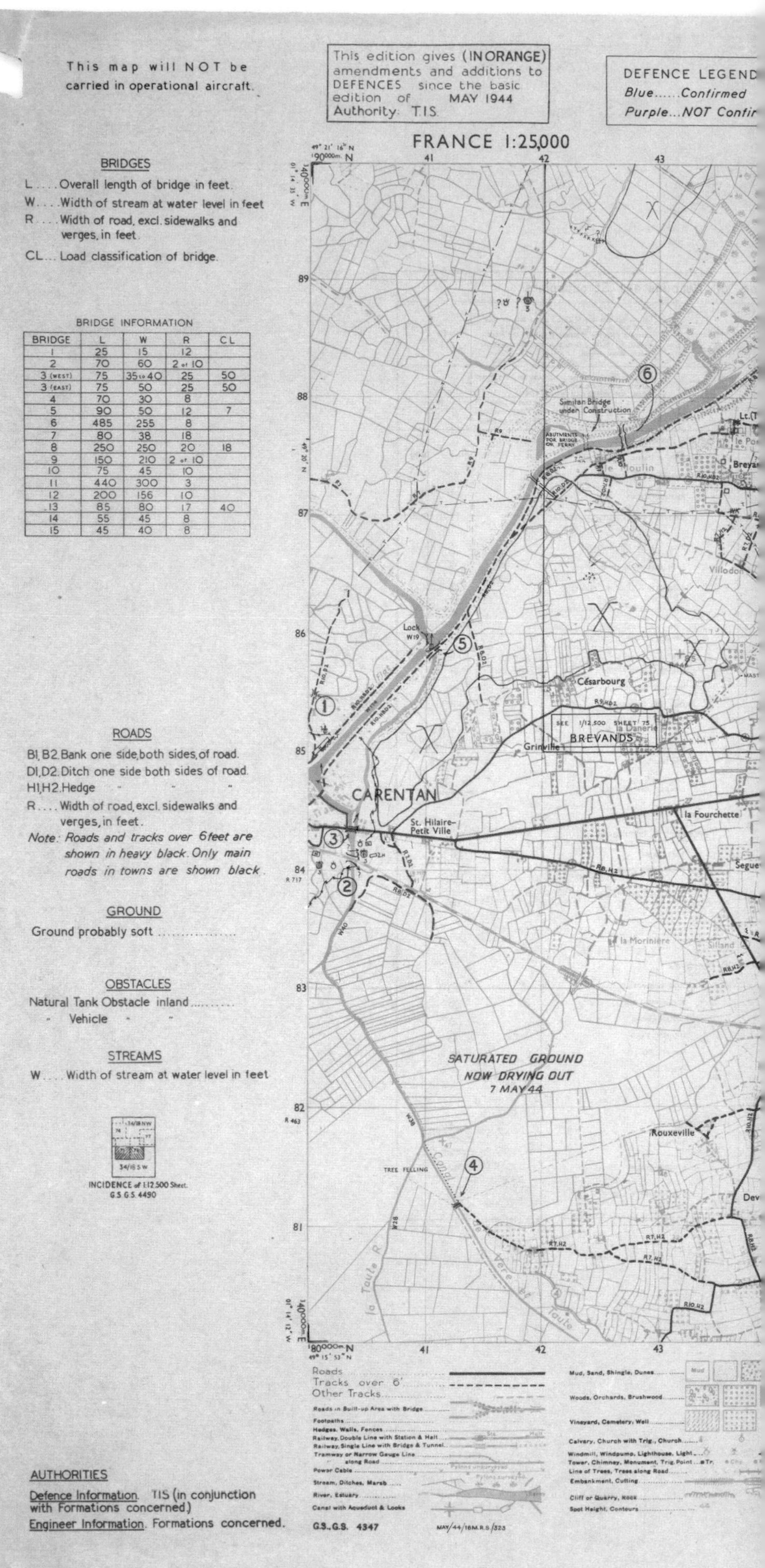

BRIDGE	L	W	R	CL
1	25	15	12	
2	70	60	2 of 10	
3 (WEST)	75	35 to 40	25	50
3 (EAST)	75	50	25	50
4	70	30	8	
5	90	50	12	7
6	485	255	8	
7	80	38	18	
8	250	250	20	18
9	150	210	2 of 10	
10	75	45	10	
11	440	300	3	
12	200	156	10	
13	85	80	17	40
14	55	45	8	
15	45	40	8	

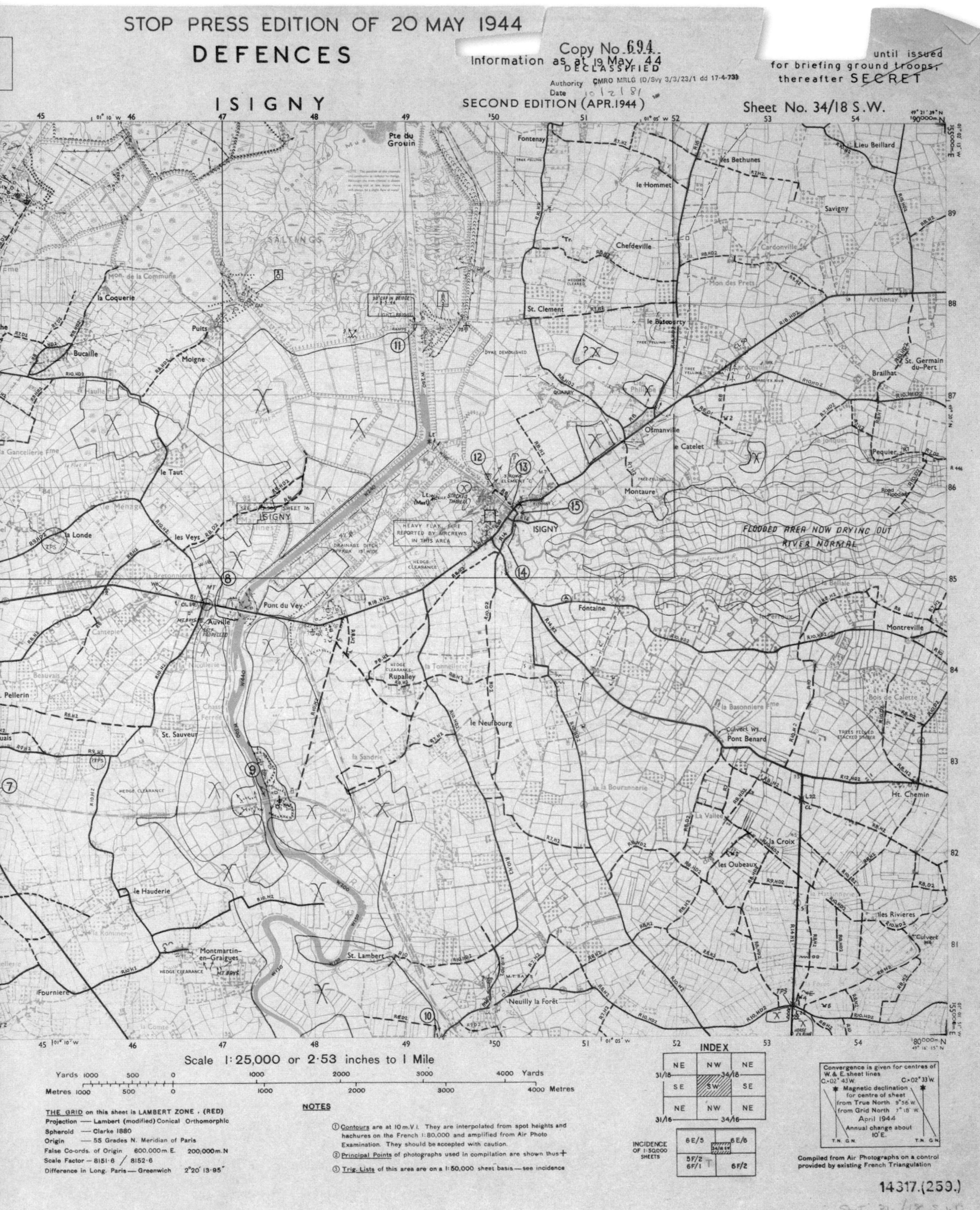

STOP PRESS EDITION OF 20 MAY 1944
DEFENCES
Copy No. 694.
Information as at 19 May 44
DECLASSIFIED
Authority CMRO MRLG (D/Svy 3/3/23/1 dd 17-4-73)
Date 10.2.81
until issued for briefing ground troops; thereafter SECRET
ISIGNY
SECOND EDITION (APR. 1944)
Sheet No. 34/18 S.W.
Pte du Grouin
Fontenay
Lieu Beillard
les Bethunes
le Hommet
Savigny
SALTINGS
Chefdeville
Cardonville
Mon des Prets
la Coquerie
St. Clement
Arthenay
le Bascourty
Puits
Bucaille
Moigne
Brailhat
St. Germain du-Pert
Ofmanville
le Catelet
Pequier
le Taut
Montaure
la Gancellerie Fme
SEE SHEET 16 ISIGNY
HEAVY FLAK FIRE REPORTED BY AIRCREWS IN THIS AREA
FLOODED AREA NOW DRYING OUT
RIVER NORMAL
la Londe
les Veys
Pont du Vey
Auville
Fontaine
Montreville
Rupalley
la Tonnellerie
le Neufbourg
la Basonniere Fme
Bois de Calette
St. Sauveur
Pont Benard
Ht. Chemin
la Vallee
la Croix
les Oubeaux
le Hauderie
les Rivieres
Montmartin-en-Graignes
St. Lambert
Neuilly la Forêt
Fourniere
Scale 1:25,000 or 2·53 inches to 1 Mile
Yards 1000 500 0 1000 2000 3000 4000 Yards
Metres 1000 500 0 1000 2000 3000 4000 Metres
THE GRID on this sheet is LAMBERT ZONE I (RED)
Projection — Lambert (modified) Conical Orthomorphic
Spheroid — Clarke 1880
Origin — 55 Grades N. Meridian of Paris
False Co-ords. of Origin 600,000 m. E. 200,000 m. N
Scale Factor — 8151·6 / 8152·6
Difference in Long. Paris — Greenwich 2°20' 13·95"
NOTES
① Contours are at 10 m.V.I. They are interpolated from spot heights and hachures on the French 1:80,000 and amplified from Air Photo Examination. They should be accepted with caution.
② Principal Points of photographs used in compilation are shown thus +
③ Trig. Lists of this area are on a 1:50,000 sheet basis — see incidence
INDEX
INCIDENCE OF 1:50,000 SHEETS
Convergence is given for centres of W. & E. sheet lines
Magnetic declination for centre of sheet from True North 9°36' W. from Grid North 7°18' W. April 1944 Annual change about 10' E.
Compiled from Air Photographs on a control provided by existing French Triangulation
14317.(259.)

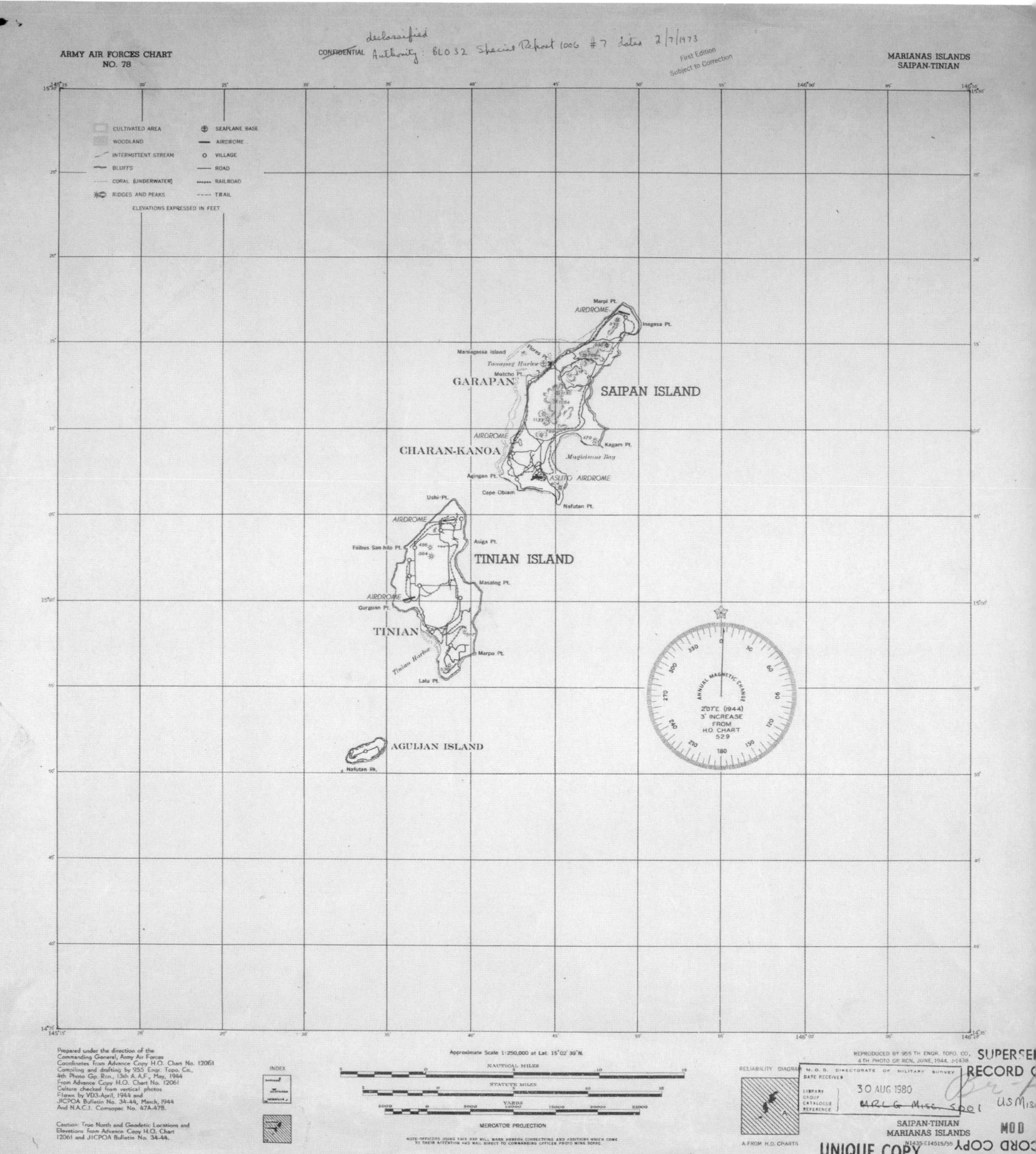

ARMY AIR FORCES CHART
NO. 78
CONFIDENTIAL
MARIANAS ISLANDS
SAIPAN-TINIAN
First Edition
Subject to Correction
SAIPAN ISLAND
GARAPAN
CHARAN-KANOA
TINIAN ISLAND
TINIAN
AGULJAN ISLAND
MERCATOR PROJECTION
UNIQUE COPY

THE MARIANA ISLANDS

955th Army Engineer Topographical Company, map of 'Marianas Islands Saipan-Tinian', June 1944.

The Mariana Islands are a group of 15 islands lying in the Pacific Ocean halfway between New Guinea and Japan. This map, produced by engineers attached to the Thirteenth Air Force, represented the most up-to-date knowledge of Saipan, which was one of the four biggest islands alongside Tinian, Rota and Guam. The latter, acquired by the United States from Spain in 1898, had fallen to Japan just two days after Pearl Harbor, and Saipan was an important administrative centre for the Japanese. The need to capture the Marianas was enhanced by the loss of Chinese airbases to Japanese attack in 1944. In addition, from the Marianas – unlike from China – all of Japan's major cities lay within reach of US strategic bombers.

Having badly hit Japanese naval power in the Battle of the Philippine Sea on 19–20 June, and thus denied Japanese island garrisons the possibility of being resupplied or reinforced, the Americans attacked the Mariana Islands, beginning with Saipan, which was the best of the islands to use for a bomber base against Japan. The engineers' map records the protective coral reefs, which had to be tackled by underwater demolition teams before the US Marines could attempt their amphibious landings.

The determination of Japanese resistance and their defensive fortifications ensured that it took three weeks (15 June to 9 July) to capture Saipan. The terrain, especially coastal cliffs, mountains and dense jungle, made the advance difficult. Nearly the entire garrison – 27,000 men – died in the strong defence and in ferocious frontal counter-attacks. Heavy casualties in the latter, however, greatly hit Japanese reserves and made the American task less difficult than it would otherwise have been had the Japanese rested on the defensive, as they were to do on Iwo Jima and Okinawa in 1945. Saipan's fall led to the resignation of the Japanese Cabinet on 18 July.

North American B25 Mitchells of the Seventh Air Force supported the landings on Tinian on 23 July. The conquest of the Marianas provided the United States with valuable bases for the aerial offensive against Japan. In October the first Boeing B29 Superfortress arrived on Saipan and from 24 November B29s were mounting air attacks on Tokyo, which was around 1,200 miles (under 2,000 kilometres) away and well within the aircraft's range.

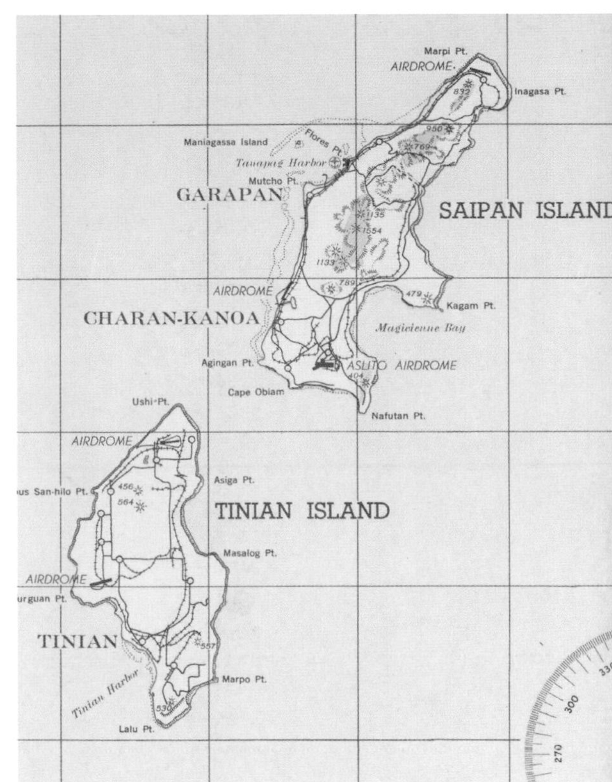

HOLLAND 1:10,000 DEFENCE OVERPRINT KOUDEKERKE - VLISSINGEN (FLUSHING)
EDITION OF 28 OCT 44
SECRET
FIRST EDITION (OCT 1944)
Sheet No 1
KOUDEKERKE
OOST
MOLENHOEK
Anteunshoef
Bergzicht
Broedershoek
Westhoek
Westerwijk
Der Boede
Tryton
DUIVENKEETENBLOK
Hoesbosch
RED CROSS
Vrijburg
Noordbeek
Huis ter Schelde
SHALLOW
De Lange Pacht
Plantlust
Zuidbeek
Papagaaienburg
Bonrepos
Klein Lammerenburg
Lammerenburg
OLD FORT
AIRFIELD
DITCHED
De Vijgeneter
Paauwenburg
Torenzicht
Zwanenburg
MAY BE HABITABLE
SATURATED
BRIDGES OUT
WIRE ALONG TOP AND FOOT OF DUNES
W3
Bonnedijke
Weewijk
HOSPITAL
KANAAL DOOR WALCHEREN
SARDINGEUL
HET DOK
VLISSINGEN
OVERPRINT REFERENCE
AIR PHOTO INFORMATION AS AT 24 OCT 44
GROUND INFORMATION (GS) AS AT 21 OCT 44
3 ROWS PILES AT FOOT OF QUAY WALL
BEACON ON TOWER
3 ROWS STAKES AT BASE OF DYKE
PERPENDICULAR SEAWALL 15'-20' HIGH
2 ROWS STAKES
3 ROWS STAKES
WESTERSCHELDE
STAKES 30-40 APART
Magnetic Declination and Grid Convergence
INDEX
REFERENCE
COMPILATION NOTE
RELIABILITY DIAGRAM
Detail interpretation in blank area is less complete than that in shaded area.
SUPERSEDED
RECORD COPY
UNIQUE COPY
DIRECTORATE OF MILITARY SURVEY MAP LIBRARY
21 APR 1955
REFER TO THIS MAP AS - HOLLAND 1:10,000 CA 21 SHEET NO. 1 DEFENCE OVERPRINT 28 OCT 44
Scale 1:10,000 or 6.34 inches to 1 Mile
HOLLAND CA 21

WALCHEREN

Map of Koudekerke–Vlissingen, 28 October 1944.

On 4 September 1944 British forces captured the docks of Antwerp, intact, on the River Scheldt. The seizure represented a key opportunity for providing effective logistical support for Allied forces advancing on Germany. The Allied need to capture Walcheren, the island that dominated the entrance to the Scheldt, was underappreciated due to the hope of pressing on northward into the Netherlands and capturing Rotterdam; however, the failure of Operation Market Garden, the attempt to cross the Rhine, defeated at Arnhem, refocused attention on the Scheldt. Walcheren was a difficult target and Canadian attempts to storm the exposed causeway that led to Walcheren from South Beveland were beaten back, and the focus instead shifted to amphibious assaults.

As the map makes clear (note the detailed annotation in the dock areas referring to '3 rows piles at foot of quay wall', '2 rows tetrahedra', '3 rows stakes at base of dyke', and so on), the defences at the main port of Vlissingen were well prepared. The map reflects the ability to use air photography, both near-vertical and oblique, to provide a good coverage. The Allies had acquired great expertise in this field. The prospect for an airborne assault on the airfield at Vlissingen (Flushing) was greatly lessened by its having had ditches dug across it. Moreover, much of the low-lying island had been flooded, in part as a defensive precaution but also as a result of the impact of bombing on the coastal dykes.

Early on 1 November, British commandos launched assaults against Vlissingen (Flushing), the port at the southern end of Walcheren, and at Westkapelle at the western end. They used landing craft supplied by the Americans, including the DUKW (boat-shaped amphibious trucks) and the LVT (landing vehicle, tracked), which carried machine guns and, given a level landing area, could climb straight out of the water and move inland. This was effectively an amphibious fleet capability within the Royal Navy.

Bad weather hindered the covering air support but a naval bombardment helped, and after landing before dawn the commandos gained a position in the port. Infantry from the 52nd Division followed, and there was two days of streetfighting before the Germans in the town surrendered. Amphibious vehicles were used to transport troops to Walcheren's inland capital of Middleburg, where the Germans surrendered on 5 November, the last on the island doing so by 8 November. No longer under threat, Allied minesweepers were able to clear the approaches to Antwerp where the first convoy arrived on 28 November.

PRISONER OF WAR MAPS

The Prisoners' Press, maps produced at Oflag 79, 1944.

These maps were printed secretly by British prisoners-of-war held in Oflag 79, a camp at Waggum near Braunschweig (Brunswick) that held 2,500 British and Commonwealth officers from 1943 to 1945.

Camp inmate and Royal Artillery officer Philip Radcliffe-Evans, a student of printing before the war, thought escapees from Oflag 79 could use location-specific maps to orient themselves. The idea arose in August 1944 when the camp sustained collateral damage during an American raid against a nearby target. The prisoners had to use roof tiles as plates and Radcliffe-Evans noticed when cleaning them that an image formed after an interaction between grease, sand and water. He realised that the tiles could be used in the lithographic process, which meant the men could try to build a clandestine press to print escape maps. This they did ingeniously, using various oddments, including a roller from a window bar covered with flying jacket leather, and inks out of bitumen and pencil lead. The prisoners could print 60 maps an hour working two shifts (right, below). Four different maps were made, with valuable details of settlements and communications, all to scale, before the press was found in January 1945 and destroyed.

Information came from local maps, details about the vicinity from temporary escapees who did reconnaissance, and from British maps of northwest Germany. The maps were made to three different scales, with the largest a 1:10,000 scale map of Oflag 79 and its perimeter (right, above), built up from a local map obtained through a local 'fixer' and therefore able to use a grid; 100 copies of it were printed. The camp is shown in relation to the railway line (to the north) between Hanover and Magdeburg, with the River Schunter to the south and the village of Querum to the southeast.

Next biggest is the 1:100,000 scale, three-colour map of the entire Braunschweig area (opposite), again gridded with coordinates, a kilometre scale bar and a legend at the bottom, including railways, roads and autobahns. The source for this was probably a smuggled-in silk map, but rather than having to laboriously copy it by hand the creation of one printing plate enabled multiple copies to be made.

The prisoners printed 1,200 copies in all on thin paper stolen from the camp administration offices. It is not known how many men actually put the maps to their intended use and escaped. The US Ninth Army liberated the camp on 12 April 1945.

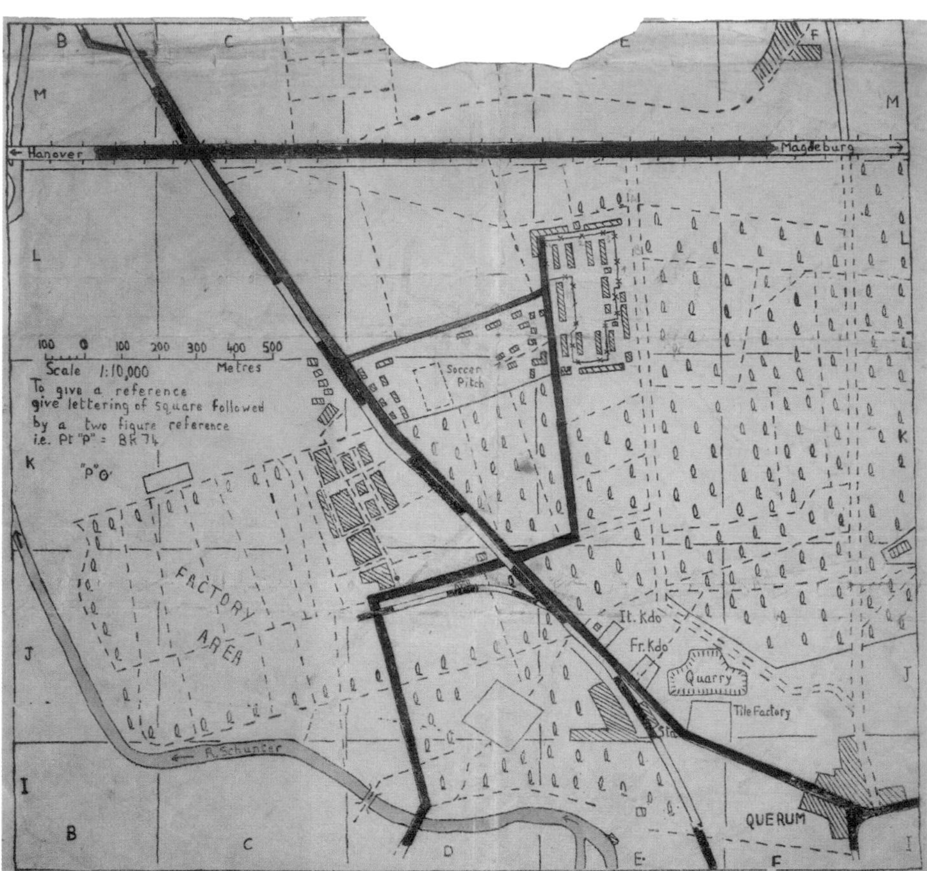

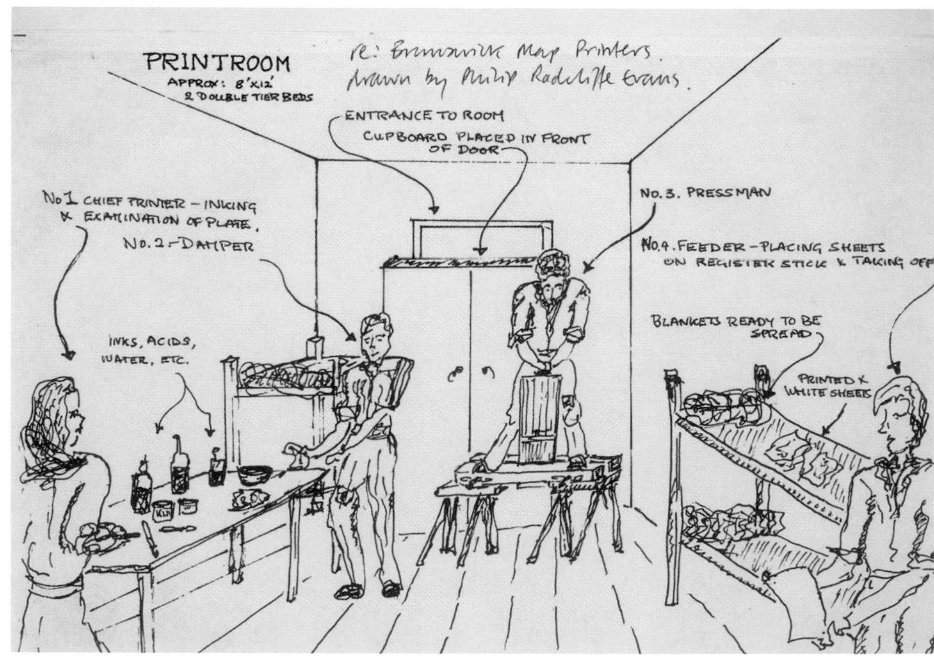

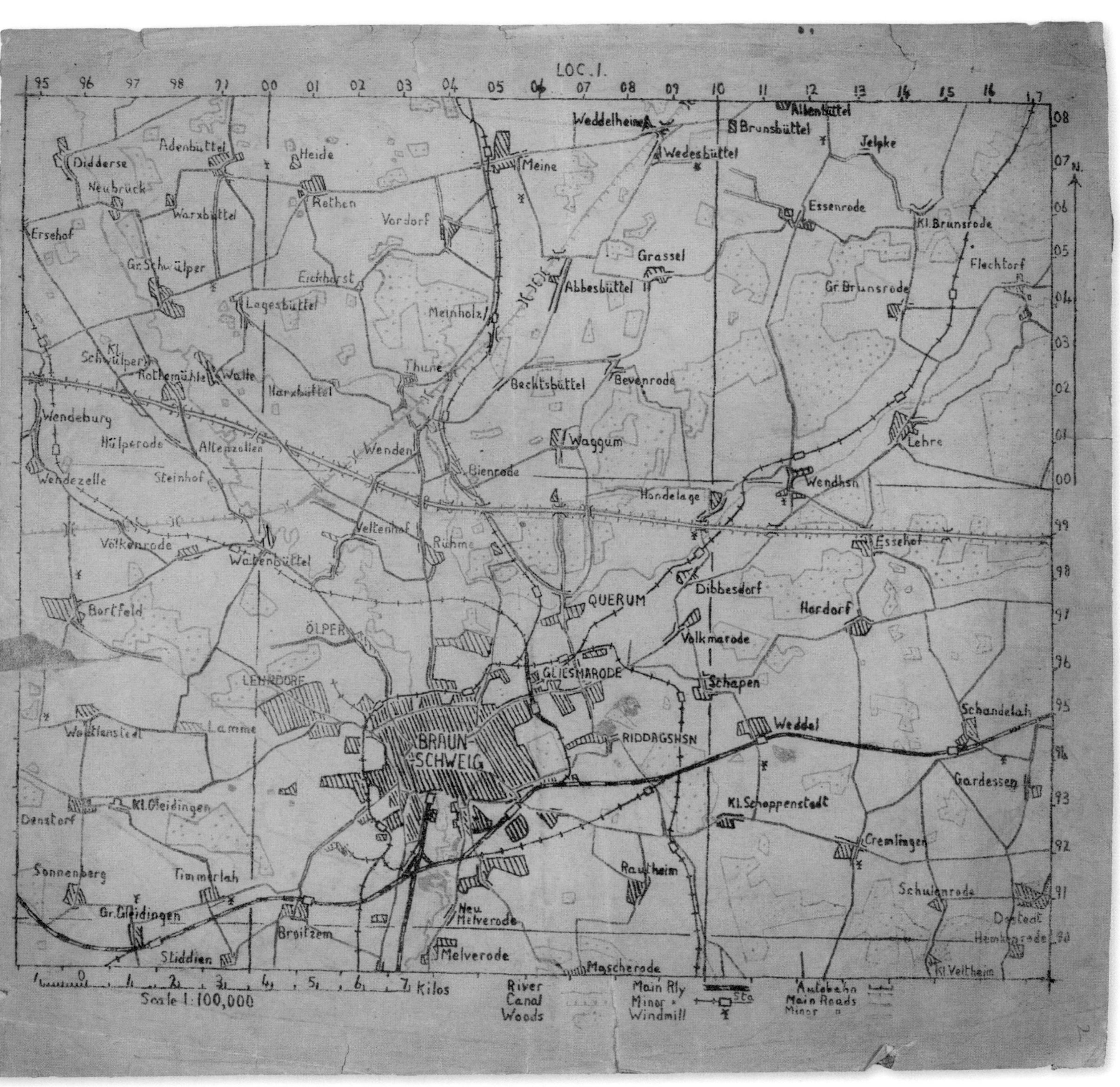
LOC. I.
BRAUN-SCHWEIG
QUERUM
GLIESMARODE
RIDDAGSHSN
Meine
Weddel
Lehre
Cremlingen
Scale 1:100,000
Kilos
River
Canal
Woods
Main Rly
Minor
Windmill
Autobahn
Main Roads
Minor

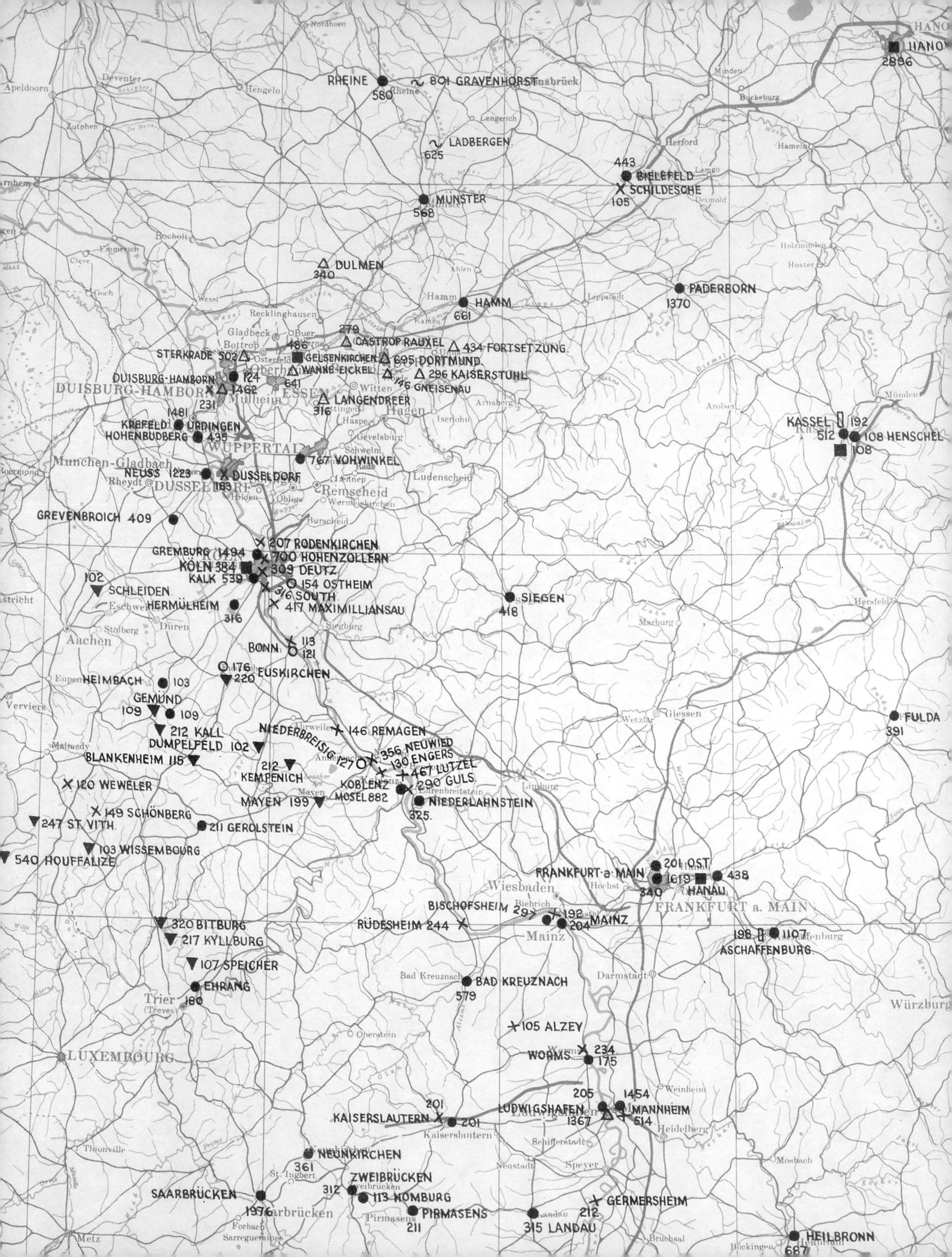

Nordhorn
HANOVER 2896
RHEINE 580
801 GRAVENHORST
Osnabrück
Deventer
Apeldoorn
Hengelo
Minden
Buckeburg
Zutphen
Lengerich
LADBERGEN 625
Herford
Hameln
443 BIELEFELD
SCHILDESCHE 105
Lemgo
Detmold
Arnhem
MÜNSTER 568
Bocholt
Emmerich
Cleve
Holzminden
Hoxter
DULMEN 340
Ahlen
Goch
Wesel
Hamm
HAMM 661
Lippstadt
PADERBORN 1370
Recklinghausen
Gladbeck
Buer
Bottrop
Kamen
279 CASTROP RAUXEL
486 GELSENKIRCHEN
434 FORTSETZUNG
STERKRADE 502
695 DORTMUND
DUISBURG-HAMBORN 124
WANNE-EICKEL 641
296 KAISERSTUHL
146 GNEISENAU
DUISBURG-HAMBORN
1462
231
ESSEN
Mülheim
Witten
LANGENDREER 316
Hagen
Arnsberg
Arolsen
Münden
1481
KREFELD
URDINGEN
HOHENBUDBERG 435
Iserlohn
Haspe
Gevelsberg
Schwelm
KASSEL 192
512
108 HENSCHEL
108
WUPPERTAL
Munchen-Gladbach
767 VOHWINKEL
Roermond
NEUSS 1223
DUSSELDORF
163
Rheydt
DUSSELDORF
Lennep
Remscheid
Ludenscheid
Hilden
Ohligs
Wermelskirchen
GREVENBROICH 409
Burscheid
207 RODENKIRCHEN
GREMBURG 1494
700 HOHENZOLLERN
KÖLN 384
309 DEUTZ
102
SCHLEIDEN
KALK 539
154 OSTHEIM
316 SOUTH
417 MAXIMILLIANSAU
HERMÜLHEIM 316
SIEGEN 418
Maastricht
Eschweiler
Stolberg
Düren
Siegburg
Hersfeld
Marburg
Aachen
BONN 113
121
176
220 EUSKIRCHEN
Eupen
HEIMBACH 103
GEMÜND
109
109
Verviers
Giessen
Wetzlar
FULDA 391
212 KALL
146 REMAGEN
NIEDERBREISIG
DUMPELFELD 102
Malmedy
356 NEUWIED
BLANKENHEIM 115
130 ENGERS
212
127
467 LUTZEL
KEMPENICH
KOBLENZ
290 GULS
120 WEWELER
MOSEL 882
MAYEN 199
NIEDERLAHNSTEIN 325.
Ehrenbreitstein
Limburg
149 SCHÖNBERG
247 ST. VITH.
211 GEROLSTEIN
103 WISSEMBOURG
540 HOUFFALIZE
201 OST
FRANKFURT-a-MAIN
1619
HANAU
438
340
Wiesbaden
Hochst
Biebrich
FRANKFURT a. MAIN
BISCHOFSHEIM 297
192
204 MAINZ
320 BITBURG
RÜDESHEIM 244
Mainz
198
1107
ASCHAFFENBURG
217 KYLLBURG
107 SPEICHER
EHRANG
180
Trier
(Treves)
Bad Kreuznach
BAD KREUZNACH 579
Darmstadt
Würzburg
105 ALZEY
Oberstein
LUXEMBOURG
234
WORMS 175
Weinheim
205
1454
201
LUDWIGSHAFEN
MANNHEIM
KAISERSLAUTERN
201
1367
514
Heidelberg
Kaiserslautern
Schifferstadt
Thionville
NEUNKIRCHEN 361
Neustadt
Speyer
Mosbach
St. Ingbert
ZWEIBRÜCKEN 312
SAARBRÜCKEN 1976
113 HOMBURG
GERMERSHEIM 212
PIRMASENS 211
315 LANDAU
Forbach
Sarreguemines
Metz
Bruchsal
Böckingen
HEILBRONN 687

STRATEGIC BOMBING

British Staff, 'Attacks of 100 tons & over by Eighth Air Force & RAF Bomber Command during January 1945: Sheet 1 – Strategic Targets', 1945.

Produced by the British Staff, this map (also overleaf) consists of data overprinted on a standard map. The strategic bombing covered here reveals a focus by the Anglo-American air assault in January 1945 on communication centres to the rear of the Battle of the Bulge, as well as a continuation, despite the winter weather, of the assault on German industry as part of the total war being waged by both sides: oil targets, factories, bridges, canals, communication centres and airfields are all included in the map legend.

On 18 January 1945, Churchill told the House of Commons:

> *'I am clear that nothing should induce us to abandon the principle of unconditional surrender, or to enter into any form of negotiation with Germany or Japan, under whatever guise such suggestions may present themselves, until the act of unconditional surrender has been formally executed.'*

Heavy bombing gave force to that stance and showed that Hitler had no viable plan.

At the Casablanca Conference in January 1943, Britain and the United States had agreed on the Combined Bomber Offensive in order to show Stalin that they were doing their utmost to weaken Germany. It was agreed that the bombing should destroy the German economic system and so damage German morale that the capacity for armed resistance would be fatally weakened. Because most German factories were in cities, these goals were linked.

Strategic bombing was made more feasible by four-engined bombers, such as the British Lancaster, as well as by heavier bombs and developments in navigational aids and training. The Allies also developed accurate radio navigation systems, which ensured that weather, darkness and smog were less of an obstacle to bombing. Pathfinder aircraft, which used the Oboe targeting system, H25 ground-looking navigation radar and target markers made a big difference, providing a valuable combination of technology and tactics.

The bombing assault on Germany was markedly strengthened as a result of the unexpectedly strong and successful German resistance in late 1944, as Allied forces advanced to near the frontier. As well as continuing to demonstrate the ability to counter-attack, shown in the Battle of the Bulge, Germany continued its V2 rocket attacks on London and released new, advanced German weaponry, notably the Me-262 jet fighter and the prospect of more, in the form of the Type XXI submarine. Luftwaffe resistance was reduced, notably after the failure of Operation Bodenplatte, an unsuccessful attempt, launched on 1 January, to gain air superiority in the Low Countries that led to the loss of many pilots. Serious losses were inflicted on Allied aircrews but the size of the bombing force was so large that the percentage of casualties was relatively low, whereas the percentage for the Luftwaffe was less favourable.

As the wartime techniques of command, control and navigation had improved, bombing became more effective. Moreover, by increasing the target area, area bombing made the task of the defence more difficult. 'V' raids, in which aircraft came in a succession of horizontal 'V' formations to sweep a continuous path across cities, proved especially destructive.

Despite the limited precision of bombing by high-flying aircraft dropping free-fall bombs, strategic bombing was crucial to the disruption of German communications and logistics, largely because it was eventually done on such a massive scale, and because the targets could not be attacked by any other means. Attacks on communications seriously affected the rest of the German economy, limiting the transfer of resources and the process of integration that is so significant for manufacturing. The reliance of industry on rail increased its vulnerability to attack, because rail systems lack the flexibility of their road counterparts, being less dense and therefore less able to switch routes. Air attack in 1945 brought the Reichsbahn (rail system) to collapse with damage extensive enough to preclude effective repairs. Bombing also acted as a brake on Germany's expanding production of weaponry, which had important consequences for operational strength. This was total war.

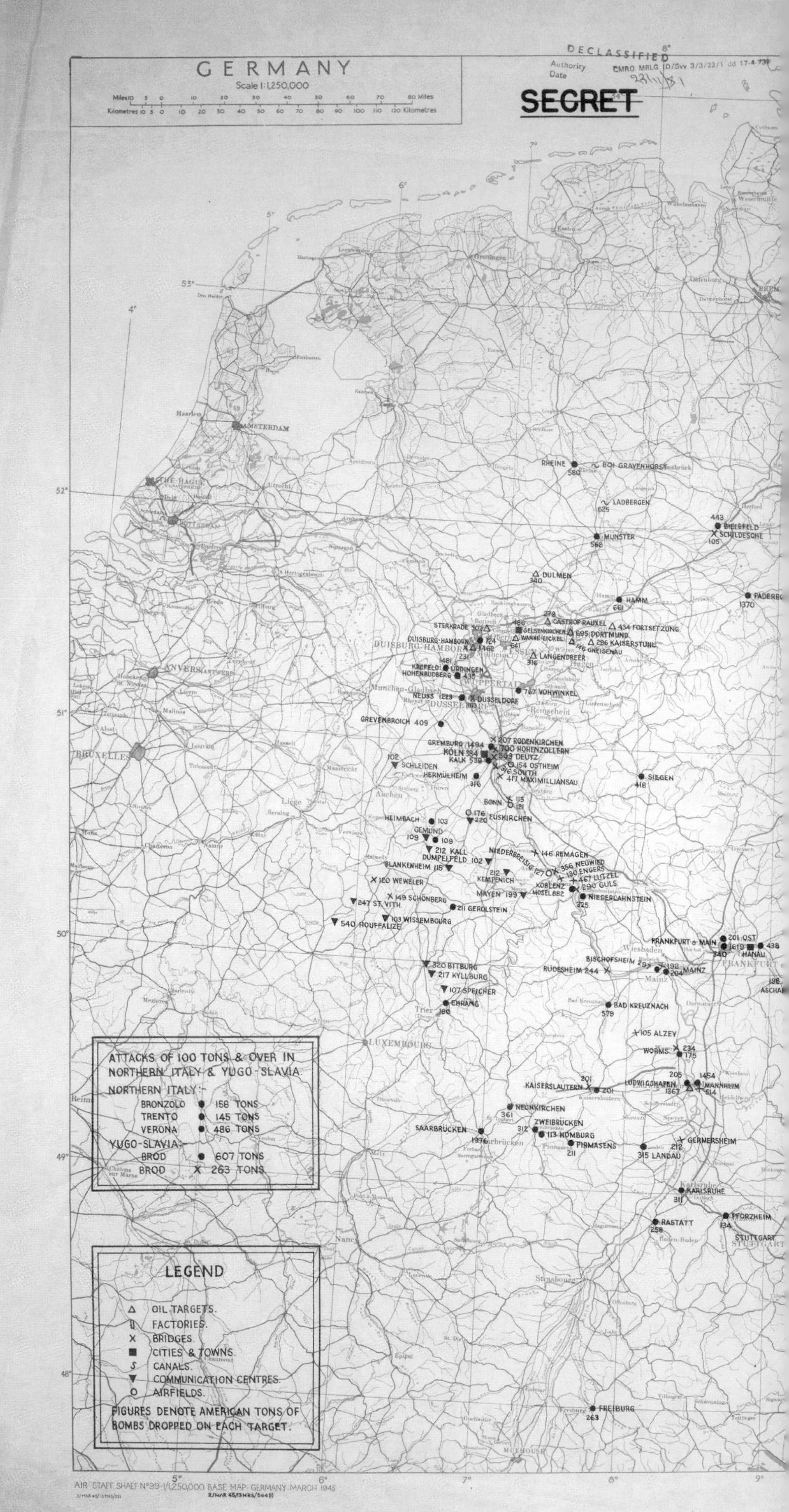
GERMANY
Scale 1:1,250,000
DECLASSIFIED
SECRET
ATTACKS OF 100 TONS & OVER IN NORTHERN ITALY & YUGO-SLAVIA
NORTHERN ITALY:-
BRONZOLO 156 TONS
TRENTO 145 TONS
VERONA 486 TONS
YUGO-SLAVIA:-
BROD 607 TONS
BROD 263 TONS
LEGEND
OIL TARGETS.
FACTORIES.
BRIDGES.
CITIES & TOWNS.
CANALS.
COMMUNICATION CENTRES.
AIRFIELDS.
FIGURES DENOTE AMERICAN TONS OF BOMBS DROPPED ON EACH TARGET.
AIR STAFF, SHAEF N°99-1/1,250,000 BASE MAP: GERMANY-MARCH 1945

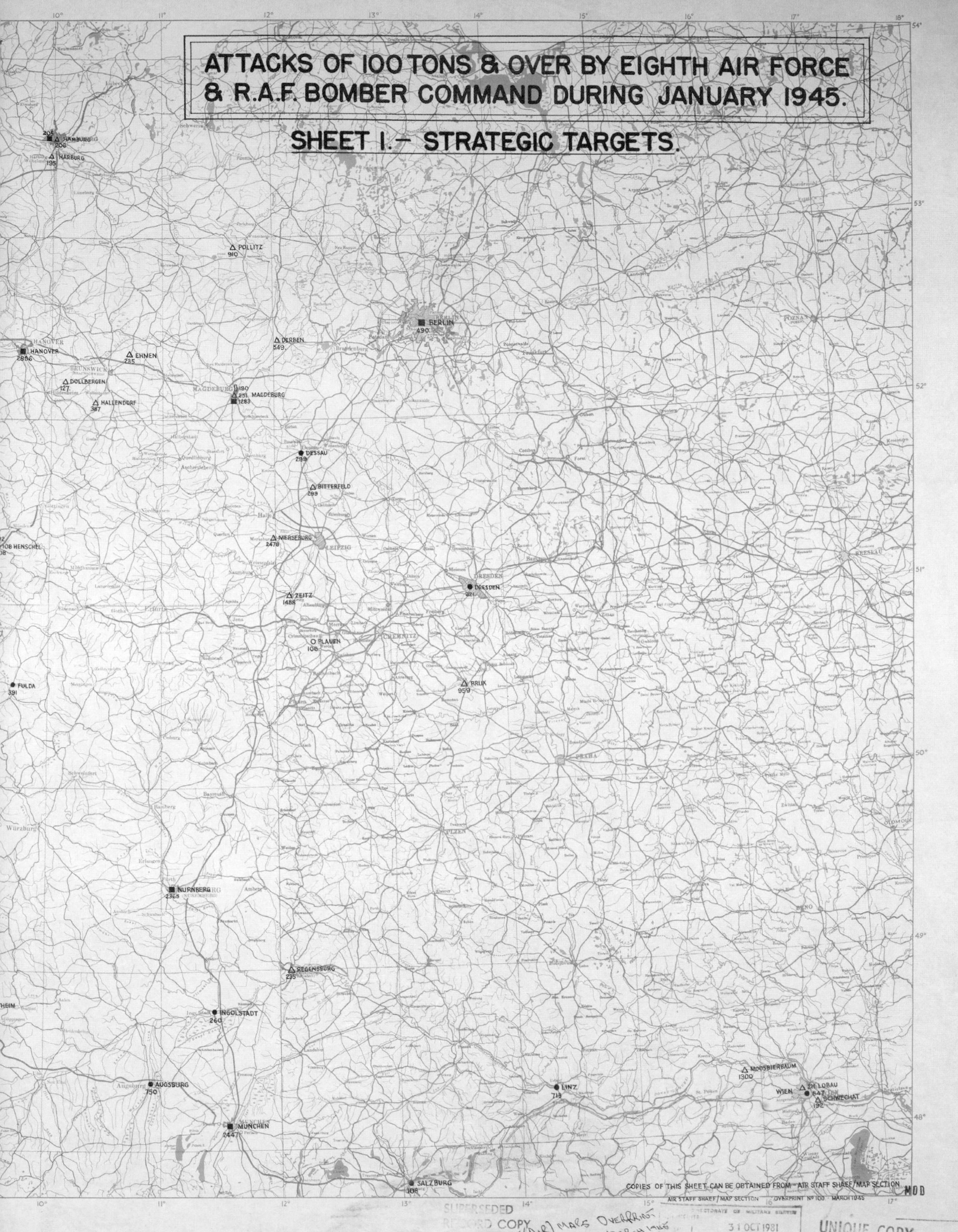
ATTACKS OF 100 TONS & OVER BY EIGHTH AIR FORCE & R.A.F. BOMBER COMMAND DURING JANUARY 1945.
SHEET I. – STRATEGIC TARGETS.
HAMBURG 205 209
HARBURG 195
POLLITZ 910
BERLIN 490
DERBEN 549
HANOVER 2856
EHNEN 235
DOLLBERGEN 127
HALLENDORF 387
MAGDEBURG 190 251 1283
DESSAU 299
BITTERFELD 299
MERSEBURG 2476
ZEITZ 1488
DRESDEN 321
PLAUEN 106
BRUX 959
FULDA 391
NURNBERG 2368
REGENSBURG 235
INGOLSTADT 260
AUGSBURG 750
MUNCHEN 2447
SALZBURG 108
LINZ 719
MOOSBIERBAUM 1300
WIEN
FL. LOBAU 647
SCHWECHAT 192
COPIES OF THIS SHEET CAN BE OBTAINED FROM - AIR STAFF SHAEF/MAP SECTION
AIR STAFF SHAEF/MAP SECTION OVERPRINT No 100 MARCH 1945
MOD

BATTLE OF THE BULGE

US Army and German Army, 'Das War der Plan' and 'Westkämpfer! Das habt Ihr geschafft!', January 1945.

As a result of the dominance of Allied air power, on 16 December 1944 Hitler sought the cover of bad weather, which would initially thwart Allied aircraft, when he launched the counter-offensive Ardennes campaign, better known as the Battle of the Bulge. The campaign was contested not only on the ground, but also in propaganda. 'Das War der Plan' ('That Was the Plan') is an American propaganda map distributed early in January 1945 that emphasised to German troops that the Bulge offensive had failed and argued, with reason, that further struggle was pointless: '...the map illustrates the breakthrough plan... It shows the German objectives. What the map doesn't show is where the Allied armies are located.' In contrast, the German leaflet 'Westkämpfer! Das habt Ihr geschafft! ('Western Front soldier! This is what you have done!') congratulates the troops and uses the backdrop of a map to claim that they have blocked Allied plans to advance via Köln (Cologne) and Strassburg (Strasbourg).

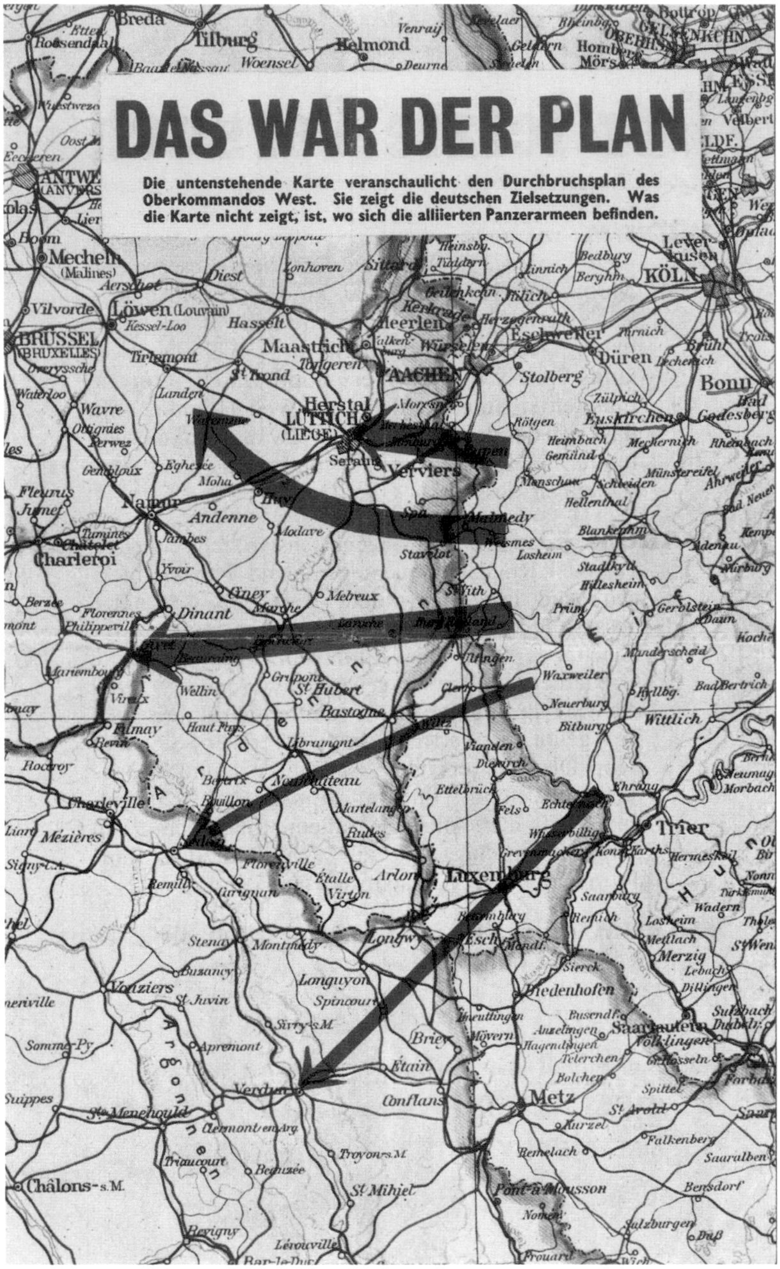

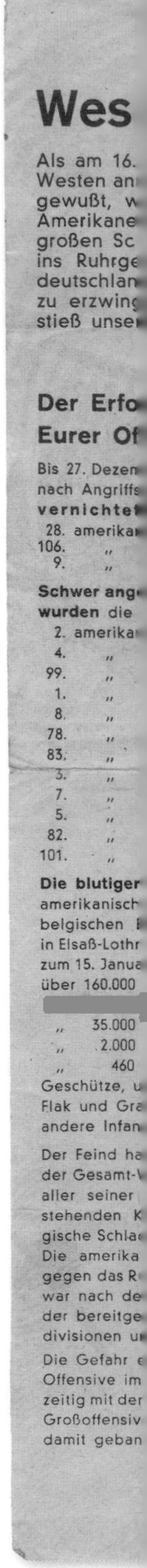

The counter-offensive also revealed a marked difference in the use of maps. The US Army was very well resourced at all levels with maps of all scales. Some were new surveys made with the benefit of air reconnaissance. All contained up-to-date overprints with accurate information of German dispositions. By contrast, the Germans had almost no mapping and little accurate idea of where the Americans were. The Germans had little chance for aerial reconnaissance by this stage of the war. They generally used Michelin 1:200,000 scale maps and tried to capture Allied maps whenever possible. Their only Intelligence was signals intercepts, prisoner interrogations and patrols. SS Kampfgruppe Peiper kept kidnapping passing civilians – to act as guides – whom they subsequently shot. In their quest for fuel, the Germans missed several important dumps not marked on maps.

The Germans were successful in the early stages of the battle. Benefitting from Germany's central position and still serviceable railway system, Hitler moved the 5th and 6th Panzer armies, core units facing the Soviets, in order to attack the Americans, hoping that a surprise offensive would lead to their defeat and the collapse of their will. With the Western Allies defeated and, hopefully, out of the war with Germany, he would then be able to focus on the Soviet Red Army in a war on one front.

Although such a development was feared by Japan, which did not want the United States and

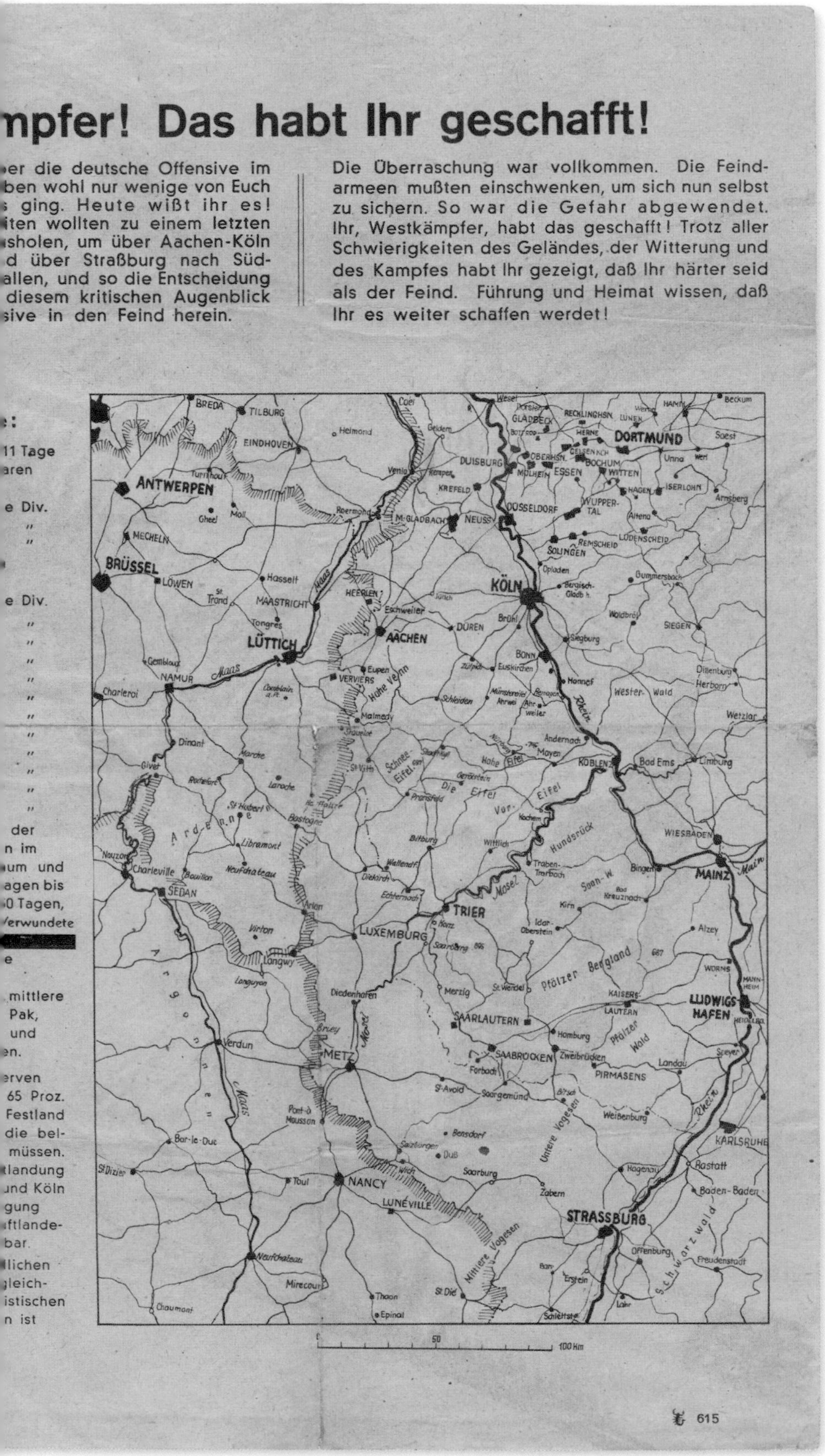
mpfer! Das habt Ihr geschafft!

er die deutsche Offensive im ben wohl nur wenige von Euch ging. Heute wißt ihr es! iten wollten zu einem letzten sholen, um über Aachen-Köln d über Straßburg nach Süd- allen, und so die Entscheidung diesem kritischen Augenblick sive in den Feind herein.

Die Überraschung war vollkommen. Die Feindarmeen mußten einschwenken, um sich nun selbst zu sichern. So war die Gefahr abgewendet. Ihr, Westkämpfer, habt das geschafft! Trotz aller Schwierigkeiten des Geländes, der Witterung und des Kampfes habt Ihr gezeigt, daß Ihr härter seid als der Feind. Führung und Heimat wissen, daß Ihr es weiter schaffen werdet!

:
11 Tage
aren
e Div.
"
"
e Div.
"
"
"
"
"
"
"
"
"
der
n im
um und
agen bis
0 Tagen,
Verwundete
e
mittlere
Pak,
und
en.
rven
65 Proz.
Festland
die bel-
müssen.
landung
und Köln
gung
ftlande-
bar.
lichen
gleich-
istischen
n ist

615

Britain to concentrate on her, Germany's assessment was a foolish one strategically, based on a flawed geopolitical analysis. There was a parallel with the German Spring Offensive in 1918, when Germany was helped by Russia already having been defeated, but that was not the situation in 1945. Moreover, in 1918 the German forces in the West were far closer to the English Channel and numerically stronger than their opponents when they attacked. In 1944, as in 1918, the opposition did not collapse as had been anticipated, and the Germans ended up, as in 1918, with a more extended front line and losses that weakened them when the Allies in turn attacked.

The aggressive nature of the surprise German assault against overstretched defences which lacked adequate fortifications and reserves was clear. Individual German units did not need to be directed from on high, the German tank forces fought well and the shock for most of the Allied troops faced with massed German armour helped to ensure that a bulge 45 miles (72 kilometres) in length was smashed in the Allied front and exploited with a German advance to close to the River Meuse (Maas).

At the same time, these German tactical and operational advantages were lessened by some deficiencies – notably a lack of fuel, supply problems, confused command structures and some poor training, especially on the part of infantry units – as well as the strength of the Americans, which included impressive artillery. Initial German successes could not be sustained, in part because of the swiftness with which the Americans deployed reinforcements, and due to the superb American defences of the towns of Bastogne and St Vith. The Germans could neither sustain their advance nor overcome fierce American resistance in flank positions. Once the weather improved, ground support air attacks proved particularly important against German tanks. The Germans were unable to reach any of the bridges over the River Meuse.

IWO JIMA

An American artillery spotter's map, Iwo Jima, 1945.

This map is a special air and gunnery target map corrected to November 1944 (three months before the US Marines' assault waves began on 19 February 1945) and it has a pencilled grid on it, indicating how the spotter might have used it to call in fire on Japanese positions. The detailed grid of numbered 1,000-yard target areas and lettered 200-yard target squares would have guided the pre-invasion bombardment, which lasted days and attempted to pulverise the entire island.

The map reflects the value of American Intelligence, which was largely based on aerial reconnaissance photographs. Marked 'secret', a classification lowered to 'restricted' within the combat area itself, the map is at a scale of 1:20,000 and printed on wet-strength paper (which was highly durable and water resistant). The military installations, which are noted in considerable detail, from radio towers to rifle pits, include probable obstacles the Japanese have placed to obstruct the likely landing beaches, consisting of tank barriers and minefields.

By the end of 1944, strategic thought on the part of the Japanese was in breakdown; once the decisive naval battle in the Leyte Gulf had been lost in late October the Japanese seemed unable, in the face of Allied power, to think through any option. Successes in China could not be translated into a broader strategic achievement, while the destruction of naval assets made it difficult to think of any further large-scale action. Japan was reduced to a defensive-offensive predicated on a tenacious defence coupled with destructive suicide missions, both of which were designed to sap their opponent's will. Neither did so and the weak government had no real strategy left other than for the military to die heroically.

The volcanic island of Iwo Jima lay midway between the Mariana Islands and those of the Japanese homeland. The Americans seized it in order to use the airstrips along the island's central area for attacks on Japan and to provide rescue bases for damaged B29s that otherwise would have had to ditch at sea. However, the battle to seize the island involved defeating well-positioned Japanese forces which fought to the death in the face of heavy pressure from the attacking US Marines. Although the Americans had massive air and sea support, the Japanese ability to exploit the natural terrain, notably by tunnelling, ensured that the bombing and shelling that preceded the landing on 19 February inflicted only minimal damage. As a consequence, the conquest of the island was slow and bloody, and much of the fighting was intense and at close quarters. The Japanese had created a dense network of underground fortifications, which not only blunted the effects of American firepower, both artillery and aircraft, but also made a fighting advance on foot difficult, not least because the Japanese had an extensive complex of interconnected and often well-concealed positions with interlocking fields of fire. There were 11 miles (18 kilometres) of fortified tunnels connecting 1,500 rooms, artillery emplacements, bunkers and pillboxes. The Japanese had sufficient artillery, mortars and machine guns to make their defences deadly, while their use of reverse-slope positions lessened the impact of American firepower.

Although much of Iwo Jima had fallen within two weeks, the time anticipated for its capture, it took 36 days of brutal combat to conquer the island, and more than one-third of the US Marines employed were killed (5,931) or wounded (17,372). The overwhelming majority of the 22,000 Japanese defenders were killed. The highest point on Iwo Jima is Mount Suribachi, visible in the map on the island's southern tip, where the iconic photograph was taken of the moment US Marines raised the US flag on D+4, the morning of 23 February.

The determination of the Japanese to fight to the last man made clear the likely cost of trying to invade the Japanese home islands and it helped to shape the later decision to use the atom bomb – to save lives on both sides.

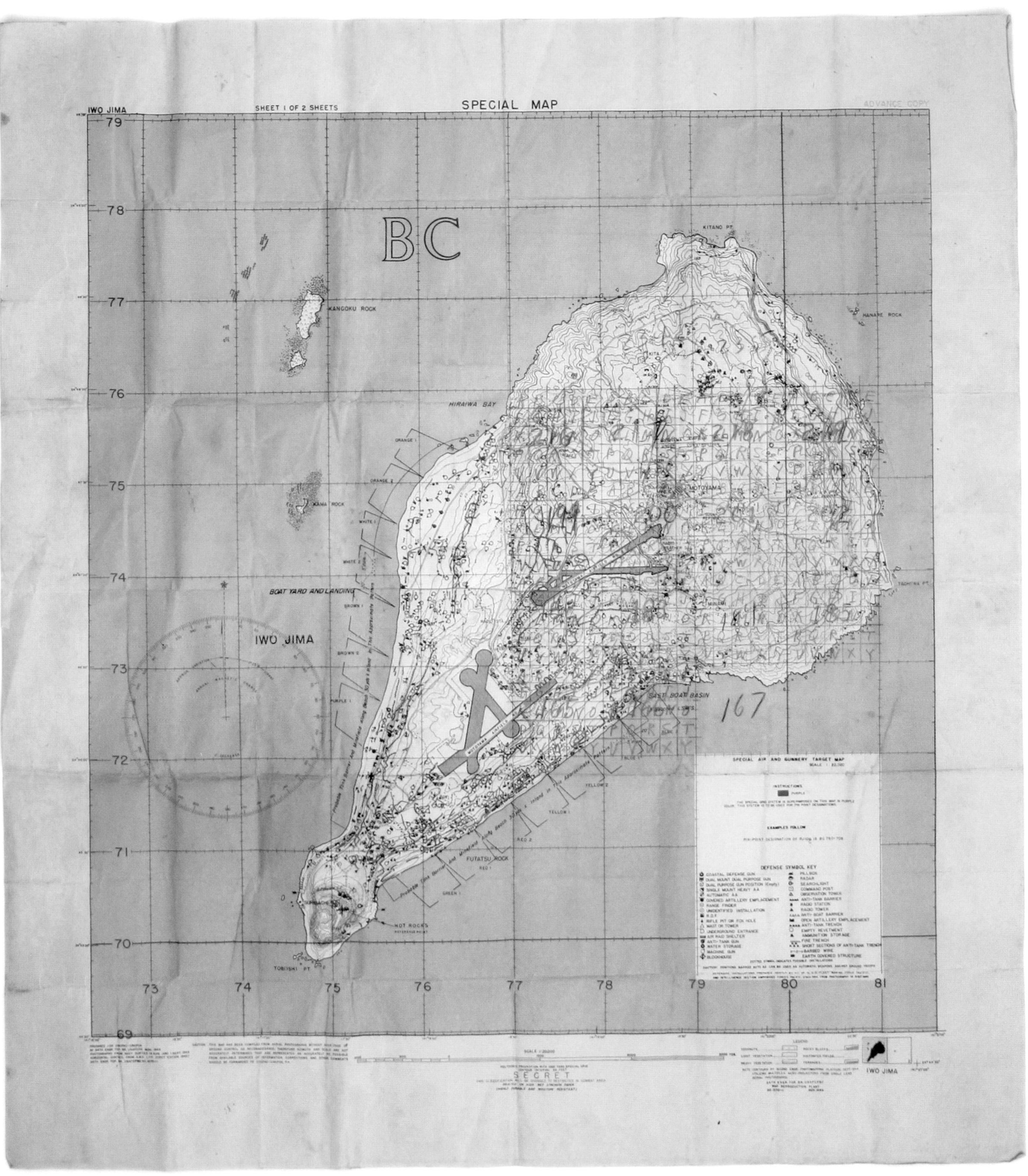
IWO JIMA
SHEET 1 OF 2 SHEETS
SPECIAL MAP
BC
KITANO PT
KANGOKU ROCK
HANARE ROCK
HIRAIWA BAY
MOTOYAMA
KAMA ROCK
BOAT YARD AND LANDING
IWO JIMA
EAST BOAT BASIN
167
FUTATSU ROCK
HOT ROCKS
TOBISHI PT
SPECIAL AIR AND GUNNERY TARGET MAP
INSTRUCTIONS
EXAMPLES FOLLOW
DEFENSE SYMBOL KEY
SECRET

CHUNGKING
SHANGHAI
KOREA
SEOUL
OMINATO
SENDAI
NAGASAKI
MOJI
OSAKA
TOKYO
YOKOHAMA
YOKOSUKA
2,580 MILES
J A P A N
KAGOSHIMA
KINHWA
NANCHANG
BURMA
FR INDO-CHINA
HANOI
CANTON
AMOY
FOOCHOW
HONG KONG
HAINAN
TAIWAN (FORMOSA)
MITSUGO JIMA
MUKO JIMA
OGASAWARA
BONIN IS.
KAZAN RETTO
MARCUS
URUCAS
MARIANAS ISLANDS
PAGAN
SAIPAN
ROTA
GUAM
WAKE
BATAAN
MANILA
SAIGON
PHILIPPINES
CEBU
DAVAO
YAP
NGULU
PALAU
SONSOROL
MERIR
TARAKAN
BORNEO
HALL IS.
NAMONUITO IS.
TRUK IS.
LOSAP
NOMOI IS.
UJELANG
PONAPE
KUSAIE
CAROLINE ISLANDS
HALMAHERA
135°
MAKASSAR
SURABAYA
BALI
TIMOR
AMBOINA
DILI
DARWIN
NEW GUINEA
MADANG
LAE
SALAMAUA
150°
KAVIENG
NUKUORO
GREENWICH
1,200 MI.
RABAUL
165°
GASMATA
BUKA
KIETA
BUNA
BUIN
CHOISEUL
KOKODA
REKATA BAY
STA ISABEL
CAPE YORK
PORT MORESBY
WYNDHAM
MILNE BAY
TULAGI

Reportage

5.

'Yanks Smash Ahead on Okinawa...'

Map of 25 April 1945 in the *Chicago Tribune*.

On 9 September 1939, Rand McNally announced that more maps had been sold at its New York store in the first 24 hours of the war than during all the years since 1918. Mapmakers responded to such demand by printing existing and new maps, and particularly so in the United States, where purchasing power was strongest. Thus, in May 1940, the National Geographic Society (NGS) brought out a map of Europe. There was another upsurge in map sales after Pearl Harbor. Moreover, popularising geopolitics in a radio speech, Roosevelt's address to the nation on 23 February 1942 made reference to a map of the world to explain American strategy, explaining that the war was different to wars of the past: '...not only in its methods and weapons but also in its geography. It is warfare in terms of every continent, every island, every air lane in the world.' He had earlier suggested that potential listeners obtain such a map, which led to massive demand and also to increased publication of maps in newspapers.

Maps were used both to convey news and for propaganda. The task of explaining engagement with distant regions posed a problem, but it also offered opportunities for innovation in conception and presentation. As in the First World War, newspapers printed large numbers of maps because readers expected maps to accompany news stories, although the mapmakers faced the difficulties posed by strict deadlines and minimal resources. In the United States, the *New York Herald Tribune*, *The New York Times*, New York's *Daily News*, *The Christian Science Monitor*, *Chicago Tribune*, *Los Angeles Times* and *The Milwaukee Journal*, all had their own cartographers, and their maps were reproduced in other newspapers. Maps were used to explain the significance of particular locations. Thus, 'Japan Pushes into Indo-China' in the *New York Journal-American*'s *Pictorial Review* in the *Seattle Post-Intelligencer* in August 1941, noted the proximity of Singapore to the new Japanese base at Saigon, as well as the tin, rubber and oil in the areas now exposed to Japanese attack.

Chief creative mappers

Leading figures included Emil Herlin, whose maps were reproduced in Francis Brown's *The War in Maps: An Atlas of* The New York Times *Maps* (1942), and Robert Chapin Jr in *Time*, whose war maps were also separately published by the magazine in much enlarged versions. Chapin's maps made sense of developments, as in 'Algeria' (23 November 1942), 'Roads to Warsaw' (24 January 1944), 'Winter Projection' (22 November 1943), a striking depiction of the direction of Soviet attack, and 'I have Returned' (30 October 1944), which illustrated Douglas MacArthur's return to the Philippines by offering a general map as well as a detailed enlargement of the American invasion sites (see pages 196–197).

Chapin used bold perspectives in his maps. 'Routes to Berlin', published in *Time* on 23 June 1943, presented the Western Allies as avoiding the heaviest defences in northern France and the Low Countries, instead probably advancing on Berlin either via landings in Schleswig-Holstein and Mecklenburg, or via southwest France and Languedoc, or the head of the Adriatic or Salonika, the last offering a base for the implausible depicted route via Belgrade, Budapest, Vienna and Prague. Chapin also liked the 'Roads' idea, as in 'Roads to Warsaw' (1944). His 'Design for Blockade', in *Time* on 2 July 1945, made clear the ability of American aircraft, based on carriers or Okinawa, and American submarines, to hit Japanese supply links from China and Korea, as shipping passed into the Kanmon Strait – a point he demonstrated by depicting a funnel. In that map he dramatised the 'Typhoon Season' by using, as in historic Japanese art, an illustration of a human-faced monster blowing wind. There were plentiful newspaper maps elsewhere, notably in Australia and Britain.

In Japan newspaper readers, too, were provided with maps explaining the situation in the Pacific, for example by Osaka Mainichi Shimbun in 1941. In Italy, the illustrated weekly magazine *Cronache della Guerra* carried a map in April 1942 of Malta and its defences. As with Allied maps, Axis ones emphasised the strength of their position. Thus, 'Kriegskarte der Nordsee' (1940) depicted the North Sea and British Isles partly in terms of the distance from German airbases, with the threat to Britain driven home by depicting German aircraft.

The rise of visual infographics

Many maps were used to locate areas of conflict. They provided a more valuable addition to the text than photographs, and were especially useful for the detail unavailable in atlases. Colour photography was not yet an established part of newspaper publishing, which meant black-and-white maps were not reduced in importance. The newspapers' mapping of war subordinated all spatial considerations to the front line; indeed, the key element

'THE PACIFIC—VITAL AREA FOR UNITED STATES AND JAPAN' by Howard Burke, *Los Angeles Examiner*, November 1942.

CHARLES HAMILTON OWENS

For the *Los Angeles Times*, Charles Hamilton Owens, who was adept at bird's-eye views, produced close to 200 full-page colour maps during the war. His perspective was very much a global one. He was skilled at producing maps that linked different areas and used colour and cartographic devices, such as arrows, to give his maps dynamism and interest. In the issue of 11 May 1942, his map 'Japs Fight to End Chinese Threat at Rear Before Drive on India' highlighted the significance of Burma to the Allied ability to support China. Relief was shown pictorially, oil fields were indicated and the strategic situation was explained in the accompanying text. On 28 August 1944, 'Day of Reckoning Near for Jap Conquerors of Philippines' was published, which used arrows to show the routes of the Japanese advance on Manila in 1941–42, as well as the location of prison camps; an inset map showed the proximity of the southern Philippines to New Guinea, where the Allies were already operating. A sense of the importance of such maps had been captured at the outset in the map of Europe in the issue of 10 September 1939, with its caption 'Save this Map! It shows the Theater of War' (see page 165), and a note on 'how to preserve this map.'

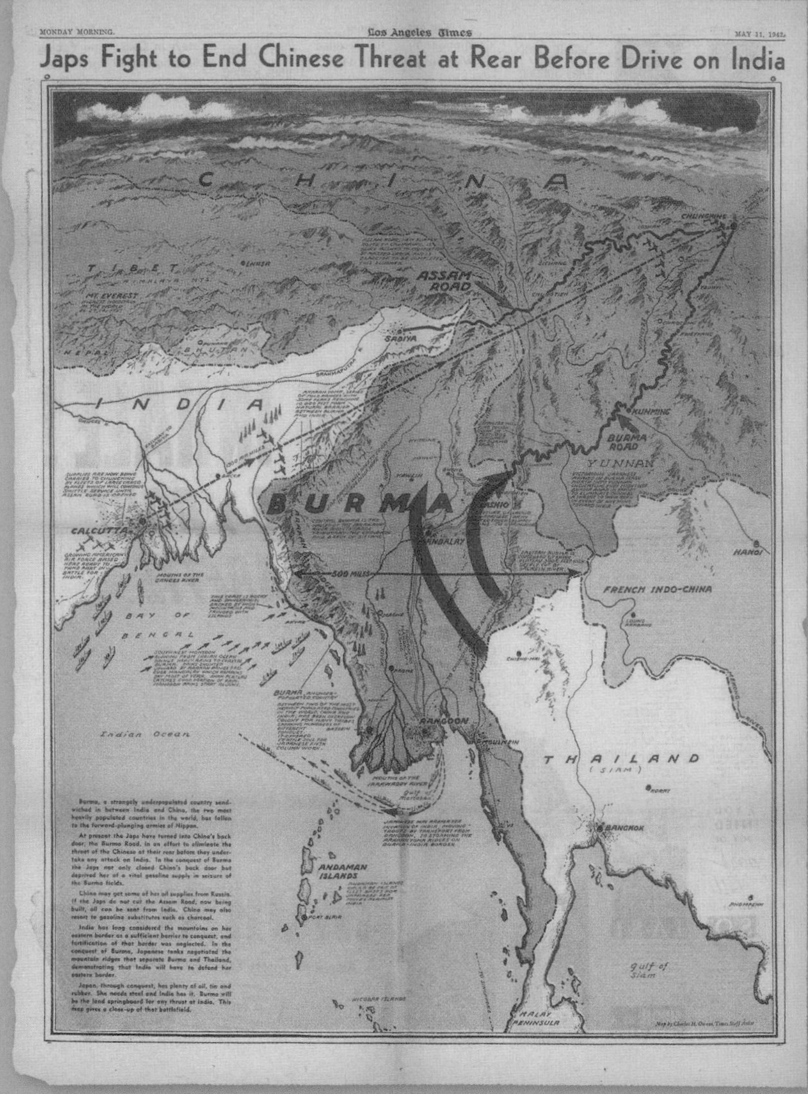

CHARLES H. OWENS, LOS ANGELES TIMES, 'JAPS FIGHT TO END CHINESE THREAT AT REAR BEFORE DRIVE ON INDIA', 11 MAY 1942.

Charles H. Owens' map explains the strategic challenge to China and India of the Japanese invasion of Burma. Communication routes are the key elements, distances are provided and oilfields are also shown. The large text block explains this effective map and the text and visuals are an excellent combination. The aerial perspective works, not least because the map is designed for American readers who are interested in the developing American air supply system from Bengal to China. Indeed, at a strategic level, the transport capabilities of aircraft were seen in the Anglo-American delivery, from July 1942 to 1945, of nearly 650,000 tons of materiel and 33,000 people from northeastern India over the 'Hump', the eastern Himalayas, to the Kuomintang (Nationalist Chinese) forces fighting the Japanese. The volume of goods moved increased significantly at the end of 1943 and, more particularly, from the summer of 1944. This achievement, which encompassed both the 'High Hump' and the 'Low Hump' routes, represented an enormous development in air transport, and in extremely difficult flying conditions, which led to the death of 1,200 pilots, even though the individual load of aircraft was low by later standards.

The map obviously says nothing about the deficiencies of America's British allies. Burma fell rapidly, with the mismanaged defence being exploited by the Japanese, who conquered the country at the cost of fewer than 2,000 dead. As in their conquest of Malaya in the winter of 1941–42, the Japanese proved adroit at outflanking manoeuvres and at exploiting the disorientating consequences for the British of their withdrawals, and, again, the Anglo-Indian force proved untrained for jungle warfare.

Invading in strength from 19 January, the Japanese captured Rangoon on 8 March. Mandalay fell on 1 May and the Japanese advanced in northern Burma and towards India, capturing Myitkyina on 8 May. Chinese intervention in Burma from March helped the British but it was not sufficient to sway the outcome.

Opened only in 1938, the Burma Road ('China's back door', as the map text puts it) was cut, which led to pressure to develop a new supply route further north beyond the range of Japanese advance. This was to be the Ledo Road, and, more immediately, the 'Hump'. In the meantime: 'Japan, through conquest, has plenty of oil, tin and rubber. She needs steel and India has it. Burma will be the land springboard for any thrust at India.'

was the location of the line. Linked to that came the employment of additional lines and shading to display the change in the front, and the use of arrows to show the direction of attacks and their possible follow-throughs. This approach ensured an operational focus in the cartographic presentation of the war, which was probably the focus that most interested contemporaries. Tactical-scale details and/or examples were not generally presented in newspaper maps, while the strategic dimension, although often advanced, received less attention than its operational counterpart.

The maps introduced readers to areas about which they knew little, notably – for the American and British publics – the Eastern Front in the Soviet Union. Accompanying news about the war with graphic information that mapped the war represented an additional visual dimension to add to those already provided by newsreels or – for those familiar with it – television, which might have prepared mass audiences for the post-war age.

However, making the newspaper maps genuinely helpful was not easy, because these simple maps reflected little of any tricky terrain or communication difficulties. For example, the detrimental effects of autumn rains, winter ice and spring thaws on the roads on the Eastern Front could not be represented. So also with other weather features.

Moreover, the rather two-dimensional notion of an easily rendered front line was not always appropriate. This was especially true of the Pacific, where a number of important Japanese bases, such as Rabaul and Truk, were simply bypassed by the Americans in 1943–45 thanks to superior air and sea power. Such a fluid and fast-moving situation, with its dependence on aerial mastery and range, was very difficult to capture adequately on a map.

Perspectives and projections

As in the First World War, commercial atlases were produced. In Britain there was *Atlas of the War* (1939) and *The War in Maps* (1940), and the Serial Map Service based on Philips' *International Atlas* produced monthly maps of the war. In turn, these monthlies were collected into an atlas. American atlases included the Los Angeles Examiner *War Atlas: Complete Book of Color Maps of All Battlefronts* (1942) and *The War in Maps: An Atlas of* The New York Times *Maps* (1942). Similarly, German firms, such as Justus Perthes and Ravenstein, produced detailed war maps for the German public.

The aerial dimension to the war encouraged the use of particular perspectives and projections for maps. The innovative cartographer Richard Edes Harrison (1911–94) had introduced the perspective map to American war journalism in 1935 when he produced maps to explain the Italian-Ethiopian war. So also in Britain with the 'Colour Relief Map of Abyssinia and War Zone' produced in 1938 by S.J. Turner for the *Daily Herald*, which adopted a bird's-eye view from over the Indian Ocean, looking north up the Red Sea, and with a note that 'This map is divided into 200-Mile Squares shown in perspective' (see pages 20–21).

The British and Commonwealth tradition of aerial maps, often in a pictorial form, continued during the war. On 9 January 1940, *The Argus* newspaper (see pages 168–169) in Melbourne offered its readers 'An Airman's View of War Fronts':

> *'Imagine yourself flying high over London in one of the fast planes of the Royal Air Force fighter squadrons, awaiting the approach of an enemy bombing formation. The map has been drawn from that centre of preparedness to represent a mental picture of the air attacking problem. ...'*

In the issue of *The Illustrated London News* of 8 February 1941, under the title 'Our Indomitable Stronghold in the Mediterranean: Malta, the Target of German Air Bombers,' George Davis produced a 'contour map of Malta' as viewed from the Axis air bases in Sicily. Davis's 'D-Day' in the same magazine adopted a similar approach.

Globalisation of viewpoints

The role of aircraft, dramatically demonstrated to American civilians (and everyone else) by the Japanese surprise attack on Pearl Harbor in December 1941, led to a new sense of space – one that reflected both vulnerability and the awareness of new geopolitical relationships. Osborn Miller's *Esso War Map* (September 1942) emphasised the wartime value of petroleum products. The theme of the text was 'Transportation – Key to Victory', and the map also provided an illuminated section, 'Flattening the Globe', showing how the globe becomes a map. North America was positioned centrally, which both made it the central target and let Germany and Japan appear as menaces from the east and west, each more threatening thanks to the other. The map included sea and air distances between strategic points, such as San Francisco and Honolulu, the significance of which had been increased by the attack on Pearl Harbor.

Esso drew upon pre-war commercial map production that was designed to encourage motorists and repurposed it for an audience engaged with the world war. The company, which also wanted to demonstrate its significance to the war effort as an independent company, produced a number of war maps (see pages 186–189).

Using the same technique, F.E. Manning's globular maps of October 1943, 'Target Berlin' and 'Target Tokyo' (see pages 178–179), printed and distributed by the US Army's Orientation Course, positioned the Axis capitals as vulnerable targets at the centre of the map. The maps included explanations of how they were constructed and to be used, including:

> *'This map is a photographic view of the world with the center at Berlin. Thus, with the detachable scale, distances can be measured along any line running thru Berlin. It should be noted that an inch at the center represents less mileage than an inch closer to the edges. The detachable scale has been designed to compensate for this and should be used only with the center on Berlin.'*

Manning's maps worked not only as accounts of aerial warfare, but also as ways to explain the converging goal and character of Allied strategy as a whole. 'Target Berlin' was swiftly followed by the beginning of a winter of British air attacks on Berlin.

The reporting and presentation of war – especially the dynamic appearance of many war maps, for example those in *Fortune*, *Life* and *Time*, with their arrows and general sense of movement – helped to make geopolitics present and urgent. Far from the war appearing to American readers as a static entity, existing at a distance, it was seen as in flux. The maps also made the war seem to encompass the spectator visually, through images of movement, and in practice by spreading in their direction. In turn, the Office of War Information, in its *A War Atlas for Americans* (1944), offered

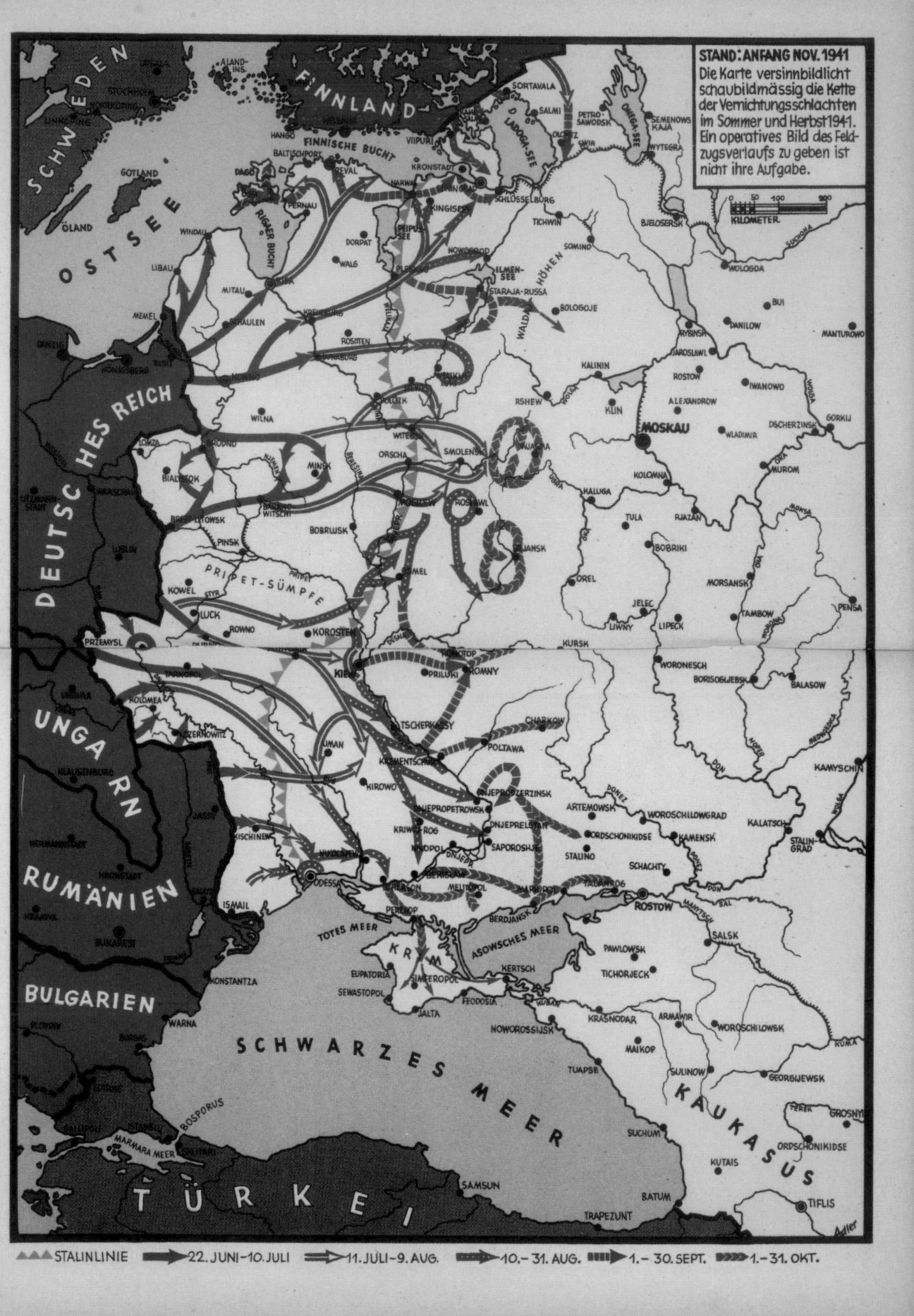

STAND: ANFANG NOV. 1941
Die Karte versinnbildlicht schaubildmässig die Kette der Vernichtungsschlachten im Sommer und Herbst 1941. Ein operatives Bild des Feldzugsverlaufs zu geben ist nicht ihre Aufgabe.
KILOMETER
SCHWEDEN
FINNLAND
OSTSEE
FINNISCHE BUCHT
RIGAER BUCHT
LADOGA-SEE
ONEGA-SEE
DEUTSCHES REICH
PRIPET-SÜMPFE
UNGARN
RUMÄNIEN
BULGARIEN
TÜRKEI
SCHWARZES MEER
ASOWSCHES MEER
TOTES MEER
KRIM
KAUKASUS
MOSKAU
LENINGRAD
KIEW
SMOLENSK
ODESSA
ROSTOW
CHARKOW
KALININ
KALUGA
TULA
OREL
KURSK
WORONESCH
STALINGRAD
SEWASTOPOL
BOSPORUS
Adler
STALINLINIE
22. JUNI–10. JULI
11. JULI–9. AUG.
10.–31. AUG.
1.–30. SEPT.
1.–31. OKT.

'STAND: ANFANG NOV. 1941' FROM DER KRIEG 1939/41 IN KARTEN, 1942.

This map, which translates as 'Status: Beginning Nov. 1941', possibly by Ernst Adler, shows German Army advances in the Soviet Union up to 31 October 1941 and it was published in 1942 in *Der Krieg 1939/41 in Karten*, which as with the 1940 book *Der Krieg 1939/40 in Karten* was under the editorship of Giselher Wirsing. The map makes the task of conquest appear nearly complete, because only part of the Soviet Union is shown, and the defence is presented in terms of the already-breached 'Stalinlinie' (Stalin Line). In practice, the Soviets were able to move much of their industry east to and beyond the Urals (not shown in this map), which was out of the range of any likely German advance and air attacks. Ministries had been evacuated from Moscow. However, the lack of Soviet collapse and the maintenance of will were also key elements, and these could not be shown in a map. Not shown either are the development of new lines of Soviet defence and the inadequacy of German supplies, complicated by weather problems and the inability of the Luftwaffe to resupply the German Army by air.

The map made clear that the Germans had advanced close to Moscow and, even more, Leningrad after what the text describes as 'the chain of annihilation battles in summer and autumn 1941'. What remained unclear to readers were the major difficulties that remained.

The propaganda dimension of German publications was obvious, but there was also a serious gap between the German assessment of the situation and the realities – a gap that related to a flawed strategy and doctrine, each of which was based on misleading assumptions of the ability to impose defeat on the Soviets.

Max Emanuel Wirsing (1907–75) was from a wealthy family and had changed his name to Giselher in 1929, becoming a writer and journalist after graduating in Economics. A nationalist and anti-Semite, he embraced Nazism and in 1943 had been appointed by Goebbels as editor of *Signal*, the armed forces magazine produced in dozens of languages by Wehrmachtpropaganda. Wirsing worked as a propaganda officer and war correspondent on the Eastern Front. He was a prominent editor in post-war West Germany.

perspective maps, part of the process by which the government produced maps for public consumption.

American public attention was being globalised, and maps helped to counter isolationism by linking distant regions to American interests. This linkage took place in a variety of ways. Emil Herlin of *The New York Times* was among those who used aerial perspectives for American audiences; another was the popular film *Casablanca* (1942), which began with a map sequence in which the zooming in moves from the entire globe to a street in Vichy-run Casablanca, Morocco. A very different film, *Victory Through Air Power* (1943), saw Walt Disney use animated maps, in which air power was depicted in terms of hurtling arrows to present the ideas of Alexander P. de Seversky's book of that name published the previous year.

At the same time as this, there were practical issues, as so often with mapping. Thus, the traditional Mercator projection was unhelpful in the depiction of air routes: great circle routes and distances were poorly presented because distances in northern and southern latitudes were exaggerated. That projection was also criticised as distorting the world – and, in particular, exaggerating the isolation of the United States from Europe while minimising the threat from Japan by making the Pacific appear bigger than it was. In August 1942, *Life* described the projection as a 'mental hazard', while in February 1943 *The New York Times* pressed for the Mercator projection's replacement by 'something that represents continents and directions less deceptively'. In *The Saturday Review of Literature* on 7 August 1943, Richard Edes Harrison wrote in favour of azimuthal projections and attacked what he presented as the misuse of the Mercator projection.

There were other problems with different types of maps. For example, maps that included pictorial images faced major issues with scale. As a result, 'The Battle of the Atlantic' (see pages 224–225) by Frederick Donald Blake (1944) suggested a crowded sphere of conflict, whereas in reality this was a large theatre of operations and there was a marked discrepancy between the roughness of the weather in the vast Atlantic, the small size of the ships and submarines, and the problems of visibility.

BASES AND BLOCKADES

Charles H. Owens, *Los Angeles Times*, 'Save this map! It shows the theater of war', 10 September 1939.

Charles H. Owens' map was designed as a means to understanding the world that was intended to be preserved for future reference: note the instructions for attaching it to heavyweight board. In keeping with the long tradition of mapmaking, it is embellished with depictions of war and is an effective map given the constraints of a lack of colour range. The focus on naval bases and blockades reflects both a strategic reality and the need to find items to fill the map, although it leads to error: Italian-ruled Rhodes is described as 'Cyprus British Naval Base'. The references to the Italian bases in Sardinia and the Balearic Islands were intended to highlight the issue of whether Germany's Italian ally would intervene. This emphasised the significance of the Mediterranean and thus of the depiction of bases there. In part, that depiction was misleading because the 'Great French Naval Base' in the Mediterranean was indeed at Toulon, but those in North Africa were at Mers-el-Kébir and Bizerta rather than Algiers and Tunis. The key British naval base of Alexandria is not marked as being such. Moreover, there was no attempt to add in the impact of air power on naval security. Malta was less secure as a 'Mighty British Base' once the prospect of Axis air attacks from nearby Sicily was considered, and, indeed, that further encouraged the focus on Alexandria. So also with the idea of a 'British Sea Dash to aid Danzig', mounted through the Skagerrak. Churchill, as First Lord of the Admiralty, had considered a naval move into the Baltic but it would have exposed the fleet to air attack in confined waters and without local bases. The rash idea was thwarted by his naval advisors.

Of greater significance at this stage was the ability of the British and the French to concentrate their forces in northeast France. With the defences against German naval, principally submarine, attack located further east in the Straits of Dover, and not where marked here, the British Expeditionary Force was safely transported to France, as were French forces from North Africa. However, the prepared defences in eastern France did not reach to the Channel because the Little Maginot Line was weak. So also with the German West Wall (Siegfried Line), which was not designed to be as strong as the Maginot Line, but was instead intended to delay attackers so that reserves could be moved up. The focus for the German line was on mutually supporting pillboxes and concrete anti-tank defences.

The Maginot Line, named after France's André Maginot, Minister of Defence from 1928 to 1930, was designed to offset Germany's larger population and therefore capacity for a larger army. It was an economy-of-force measure and a means to create jobs. The French regarded the fortifications, which covered the Franco-German frontier from Switzerland to Luxembourg, albeit being denser in particular areas, as an aspect of a force structure that could support an offensive or a defensive strategy. As intended, the Maginot Line constrained German options in 1940 and channelled the Germans into Belgium, but the French lost the mobile war as a consequence of the German exploitation of their unexpected axis of advance there.

ROTOGRAVURE SECTION — Los Angeles Times — SUNDAY, SEPTEMBER 10, 1939

SAVE THIS MAP! IT SHOWS THE THEATER OF WAR

HOW TO PRESERVE THIS MAP

• Use a good heavy-weight cardboard (lined chip board is best) and either library or flour paste. You may either cover the entire back of the map with paste or just tip the edges, but when rubbing the map onto the board be sure to cover it with another paper to avoid smearing the roto ink. If a very heavy board is not available it is best to mount a sheet of paper on the back of the board thus preventing warping of board when paste dries.

Map prepared by Charles H. Owens of The Times art staff

POLAND at WAR
KEY
Highways
Bombed Towns
Fortified Towns
Polish Fortifications
0 25 50 100 mi.
Striped Areas: 1914 Germany
SWEDEN
BALTIC SEA
LATVIA
RIGA
Libavia
Leningrad – 525 mi.
Moscow
LITHUANIA
Memel
KAUNAS
GDYNIA
(Port closed and mined)
Puck
(First German bomb fell here at 5:20 A.M.-Sept. 1.)
Westerplatte
(Munitions dump)
Hel
KÖNIGSBERG
EAST PRUSSIA
Vilija R.
WILNO
Niemen R.
Moscow 400 mi.
MINSK
Tczew
Allenstein
Grodno
U.
Baranowicze
Schnedemühl
Grudziąz
TORUN
Narew R.
BIALYSTOK
S.
Bydgoszcz
Mlawa
Pultusk
POZNAŃ
(Sugar Factories)
Vistula R.
Kutno
WARSAW
Bug R.
BRZEŚC-on-BUG
S.
Warta R.
LODZ
(Textile Centre)
Pripet Marshes
Sieradz
Pilica R.
Radom
LUBLIN
Skarzysko-Kamienna
Radomsko
R.
BRESLAU
KIELCE
LUCK
Rowne
Częstochowa
Sandomierz
Russo-Austrian Border 1914
Oppeln
C. O. P.
(Central Region of Industry)
Kiev 250 mi.
KATOWICE
CRACOW
San R.
LWOW
Tarnopol
Nowy Targ
CARPATHIAN MTS.
POLAND'S POSTERN
Only here can help come
Dukla Pass
STANISLAWÓW
Dniester R.
Jablonka Pass
SLOVAKIA
Bratislava
HUNGARY
VIENNA
Black Sea 425 mi.
RUMANIA
TIME Map by R. M. Chapin.

PENETRATION AND ENCIRCLEMENT

R.M. Chapin, *Time*, 'Poland at War', 11 September 1939.

This map, by Robert Chapin Jr in *Time* magazine, captures the challenge to Poland posed by Germany's ability to attack from north, west and south (the last thanks to the takeover of Czechoslovakia), but it does not yet capture the added challenge posed when the Soviet Union joined in on the German side on 17 September. Beginning their attack on 1 September, the Germans captured Lvov on 22 September and Warsaw on the 27th. The last Polish troops stopped fighting on 6 October. 'Poland's postern' to Romania marked on the map was to be the route of flight. In practice, no Allied help could viably have been sent via the Black Sea and Romania.

Maps for German audiences made victory appear inevitable. No role was presented for the Poles, while mapping such events in terms of the German Army did not offer any share of the success to the air force. In practice, the Poles defended the full extent of their borders, rather than concentrating in the heart of Poland to provide defence in depth and respond to German thrusts. The Polish deployment, which was done to signal the determination to defend frontier areas, actually helped the German penetration and encirclement strategy. By forcing the Poles into a one-sided war of manoeuvre, the Germans put them at a tremendous disadvantage, and Polish positions were successively encircled. The Germans were helped by seizing and using the initiative, while the flat terrain of most of Poland and the dryness of the soil and the roads following the summer all helped in tank and infantry advances. In particular, it was good tank weather. Hitler had explained in August 1939 to Count Ciano, the Italian Foreign Minister, that Germany could not long delay as otherwise rain would make the ground unsuitable for operations.

Whereas the Polish army was weak in tanks, in both quantity and quality, as well as in anti-tank guns and training, the Germans had five panzer divisions, each with about 300 tanks, as well as four light divisions, although they had far fewer tanks. The location of the panzer divisions had become an important indication of operational planning and prioritisation: most of the panzer and light divisions were in Army Group South in Silesia, which was entrusted with taking Warsaw and made the early key breakthroughs, whereas Army Group North, which cut off the Polish Corridor to the Baltic, had only one panzer division.

Inner and outer pincers created by German armoured columns were closed, isolating Polish armies and making it difficult for them to maintain supplies or launch counter-attacks. Confidence in this prospect enabled the German armour to advance ahead of the marching infantry and with exposed flanks. This, however, was a risky technique, particularly when the Poles had overcome the initial disorganisation and fear that resulted from coming under German attack. The Germans also faced problems with supplies, especially that of fuel, for which the panzer divisions had substantial requirements. Nevertheless, the Germans advanced too fast for the Poles to organise effective resistance and were able to cope with the problems that occurred and to regain momentum. Thus, a temporarily successful Polish counter-attack on the River Bzura (near Lodz) launched on 9 September was quashed by German tank reinforcements and air and artillery superiority. The Germans lost 17,800 dead, as well as 300 tanks, but the Poles had 66,300 dead and 124,000 wounded, as well as 694,000 men captured by the Germans.

The Slovaks invaded Poland alongside the Germans, but were a minor factor, unlike the Soviets who captured 240,000 Poles. The defeat of Poland was followed by large-scale murder and deportation of Poles by both the Germans and Soviets.

WAR FROM ABOVE

The Argus newspaper, 'An Airman's View of War Fronts', 9 January 1940.

This map was prepared ('No. 16' in the Melbourne newspaper's 'War Maps' series) to show the crucial role of RAF fighter aircraft, notably the Spitfire, and to present an aerial bird's-eye perspective, although, as noted, the map presents the 'mental picture' and not what the pilot could actually see. Underscored by large British investments in the aircraft industry, the British had developed two effective monoplane fighters, the Hawker Hurricane and the Supermarine Spitfire, in accordance with 1934 Air Ministry specifications for a fighter with improved speed, rate of climb and ceiling, and with eight machine guns. The new fighters entered service in December 1937 and June 1938 respectively.

In 1937, in a speech titled 'Fighter Command in Home Defence', Sir Hugh Dowding told the RAF Staff College that the major threat to British security would be an attempt to gain air superiority in which the destruction of airfields would play a key role, leaving the way clear for attacks on industry, London and the ports. A system of defence was therefore the key necessity, and alongside early-warning radar the Spitfires and Hurricanes were to help rescue Britain in 1940 from the consequences of devoting too much attention earlier to bombing.

On 23 May 1939, Hitler had informed his army commanders that it was necessary to seize France, Belgium and the Netherlands in order to provide the navy and air force with bases from which to attack Britain. The strength of Britain's economy, its maritime ability to blockade Germany and its capacity to bomb Germany's leading industrial zone, the Ruhr, which was vulnerable to aircraft based in France, all worried Hitler at this stage. Given the then capabilities of bombers, notably in bombload, he exaggerated the threat.

The map underplays the risk that was to face Britain later in 1940, as the consequence of the loss of France, Belgium, the Netherlands and Norway meant that German-controlled airfields, notably in the Pas de Calais, were closer to Britain, greatly cutting the journey time of bombers and allowing fighters to escort them.

There was also no discussion of radar. In 1904, Christian Hülsmeyer had first used radio waves to detect the presence of distant metallic objects. Thirty years later, the French CSF company took out a patent for detecting obstacles by ultra-short wavelengths, while in 1935 Robert Watson-Watt wrote *The Detection of Aircraft by Radio Methods* and the Air Ministry decided to develop radar. The following year, the American Naval Research Laboratory demonstrated pulse radar successfully.

In the Battle of Britain, the British were to benefit greatly from an integrated air-defence system founded on a chain of radar-equipped early-warning stations built from 1936 to 1939. These stations, which could spot aircraft 100 miles (160 kilometres) off the south coast, were linked to centralised control rooms where data was analysed and then fed through into instructions for fighters. No other state had this capability at this stage and the weakness of anti-aircraft defences enhanced its significance. The integrated British control system, with its plotters of opposing moves (see page 111) and its telephone lines, was an early instance of the network-enabled capability seen in the 2000s with its plasma screens and secure data links.

IMAGINE YOURSELF flying high over London in one of the fast planes of the Royal Air Force fighter squadrons, awaiting the approach of an enemy bombing formation. The map has been drawn from that centre of preparedness to represent a mental picture of the air attacking problem. The coastline is spread out in aerial bird's-eye form. Holland is shown to the left, with the German coastline at the Elbe and the Weser. Beyond is Danzig (about 750 miles), Warsaw (850 miles), and Berlin (550 miles). Belgium lies in the centre of the view, and the mouths of the Schelde, the Meuse, and the Rhine are shown, with Frankfort, Cologne, and Essen beyond. To the right is Luxembourg (overlooking the Saar valley), the Vosges, and Munich.

APART FROM a narrow "swept channel" along the British coastline, the Dover Barrage makes it impossible for enemy warships to attack in that area, and explains the extent to which the Nazis must depend upon the effectiveness of Goering's bombers if they decide to launch a large-scale air attack. So far the Nazi raiders have been shot down, one by one, by Britain's one-man fighter planes, and the small drawing on the right shows how it is done. The Spitfire fighter is a mile a minute faster than the Heinkel bomber. The secret of the 200 bullets fired every second by the Spitfire is their assortment. Following the armour-piercing bullets at given intervals are the incendiary or tracer bullets, which, beside setting fire to the enemy plane, show the British pilot, by their trail of fire, exactly where he is hitting the enemy.

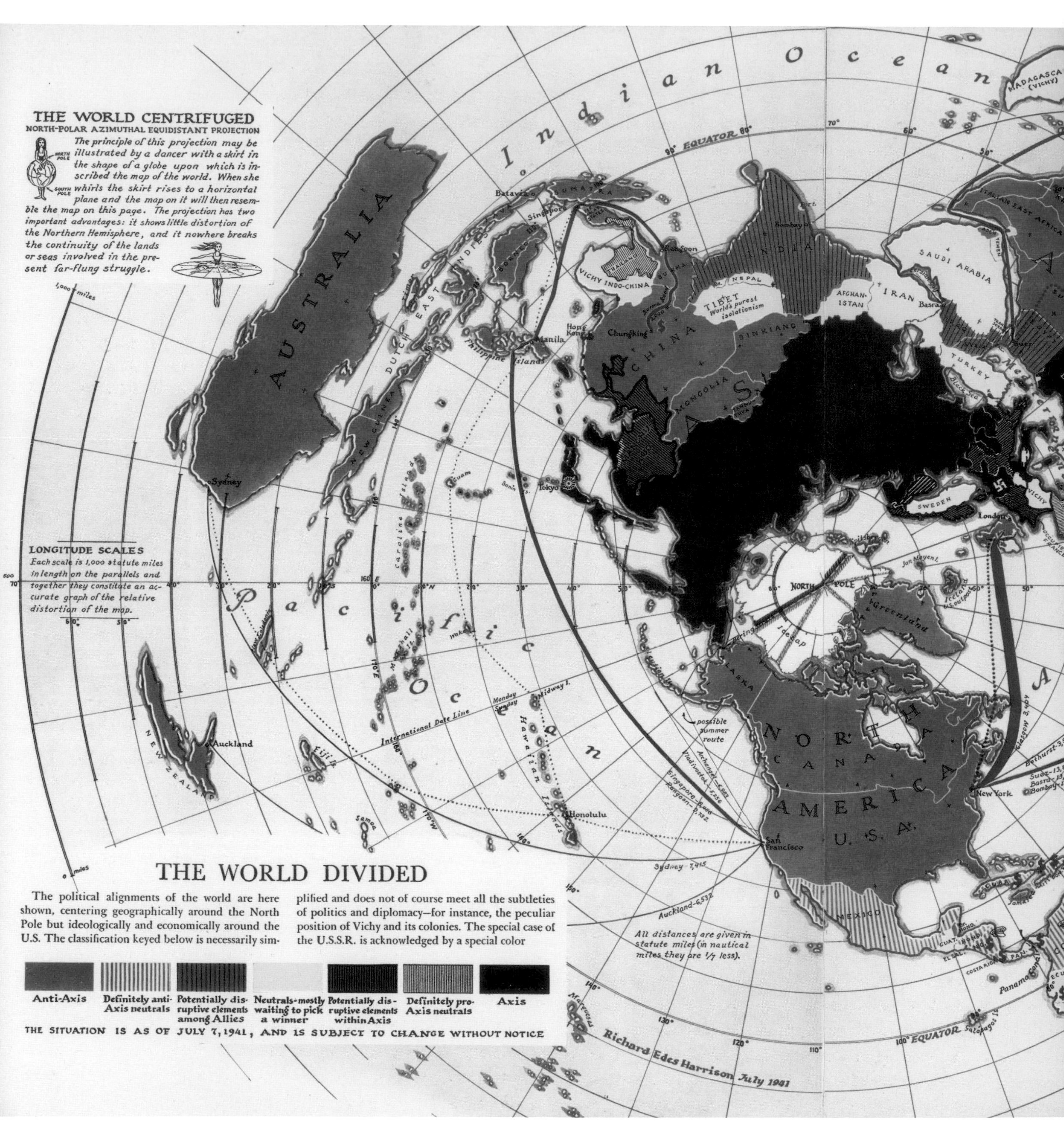
THE WORLD CENTRIFUGED
NORTH-POLAR AZIMUTHAL EQUIDISTANT PROJECTION
The principle of this projection may be illustrated by a dancer with a skirt in the shape of a globe upon which is inscribed the map of the world. When she whirls the skirt rises to a horizontal plane and the map on it will then resemble the map on this page. The projection has two important advantages: it shows little distortion of the Northern Hemisphere, and it nowhere breaks the continuity of the lands or seas involved in the present far-flung struggle.
NORTH POLE
SOUTH POLE
1,000 miles
LONGITUDE SCALES
Each scale is 1,000 statute miles in length on the parallels and together they constitute an accurate graph of the relative distortion of the map.
0 miles
Indian Ocean
Pacific Ocean
AUSTRALIA
NEW ZEALAND
Sydney
Auckland
DUTCH EAST INDIES
NEW GUINEA
BORNEO
SUMATRA
Batavia
Singapore
Manila
Philippine Islands
Hong Kong
VICHY INDO-CHINA
THAILAND
BURMA
Rangoon
Chungking
CHINA
MONGOLIA
SINKIANG
TIBET
World's purest isolationism
NEPAL
INDIA
Bombay
AFGHANISTAN
IRAN
Basra
SAUDI ARABIA
ITALIAN EAST AFRICA
TURKEY
Black Sea
MADAGASCAR (VICHY)
Tokyo
Guam
Caroline Islands
Marshall Is.
Wake I.
Midway I.
Hawaiian Islands
Honolulu
Fiji Is.
Samoa
International Date Line
Monday
Sunday
NORTH POLE
Greenland
Jan Mayen I.
Iceland
Spitsbergen
Ice Cap
Bering
Aleutian Is.
ALASKA
SWEDEN
London
VICHY
NORTH AMERICA
CANADA
U.S.A.
MEXICO
New York
San Francisco
possible summer route
Panama Canal
Galapagos Is.
EQUATOR
Sydney-7,415
Auckland-6,532
All distances are given in statute miles (in nautical miles they are 1/7 less).
Richard Edes Harrison July 1941
THE WORLD DIVIDED
The political alignments of the world are here shown, centering geographically around the North Pole but ideologically and economically around the U.S. The classification keyed below is necessarily simplified and does not of course meet all the subtleties of politics and diplomacy—for instance, the peculiar position of Vichy and its colonies. The special case of the U.S.S.R. is acknowledged by a special color
Anti-Axis
Definitely anti-Axis neutrals
Potentially disruptive elements among Allies
Neutrals-mostly waiting to pick a winner
Potentially disruptive elements within Axis
Definitely pro-Axis neutrals
Axis
THE SITUATION IS AS OF JULY 7, 1941, AND IS SUBJECT TO CHANGE WITHOUT NOTICE

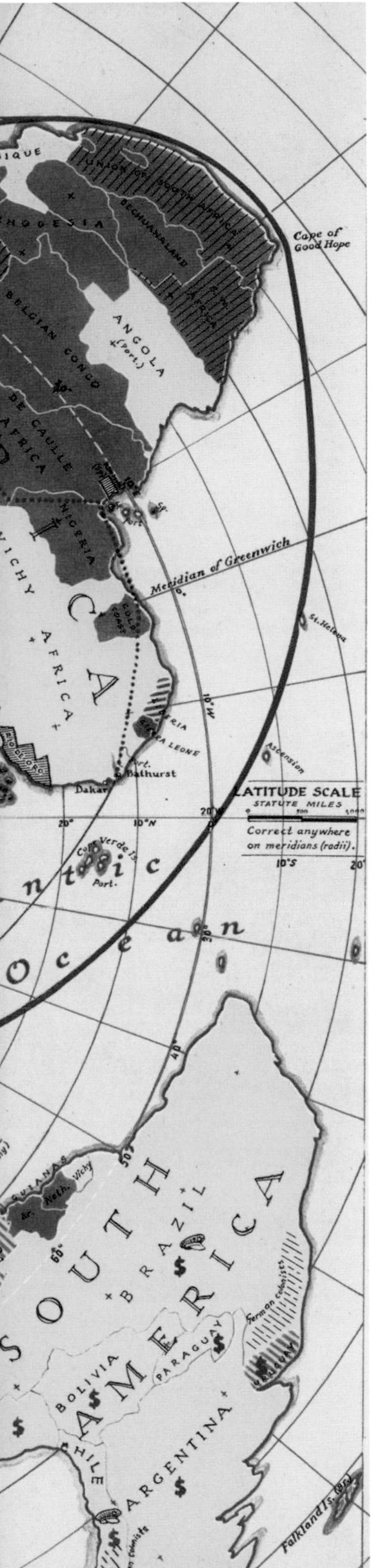

THE AMERICAN PIVOT

Richard Edes Harrison, 'The World Divided', July 1941.

Richard Edes Harrison's orthographic projections and aerial perspectives brought together the United States and distant regions with great geographical imagination and visual flair, and were part of a worldwide extension of American geopolitical concern and military intervention linked to a reshaping of the world. Like Charles Hamilton Owens in the *Los Angeles Times*, Harrison (1911–94) worked from a globe, which was a key element in mapping on a hemispheric and, even more, world scale. Colour greatly assisted the impression that was possible through visual contrasts, for example of terrain.

Harrison underlined the need for vigilance in his series of three maps, 'Three Approaches to the United States' (see pages 26–27), which appeared in *Fortune* in September 1940 alongside an article emphasising the dangers posed by isolationism and inadequate military strength. Harrison's maps showed the threat of a German attack via Canada, a Japanese one from the Northern Pacific and one on the Atlantic coast from South America, and the different aerial perspectives offered contributed to the sense of a general crisis. The first threat he describes as a 'great-circle route from Berlin' that passes through Detroit:

> *'... a pincer movement extending from Newfoundland down the New England coast and down the St. Lawrence to the continent's heart; a third arm reaching to the south shore of Hudson Bay, where the terrain permits quick construction of landing fields.'*

The industrious Harrison's output included maps centred on the North Pole. 'The World Divided,' published in *Fortune* in July 1941 and depicting the world as of 7 July, carries a note explaining the North-Polar Azimuthal Equidistant projection used:

> *'The World Centrifuged... The principle of this projection may be illustrated by a dancer with a skirt in the shape of a globe upon which is inscribed the map of the world. When she whirls the skirt rises to a horizontal plane and the map on it will then resemble the map on this page. The projection has two important advantages: it shows little distortion of the Northern Hemisphere, and it nowhere breaks the continuity of the lands or seas involved, in the present far-flung struggle.'*

The picture of the dancer (opposite, top left) has her skirt marked with North Pole and South Pole. The explanation includes: 'Longitude Scales. Each scale is 1,000 miles (1,600 kilometres) in length on the parallels and together they constitute an accurate graph of the relative distortion of the map.' The key of national alignment includes: 'The classification keyed below is necessarily simplified and does not of course meet all the subtleties of politics and diplomacy – for instance, the peculiar position of Vichy and its colonies.' The text notes the threat to the United States: 'The Western Hemisphere lies between the shores of the great land mass of the Eastern Hemisphere.'

'One World, One War' showed the war as on 15 October 1943, with the United States depicted in a key position. The preface to his *Look at the World: The Fortune Atlas for World Strategy* (1944), an atlas, published by the major New York house of Alfred A. Knopf, that reproduced his maps from the magazine *Fortune*, explained that its 'main purpose is not to locate supply lines.... Instead they try to show why Americans are fighting in strange places and why trade follows its various routes. They [the maps] emphasise the geographical basis of world strategy'.

Harrison's maps put the physical environment before national boundaries, and also reintroduced a spherical dimension, offering over-the-horizon views: an aerial perspective that did not exist for humans in nature and was not to exist until the age of rocketry, but that captured physical relationships, as in his 'Russia from the South' (1944). The first edition of the atlas sold out rapidly, and Harrison's techniques were widely copied and his maps used when training aircraft pilots. *Fortune*'s generous use of colour made the images much more vivid and added considerably to the amount of information that could be communicated. The maps adopted aerial perspectives that shrank conventional ideas of space and thus emphasised the impact of air power.

THE JAPANESE EMPIRE

Nobarasha Publishing, Daitoa Battle Bureau Map, 1942.

This Japanese propaganda map 'Dai Tōa senkyoku chizu' is thought to have been inserted into an annual and it shows pictorial relief as well as signs of Japanese influence and occupation. Made more attractive by the animals, whales and plants shown, this map exaggerates Japanese activity: imperial Japan's Rising Sun war flag flies from the Andaman Islands in the Bay of Bengal, captured in March 1942, to the western Aleutian Islands, occupied that June, but the Japanese were less active in the central and southern Pacific than suggested in this map. Moreover, the carriers that had entered the Bay of Bengal in April were swiftly withdrawn in order to be deployed for an attack on Port Moresby in New Guinea, an attack that was thwarted by the Americans in the Battle of the Coral Sea on 7–8 May. That battle was the first fought entirely between carrier groups in which the ships did not make visual contact, one that also demonstrated the failure of the Pearl Harbor attack to wreck American naval power. The map understandably did not show the American air attack on Tokyo on 18 April, the Doolittle Raid, launched from the carrier USS *Hornet*.

The Japanese strategy did not focus on continual conquest. For example, the ideas of advancing into Australia or India were opportunistic additions to the original strategy. Instead, the Japanese thought in terms of advancing to establish a strong empire in East and Southeast Asia and the western Pacific, an empire that would be protected by a powerful defensive perimeter that would inflict such heavy casualties on any attackers that they would abandon the effort, and notably so the Americans with their, in Japanese eyes, weaker will. The Japanese navy, in its strategy of Zengen Sakusen ('Great All-Out Battle Strategy'), planned on the successive use of submarines, long-range shore-based bombers, carrier-based dive bombers and destroyer night-time torpedo attacks against the advancing American fleet to weaken it before a decisive clash with Japanese warships. This strategy encouraged the resource-draining construction of the two battleships of the *Yamato* class, the largest ever built, as well as a policy of denying island bases to the Americans.

大東亞戰局地圖
北極海
太平洋
大西洋
東京
日
ベーリング海
ベーリング海峡
カムチャツカ半島
オホーツク海
ペトロパウロフスク
アリューシャン列島
鳴神島（キスカ島）
熱田島（アッツ島）
ダッチハーバー
コヂャック
アラスカ
カナダ
ハドソン湾
バフィンランド島
ラブラドル半島
ケベック
モントリオール
トロント
ボストン
ニューヨーク
フィラデルフィア
ワシントン
ハリファクス
シカゴ
デンバー
アメリカ合衆國
サンフランシスコ
ロスアンゼルス
サンヂエゴ
メキシコ
ハバナ
西印度諸島
中央アメリカ
カラカス
ベネズエラ
ボゴタ
コロンビア
リマ
ペルー
ブラジル
パラグアイ
ブエノスアイレス
モンテビデオ
アルゼンチン
マゼラン海峡
ハワイ諸島（米）
ホノルル
ミッドウェー島（米）
ジョンストン島（米）
パルミラ島（米）
カントン島（米）
ファンニング諸島（英）
ギルバート諸島（英）
ナウル島
ビスマルク諸島
ソロモン諸島
ニューヘブライヅ諸島（佛・英）
フィジー諸島（英）
サモア諸島（米）
トンガ諸島（英）
クック諸島（英）
ソシエテ諸島（佛）
タヒチ島（佛）
パウモツ諸島（佛）
マーキーズ諸島（佛）
ノーフォーク島（英）
ニュージーランド（英）
オークランド
ウェリントン
ダニーデン
ブリスベン
シドニー
メルボルン
ホバート
南洋群島

A VAST COMBAT ARENA

Howard Burke, *Los Angeles Examiner*, 'The Pacific—Vital Area for United States and Japan', November 1942.

The text attempts to explain strategy, claiming that 'Allied strategy has been to save Australia first'. 'Military and commercial air centers' emerge in the 'battle of bases' as being of significance alongside naval bases. The strategic value of Pearl Harbor for controlling the eastern Pacific and advancing across the western Pacific had been apparent to all, but the problems faced by such an advance and, in particular, supporting the American garrison in the Philippines, had proved insurmountable when Japan attacked in December 1941.

The need to plan for conflict across very large bodies of water encouraged an emphasis on a greater range for ships and aircraft and led Japan and the United States both to convert ships into carriers and to commission purpose-built carriers. There was also a focus on increasing the range, size and speed of submarines. In July 1937, the Americans authorised the *North Carolina* and *South Dakota* classes of fast battleships. However, that year, Japan ordered the *Yamato* and *Musashi*, the largest capital ships in the world, each displacing 72,000 tons and carrying nine 18.1-inch guns.

The Doolittle Raid, an American air attack on Tokyo on 18 April 1942, was mounted from carriers and took its name from the bombers' commander. The raid was essentially symbolic, a public relations event for the US Home Front, but crucially it hit at Japanese prestige and forced the relocation of fighter aircraft to home defence, which put further pressure on air resources. The raid therefore had a strategic impact and recentred Japanese naval concerns away from southwest Asia and toward the northern Pacific, where there was less of an extensive defence in depth for Japan and therefore apparently greater possibilities for an American advance.

As a result, the Japanese decided both to seize the westernmost Aleutian Islands ('into the primary defense area'), lessening American options in the northern Pacific, and to tackle the US Pacific Fleet. To that end, and to enhance the defensive perimeter, Admiral Isoroku Yamamoto, the fleet commander, proposed to seize Midway and other islands that could serve as support bases for an invasion of Hawaii, which he correctly assumed would lead to the battle he wanted. In this battle, Japanese battleships and submarines would destroy American carriers.

In part, the Japanese suffered, like Hitler's forces in the Soviet Union in 1941 and 1942, from the adoption of an offensive strategy that both contained too many goals and did not adequately address prioritisation, sequencing, timetabling and responses to opposing moves. Having operated in the Indian Ocean, the Japanese navy was expected next to obtain advantages and secure goals in the southwest Pacific, and then across the expanse of the northern Pacific to seize Midway, destroy the American fleet, and capture Attu and Kiska in the Aleutian Islands. This misconceived strategy was further weakened by poor execution, notably the loss of the Intelligence war, but also the events of the desired battle itself.

In the Battle of Midway on 4–5 June, American carrier aircraft sank four Japanese carriers. The battle demonstrated Yamamoto's insight about the significance of fleet action as opposed to seizing territory; but, due to poor planning and execution, did so to the detriment of Japan. Although there was still the damaged *Shōkaku*, the *Zuikaku* and a number of smaller carriers, Japan after Midway no longer had a significant carrier fleet nor the linked air cover, and thus Japanese plans for further expansion in the Pacific lacked traction.

Already prior to the Japanese defeat at Midway the initial Japanese ability to mount successful attacks had not deterred the Americans from the long-term effort of driving back and destroying their opponents. The American government and public were not interested in the idea of a compromise peace. Even more than Germany toward the Soviet Union, Japan, the weaker power, had gone to war with the one power that could beat it, and in a way that was calculated to ensure that it did so. As a result, helped greatly by the superior command skill that made possible the effective use of resources at a stage when they were still in short supply, the Americans were able to exercise strategic leverage, and to take the strategic initiative successfully by the end of 1942.

November 1942 was the month in which the Americans won the naval conflict off Guadalcanal, which ensured that Japan could not support the force it had landed on the island on 7 July. The Americans had landed troops in August and the defeated Japanese withdrew in January 1943. This map is successful in explaining the importance of the Solomon Islands to both sides, but it exaggerates the potential of an American advance from the Aleutians. In May 1943, an American amphibious force reconquered Attu, the westernmost of the

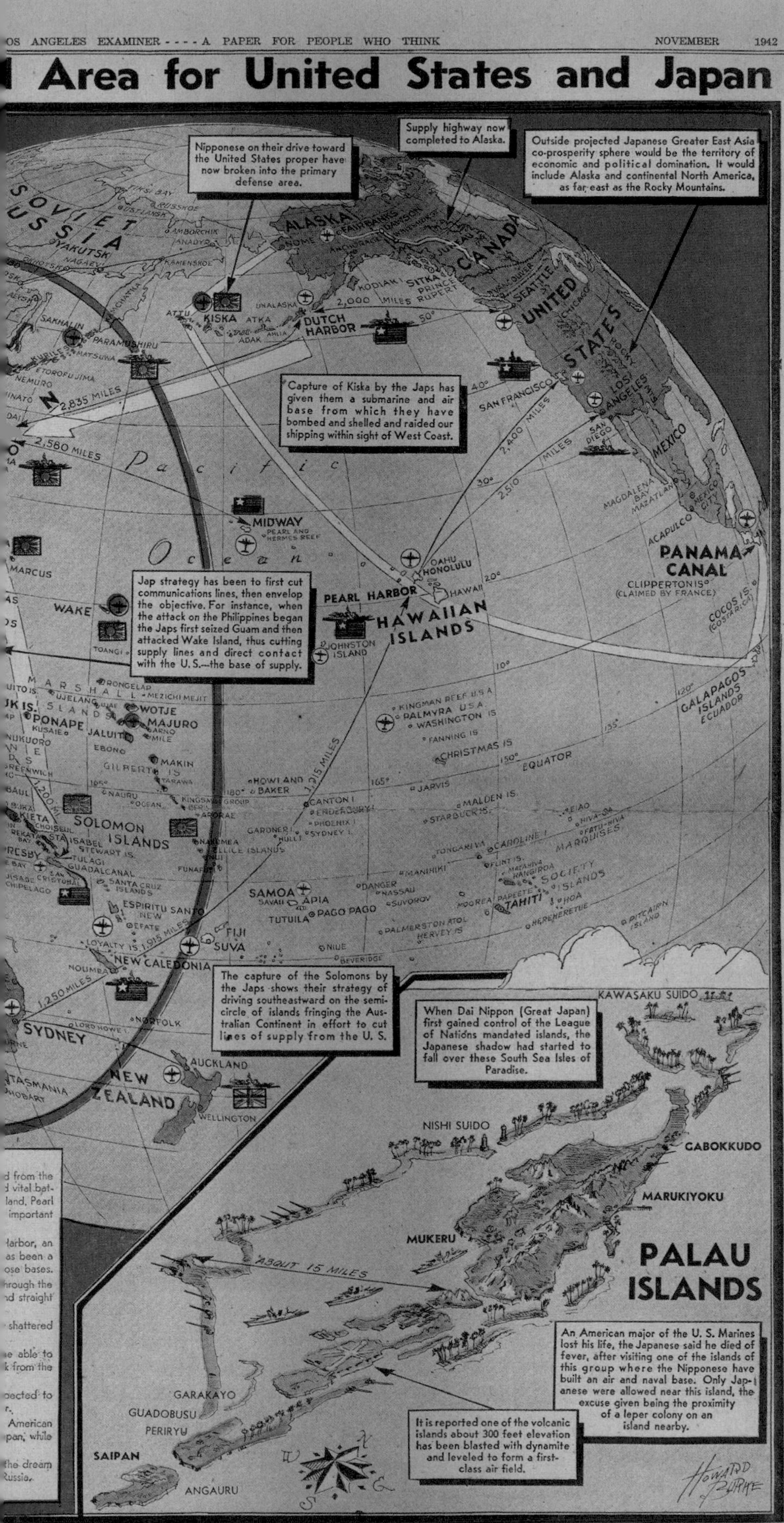

Aleutian Islands after a naval clash in the Battle of the Komandorski Islands to the west, but for geopolitical, strategic, logistical and climate reasons there was to be no North Pacific road to Tokyo. Instead, the Kurile Islands were to fall to Soviet attack in 1945.

Australia had turned to the USA. Under the threat of apparently imminent Japanese invasion, the minutes of the Australian War Cabinet make clear its anxieties about British priorities in goals and force allocation, notably of the British reluctance to release Australian forces from the Middle East, and once released a British attempt to have them used to defend India against Japanese attack. The New Zealand government shared these anxieties.

In July 1942, the Australian War Cabinet cabled Churchill: '...superior sea power and airpower are vital to wrest the initiative from Japan and are essential to assure the defensive position in the Southwest Pacific Area.' Britain was not in a position to provide either. The USA, in contrast, both could and would, as the events of 1942 made abundantly clear. After the devastating Japanese carrier-borne air attacks on Darwin on 19 February, designed to cover the successful invasions of Java and Timor, American aircraft helped to protect northern Australia as part of what became a significant deployment of American forces. In April 1942, General MacArthur established his Southwest Pacific Area (SWPA) headquarters in Brisbane.

THE SOUTHERN MEDITERRANEAN

Richard Edes Harrison, 'The Not-So-Soft Underside', January 1943.

Presenting the land from above both invited the viewer into the drama of the map, communicating excitement, and – by bringing the topography into a high relief thanks to a vertical scale that was out of proportion to that on the ground – emphasised the geographical factors stemming from the topography. This can be seen clearly in Harrison's map of Europe seen from Africa, 'The Not-So-Soft Underside'.

It was published in *Fortune* on 27 January 1943 as the Allies prepared to battle Axis forces in Tunisia. The magazine article drew on the lesson offered by the topography:

> *'No full-fledged military expedition since ancient times has succeeded in crossing the Pyrenees or the Alps from south to north and making the invasion stick. The great formative invasions since the time of the Romans have all come from east to west, from the Russian plains or the Anatolian plateau of Turkey. The "soft underside of the Axis," the "unprotected belly of Europe," is, then, a figure of speech that lacks geographical common sense. The mountains and sketchy roads of crippled Spain, the narrow, easily closed gap of the Rhône, the tunnels of Switzerland, the Nazi air force in Crete, pose terrifying problems of both military tactics and supply. From the communications officer's view it is thus American dollars to Italian lire that Hitler's Germany will not be invaded in force from North Africa.*
>
> *... what did we get out of the African campaign? When the Mediterranean is cleared, it will save miles of shipping. But from the positive standpoint, it spreads Hitler thin all around the margins of Europe. He must defend Italy to keep Americans from taking over airfields within easily striking distance of the Skoda works in Pilsen and Munich. He must watch Turkey, lest the United Nations, with the compliance of Ankara, bypass Crete and the Balkan mountains for a thrust up through Bulgaria. He must keep an eye on Spain and Portugal, while he is also watching Rzhev and Rostov. In short, possession of the Mediterranean south shore gives the United Nations the opportunity to deliver confusing multiple blows – and Hitler's own power of the initiative has been critically impaired.'*

While the map – which has a key to the location of important minerals (coal, copper, iron, manganese, lead, phosphate and zinc) – explains the significance of the conflict in Tunisia, providing American land forces with their first experience of combat with German and Italian units, it was very much a map dependent on the text, and vice versa, although the geography was overworked. The message of the detailed caption beneath the map is clear: 'In losing the southern shore of the Mediterranean he [Hitler] has lost the initiative and must now operate to counter Allied plans.'

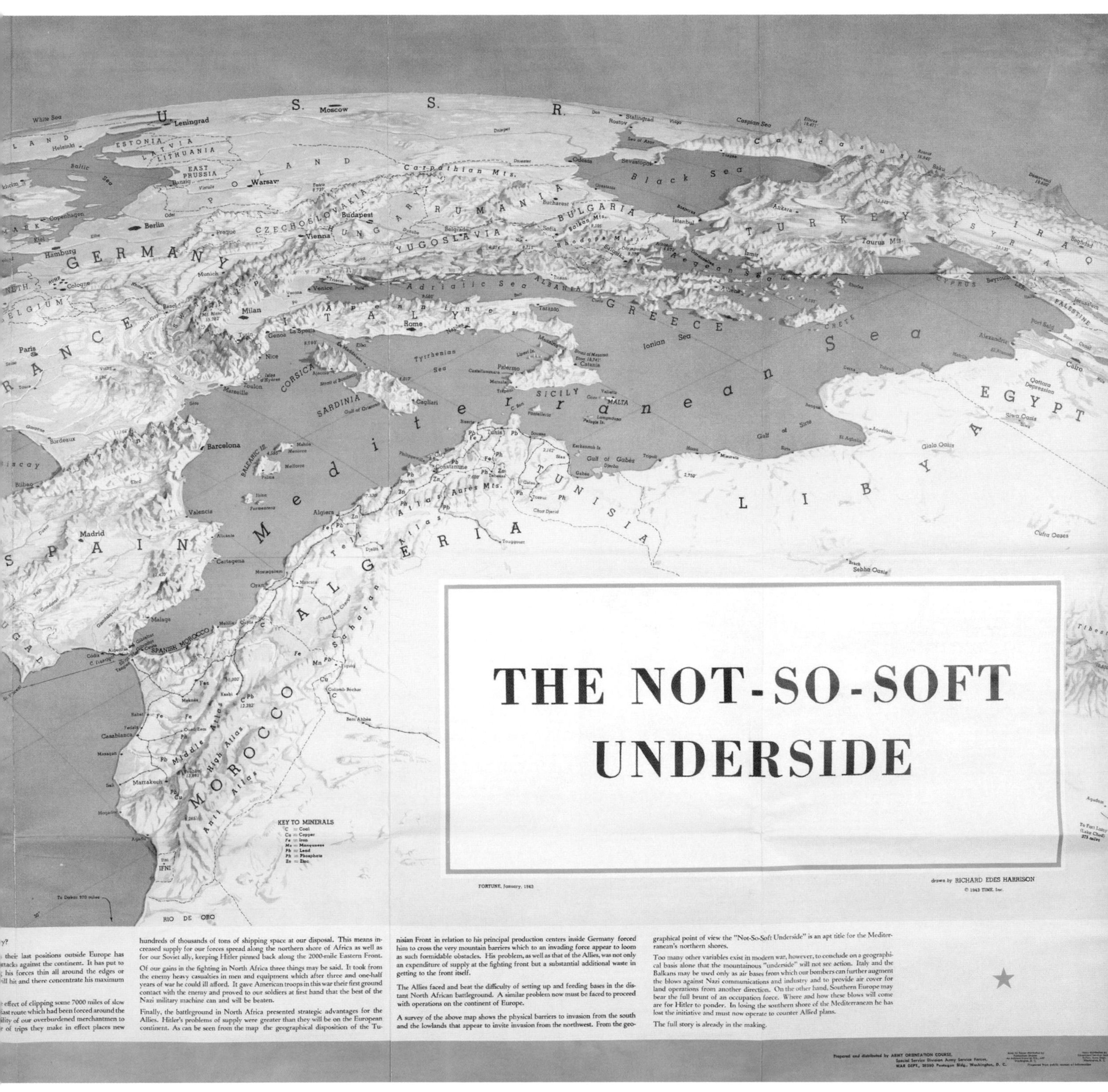
THE NOT-SO-SOFT UNDERSIDE
FORTUNE, January, 1943
drawn by RICHARD EDES HARRISON
© 1943 TIME Inc.
U. S. S. R.
GERMANY
POLAND
CZECHOSLOVAKIA
HUNGARY
RUMANIA
BULGARIA
YUGOSLAVIA
ALBANIA
GREECE
TURKEY
SYRIA
IRAQ
FRANCE
SPAIN
ITALY
CORSICA
SARDINIA
SICILY
MALTA
Mediterranean Sea
Black Sea
Adriatic Sea
Tyrrhenian Sea
Ionian Sea
Aegean Sea
TUNISIA
ALGERIA
MOROCCO
SPANISH MOROCCO
LIBYA
EGYPT
RIO DE ORO
IFNI
Atlas Mts.
Carpathian Mts.
Taurus Mts.
Caucasus
KEY TO MINERALS
To Dakar 970 miles
y?
their last positions outside Europe has
ttacks against the continent. It has put to
his forces thin all around the edges or
ill hit and there concentrate his maximum
effect of clipping some 7000 miles of slow
ast route which had been forced around the
ility of our overburdened merchantmen to
of trips they make in effect places new
hundreds of thousands of tons of shipping space at our disposal. This means increased supply for our forces spread along the northern shore of Africa as well as for our Soviet ally, keeping Hitler pinned back along the 2000-mile Eastern Front.
Of our gains in the fighting in North Africa three things may be said. It took from the enemy heavy casualties in men and equipment which after three and one-half years of war he could ill afford. It gave American troops in this war their first ground contact with the enemy and proved to our soldiers at first hand that the best of the Nazi military machine can and will be beaten.
Finally, the battleground in North Africa presented strategic advantages for the Allies. Hitler's problems of supply were greater than they will be on the European continent. As can be seen from the map the geographical disposition of the Tunisian Front in relation to his principal production centers inside Germany forced him to cross the very mountain barriers which to an invading force appear to loom as such formidable obstacles. His problem, as well as that of the Allies, was not only an expenditure of supply at the fighting front but a substantial additional waste in getting to the front itself.
The Allies faced and beat the difficulty of setting up and feeding bases in the distant North African battleground. A similar problem now must be faced to proceed with operations on the continent of Europe.
A survey of the above map shows the physical barriers to invasion from the south and the lowlands that appear to invite invasion from the northwest. From the geographical point of view the "Not-So-Soft Underside" is an apt title for the Mediterranean's northern shores.
Too many other variables exist in modern war, however, to conclude on a geographical basis alone that the mountainous "underside" will not see action. Italy and the Balkans may be used only as air bases from which our bombers can further augment the blows against Nazi communications and industry and to provide air cover for land operations from another direction. On the other hand, Southern Europe may bear the full brunt of an occupation force. Where and how these blows will come are for Hitler to ponder. In losing the southern shore of the Mediterranean he has lost the initiative and must now operate to counter Allied plans.
The full story is already in the making.
Prepared and distributed by ARMY ORIENTATION COURSE,
Special Service Division Army Service Forces,
WAR DEPT., 2E580 Pentagon Bldg., Washington, D. C.

CAPITALS IN THE CROSSHAIRS

F.E. Manning, 'Target Berlin' and 'Target Tokyo', October 1943.

Chicago Sun journalist F.E. Manning's globular maps of October 1943, 'Target Berlin' and 'Target Tokyo', positioned the Axis capitals as vulnerable targets at the centre of the map, creating the effect through the use of concentric circles. Both maps are newsmaps, dated 18 and 25 October respectively, and were printed and distributed by the Army Orientation Course Special Service Division Army Service Forces, a propaganda arm of the War Department. The wartime development of long-range aerial reconnaissance ensured that targets were increasingly well mapped.

In keeping with the purpose of the newsmaps – to relay developments in the war from an uplifting and positive perspective – the other side of each map carries summaries of recent war news on a variety of battlefronts. By late 1943 the Allies were winning in Italy, the Soviet Union and East Asia. The inclusion along the bottom of the maps of a cut-away scale to measure distance emphasised to service personnel that these cities were Allied mission objectives and they were closing in on them.

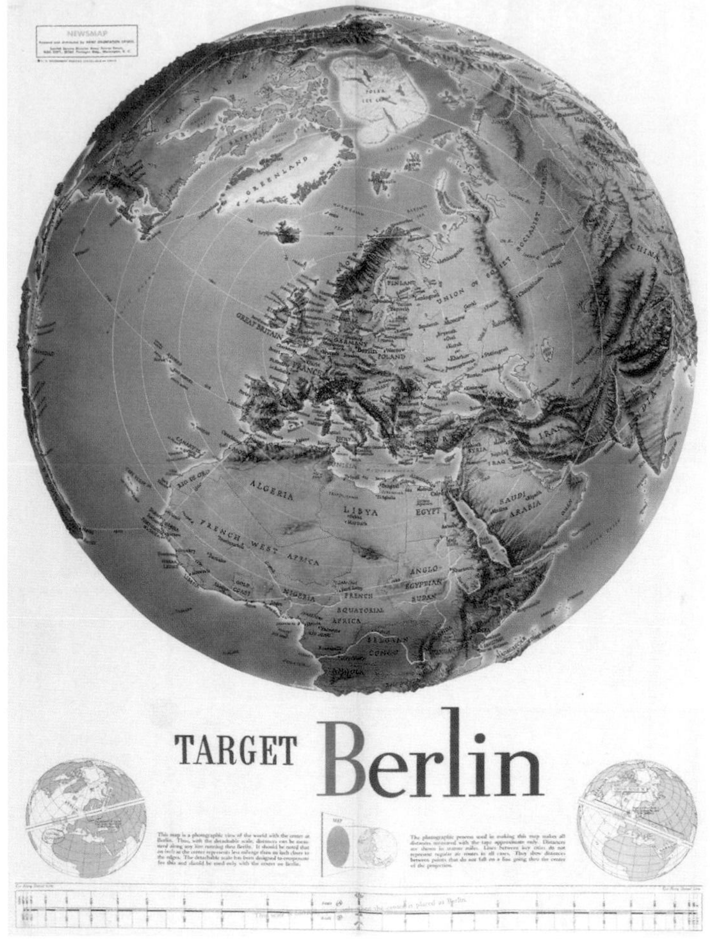

By October 1943, the Americans increasingly had victory in consideration. The strategy assessment included an attempt to educate the public on the geographical relationship between places, notably between the Philippines and the Southwest Pacific Area. The means for strategy were greatly improved by the build-up, training and deployment of much larger forces. The build-up comprised carriers, submarines and other classes of vessel, including destroyers, which played a key role in escorting, patrolling and supporting amphibious tasks. By 31 December 1943, 332 destroyers were on active service.

Moreover, the Americans developed important organisational advantages, from shipbuilding to the use of resources. Their advance would have been impossible without the ability to transport large quantities of supplies and to develop the associated infrastructure, such as harbours, oil-storage facilities and the ships of the support train. In many respects, this was a war of engineers, and the American aptitude for creating effective infrastructure was applied to great effect in the Pacific theatre.

The Americans had enough resources to follow a southern drive on the Philippines via New Guinea and a central Pacific drive westward, the choices respectively of the US Army and the US Marine Corps with the US Navy. Resources, however, were far from the only element: the Americans had also gained cumulative experience in amphibious operations, in how to coordinate them effectively with naval and air support, and in naval logistical support. Another contributor to success was superior American inter-war leadership development, based at the war colleges and focused on solving complex higher operational and strategic problems.

Naval and air capabilities meant the Americans could identify and neutralise key targets and thus bypass many of the islands the Japanese continued to hold, which was a sensible strategic and operational decision given the time, effort and casualties taken to capture Guadalcanal in 1942–43. Thanks to a growing apparent American superiority at sea and in the air, the Japanese became less able to mount ripostes, and bypassed bases such as Rabaul and Truk, both marked on the map, were isolated. Thus, the Pacific War became one that was far from linear. Bypassing at the strategic level could not be matched at the tactical level, where the frontal attack on island defences frequently led to heavy casualties.

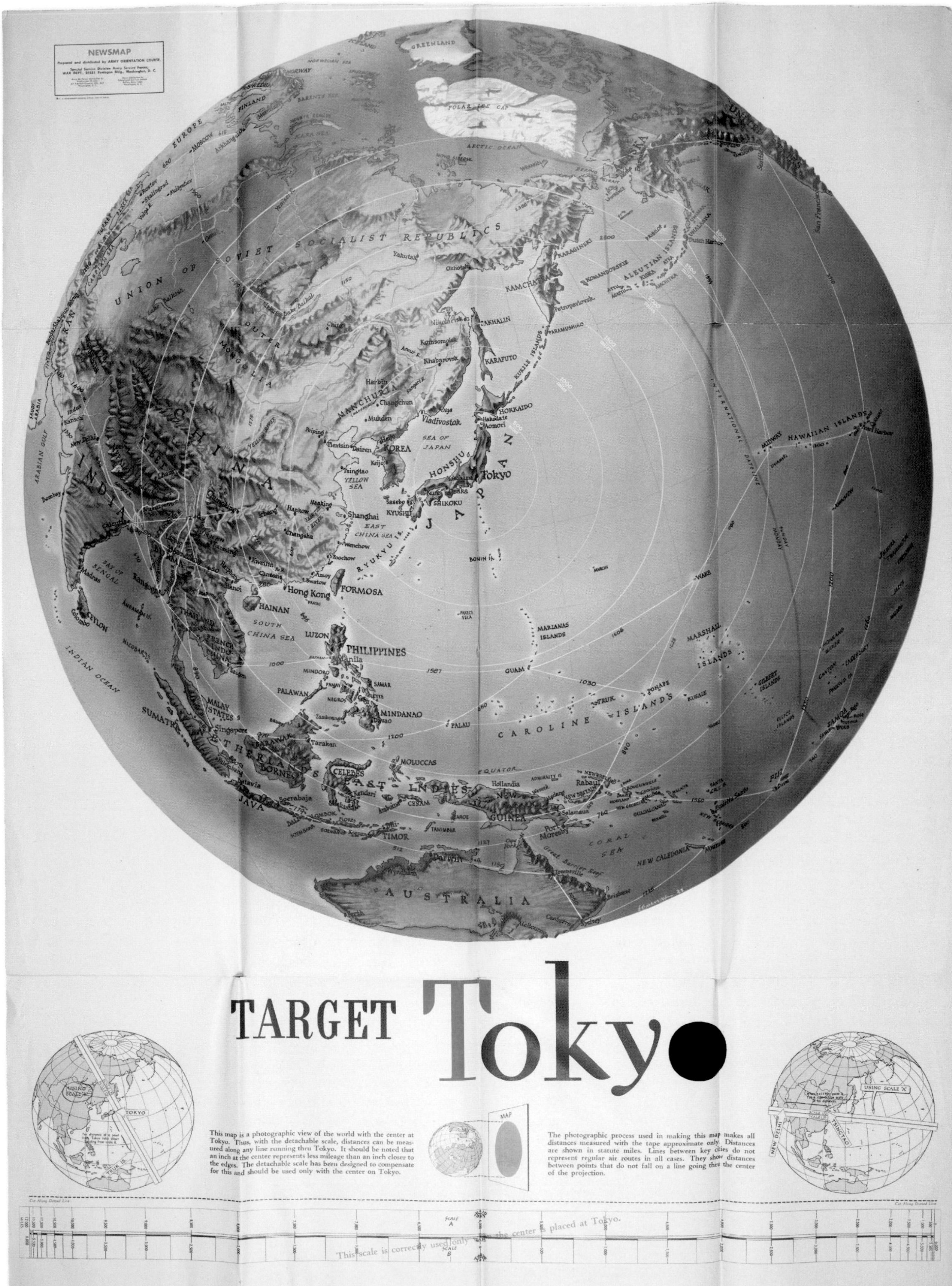

NEWSMAP

Prepared and distributed by ARMY ORIENTATION COURSE.

Special Service Division Army Service Forces,
WAR DEPT., 2E581 Pentagon Bldg., Washington, D. C.

TARGET Tokyo

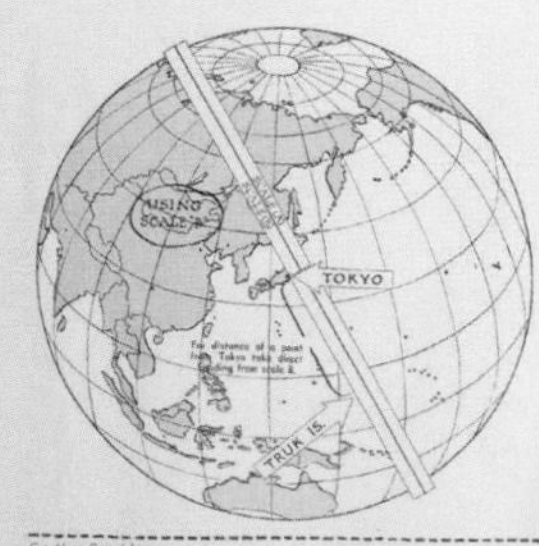

This map is a photographic view of the world with the center at Tokyo. Thus, with the detachable scale, distances can be measured along any line running thru Tokyo. It should be noted that an inch at the center represents less mileage than an inch closer to the edges. The detachable scale has been designed to compensate for this and should be used only with the center on Tokyo.

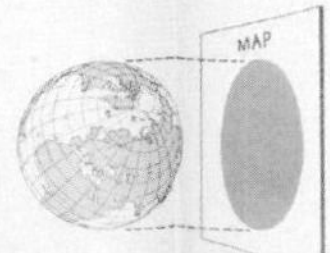

The photographic process used in making this map makes all distances measured with the tape approximate only. Distances are shown in statute miles. Lines between key cities do not represent regular air routes in all cases. They show distances between points that do not fall on a line going thru the center of the projection.

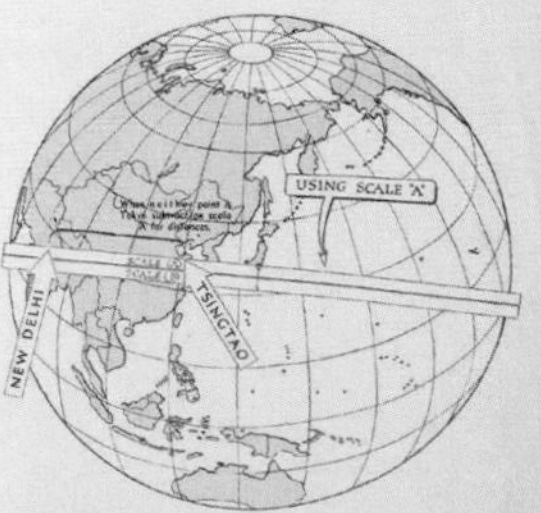

SCALE A

SCALE B

This scale is correctly used only when the center is placed at Tokyo.

White smoke from bursting shells along the Volturno River in Italy marks the start of a smoke screen laid down by American artillery during operations against the enemy defense line. Reconnaissance plane circles over the scene.

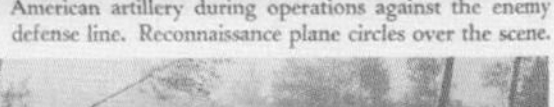

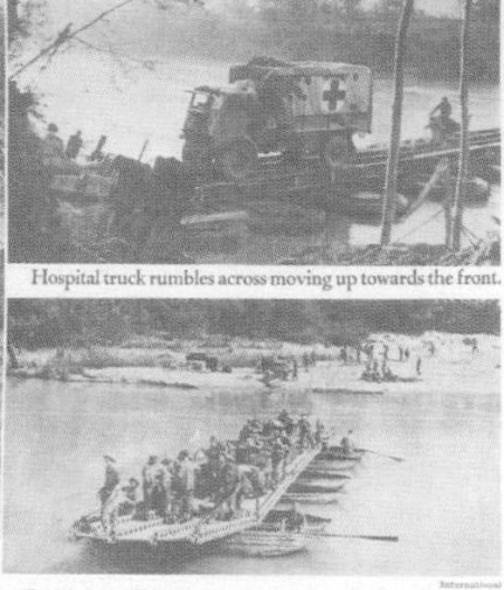

Hospital truck rumbles across moving up towards the front.

American Engineers build a ponton bridge over the Volturno after shock troops in rubber boats secured a bridgehead. Note tractor on one section of the bridge.

British Army Engineers cross on a shore-landing deck craft.

One of the first victims of the Navy's new carrier-based Hellcat fighter planes was this Jap four-engined flying boat, nicknamed "Emily." The big plane was downed while on patrol east of the Jap-held Gilbert Islands in September.

Flames flare from one engine as the Hellcat closes in.

Emily is nearly down to the water with cockpit an inferno.

Five minutes from time of sighting this marked the end.

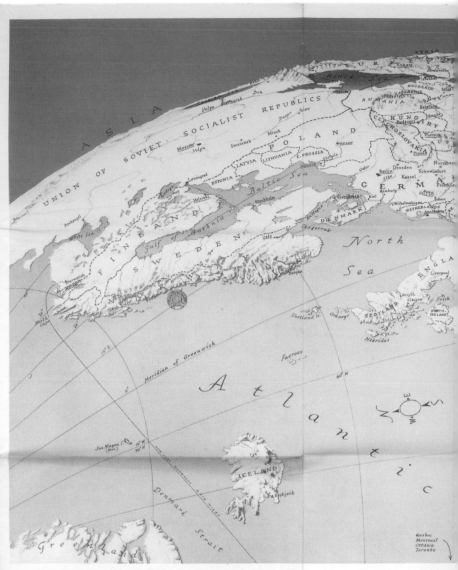

NEWSMAP

MONDAY, NOVEMBER 8, 1943

WEEK OF OCTOBER 28 TO NOVEMBER 4

217th Week of the War—99th Week of U. S. Participation

Volume II No. 29

THE WAR FRONTS

RUSSIA: All German land exits from the Crimea were blocked off last week. The Red Army drove westward from its positions north of the Sea of Azov and passed beyond the Nogaisk Steppe to gain holds on the lower reaches of the Dnepr River. Advances of unspecified depth were reported into the Crimea from the Perekop Peninsula as well as the narrow land strip below Salkovo, which lies on the only other rail line leading southward.

While no confirmation came from Moscow, the Berlin Radio announced that the Soviets seized an additional hold on the Crimea by crossing the Kerch Strait and making landings both north and south of the town of the same name.

The heaviest fighting in Southern Russia was reported 130 miles north of the Crimea before the enemy-held town of Krivoi Rog. Heavy tank and artillery engagements were reported as the Nazis tried to hold open the narrowing neck of the Soviet encirclement drive.

ITALY: Battling successfully against an enemy entrenched in rugged and rising ground and further hampered by heavy rains which made the going even more difficult, American and British infantrymen pushed forward slowly but definitely in the advance toward Rome.

Fighting on the western half of the battle line that stretched across Central Italy centered principally about Mt. Massico, on the west coast nine miles above the Volturno River, and the Matese mountain ridge which lies 35 miles inland to the northeast. The enemy-held towns of Venafro and Isernia were reported under American artillery fire.

British units of the Fifth Army were making progress along the coast while American troops were in the inland action. On the Adriatic coast the Eighth Army maintained pressure and made slight advances.

AIR OFFENSIVE: Creation of a new U. S. Fifteenth Air Force to concentrate on bombing of strategic targets in Nazi-held Europe and to operate from the Mediterranean area was announced at the same time that the first mission of the new force was reported.

Wiener-Neustadt, near Vienna in Austria, was the first target, and our heavy units were reported highly successful in the 1400-mile round trip flight which had as its target the Messerschmitt assembly plant.

Lt. Gen. Spaatz will head the new force as well as continuing as Commanding General of the Twelfth Air Force which is the American unit in the Northwest African Air Force. The Twelfth and the Fifteenth U. S. Air Forces continue to operate under the Mediterranean command directed by Air Marshal Tedder.

In the first attack on Southern France since August, our Flying Fortresses bombed the 540-foot rail viaduct on the Mediterranean coast near Antheor. The bridge carried military loads from Marseille to Genoa. Other attacks were made against Genoa, La Spezia and targets in Yugoslavia and Albania.

Eighth Air Force bombers based in England, which had previously been reported building up strength to equal the RAF, carried out a heavy attack against Wilhelmshaven. While the exact number of planes that took part in this raid was not immediately revealed, the headquarters did state that it was the greatest force of U. S. planes to attack Germany and on this basis unofficial estimates placed the number in the neighborhood of 1000 bombers and fighters.

SOUTHWEST PACIFIC: Three new Allied landings in the Northwest Solomons opened the campaign to drive the enemy from his last holdings in the strategic islands northeast of Australia.

The landings were made on Mono and Stirling Islands, of the Treasury Group on Choiseul Island, which the enemy had used as an intermediate escape stopover on the way from Vella Lavella to Bougainville, and on Bougainville itself, where reports from the Southwest Pacific estimated 40,000 enemy troops were established. The Bougainville landing was carried out at Empress Augusta Bay, on the southwest coast and landings on Choiseul by seaborne paratroops were also made on the southwest coast. American and New Zealand troops were in the Treasury Islands landing. U. S. Marines carried out the landing on Bougainville.

Prior to the operations the enemy airfields at Buin, Kahili, Kara, Ballale and Kieta were neutralized by heavy air attacks. Our warships provided cover for the landings and also bombarded enemy positions on Buka Island at the northern tip of Bougainville. Gen. MacArthur directed heavy blows at the enemy's main base at Rabaul from which the Japs would normally bring down reinforcements. An enemy attempt to oppose the landing with warships was driven off.

BURMA: American and British air attacks against targets in Burma were increasing. In addition to supporting the Chinese defense in Southern Yunnan Province where the enemy has attempted to drive across the Salween River, Allied bombers hit Akyab and Rangoon, and continued the steady campaign against enemy rail and river traffic.

Liberators dropped 40 tons of bombs on a zinc smelting plant at Kwangyen, near Haiphong, and a medium bomber attack was made against an airdrome at Fort Bayard in the former French-leased territory of Kwangchowan.

Our units also carried out air operations in support of Chinese troops fighting to block a two-pronged Jap drive from above Lake Tungting in Hunan Province.

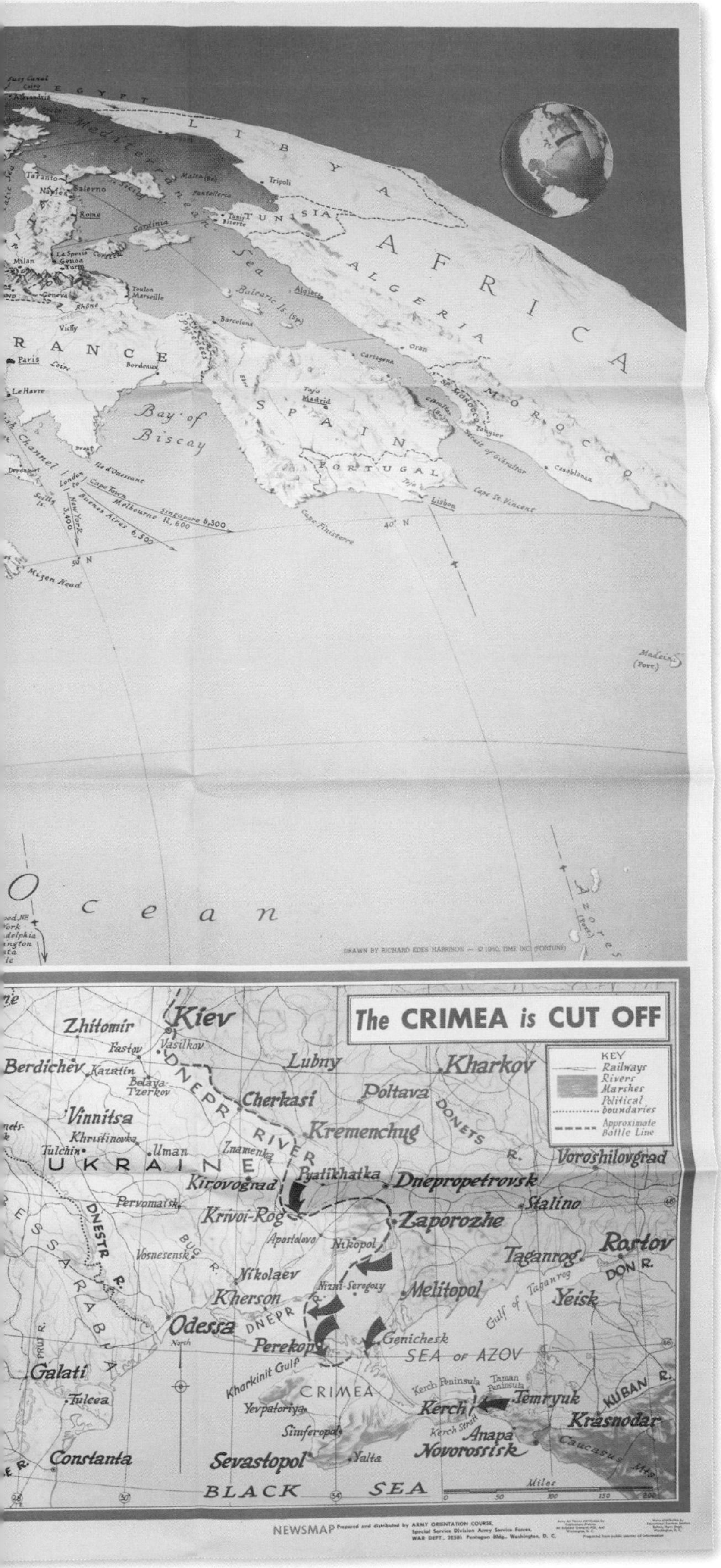

THE WAR FRONTS

Newsmap, 8 November 1943.

The 'Newsmap' series was produced weekly from April 1942 to March 1946 for the Army Information Branch and it brought together operations in different parts of the world, including maps, photographs and text. Published for US service personnel overseas by the Army Orientation Course Special Division Army Service Forces, this issue for 'Monday, November 8, 1943' provided a number of maps, including one by Richard Edes Harrison that uses an unorthodox perspective to make sense of Britain's significance for the Allied air offensive on Germany. The supporting maps record American and Soviet offensives in the Pacific and the Ukraine.

As part of General MacArthur's Operation Cartwheel, the Americans had attempted to isolate the Japanese air and naval base at Rabaul, which is shown on the map 'BOUGAINVILLE: Last Stop in the Solomons'. Rabaul, in New Britain, had fallen to the Japanese in February 1942 and become their major air and naval base in the area. The Allies had bombarded it regularly while the Americans had begun a process of island-hopping in the Solomon Islands in June 1943. As with many of the remarks in this book, that is a bland statement for what was in reality a very difficult experience of hard fighting that included attritional conflict in difficult jungles and swamps, advancing in the face of determined opponents and their machine guns and snipers, and facing night-time Japanese infiltration of American positions. Although Bougainville was invaded on 1 November 1943, and the Americans established a beachhead perimeter from which they beat off a series of Japanese attacks, much of the island was only cleared from December 1944 when Australian troops took over. But it was the capture of the major positions that was significant, and the Americans pressed on to bypass Rabaul in early 1944, landing on the Green Islands to the east, the Admiralty Islands to the west and St Matthias to the north. The large Japanese garrison in Rabaul had been taken out of the strategic equation, and the way was clear to attack the Philippines.

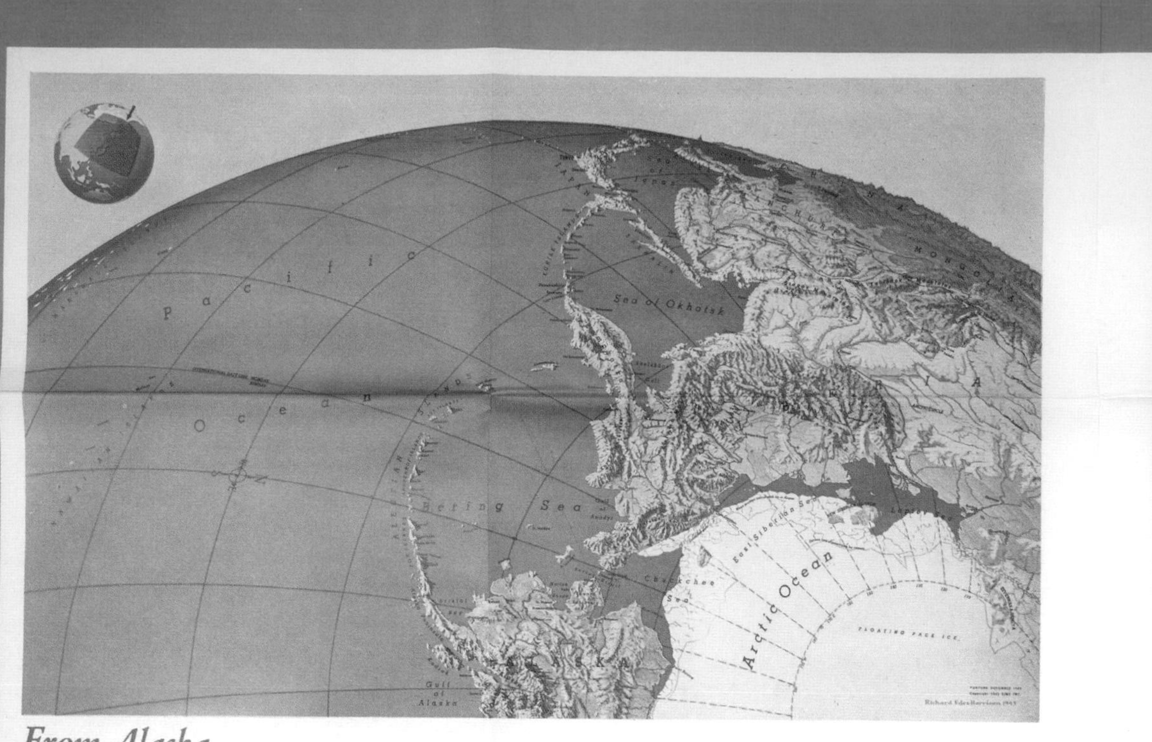

From Alaska

FOUR APPROACHES TO

From China-Burma

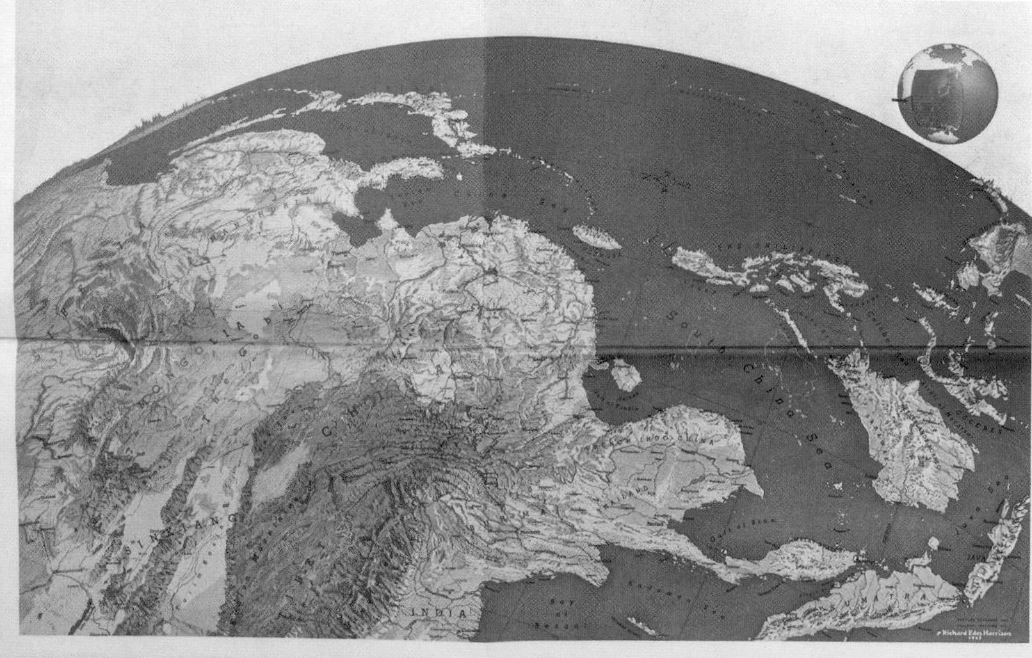

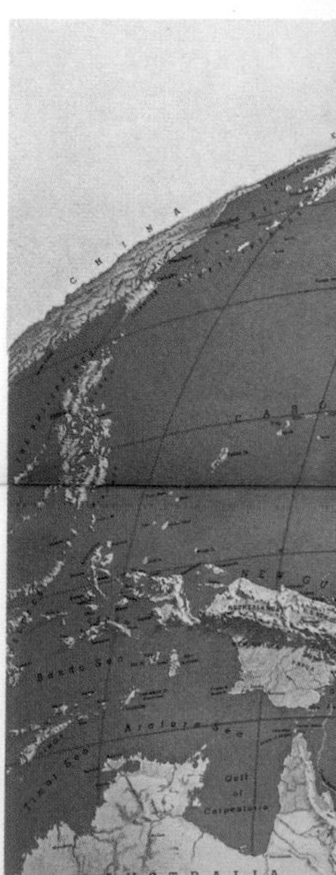

NEWSMAP Prepared and distributed by ARMY INFORMATION BRANCH
ARMY SERVICE FORCES
205 E. 42nd Street, NEW YORK 17, N. Y.

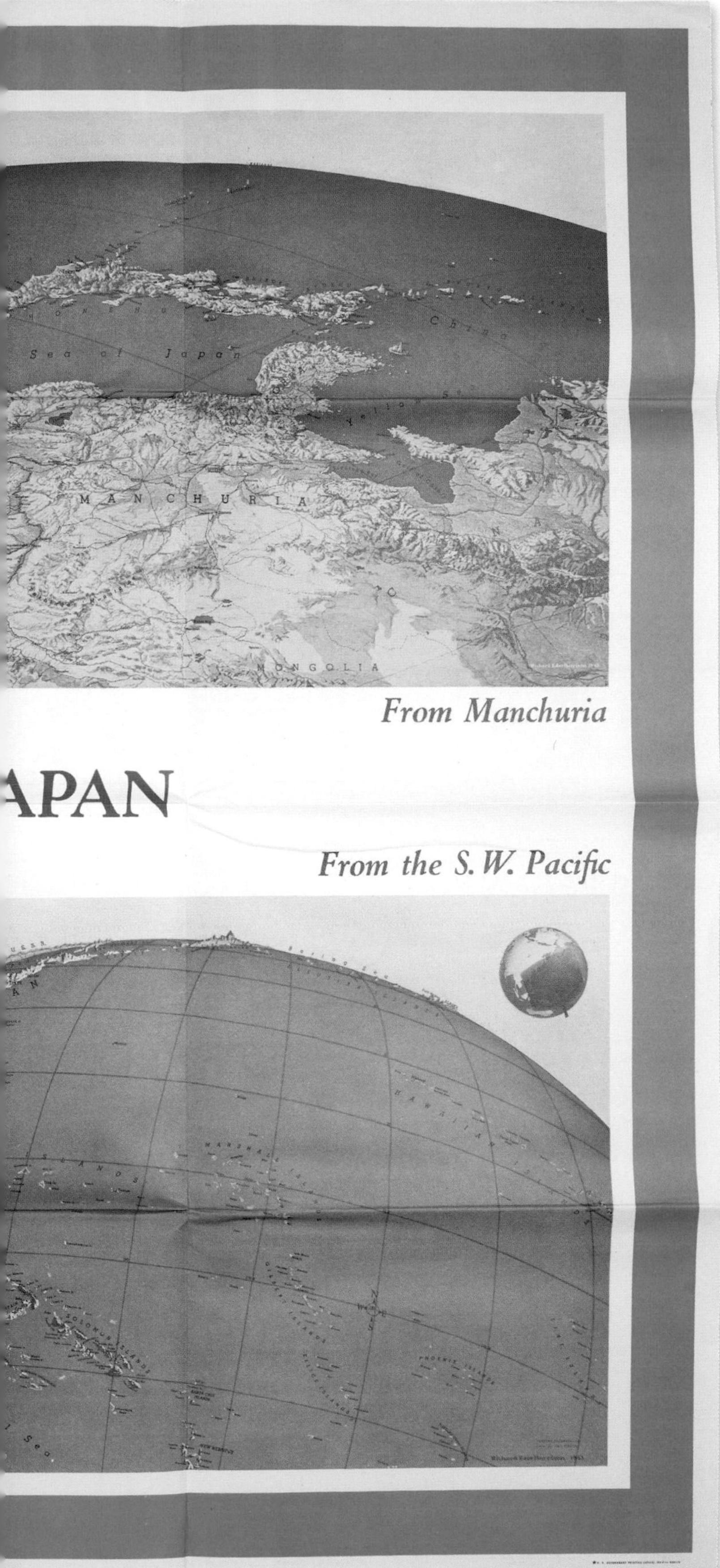

STRATEGIC CHOICES

Richard Edes Harrison, 'PRE-INVASION Air Offensive' and 'Four Approaches to Japan', 8 May 1944.

In this double-sided newsmap for the American armed forces, Richard Edes Harrison uses maps to indicate the advance of forces in New Guinea (overleaf), show the extent to which occupied Europe is under Allied air attack (overleaf) and present the options for dealing with Japan (left).

In part, the Germans had lost the air war when the Allied air attack was stepped up as a result of the introduction of American long-range fighters: Lockheed P-38 Lightnings, Republic P-47 Thunderbolts and North American P-51 Mustangs, the latter two with drop fuel tanks. These aircraft provided necessary escorts for the bombers and enabled the fighters to take part in dogfights over Germany. In late February and March 1944, major American raids in clear weather on German sites manufacturing aircraft and producing oil led to large-scale battles with German interceptors which contributed to the Allies gaining air superiority.

There was also a marked improvement in the operational effectiveness of Allied bombing in 1944. The map shows northern Italy and the Balkans, including the oilfield at Ploesti, as exposed to American attacks from bases in southern Italy. On the whole, American planners did not fully appreciate Hitler's paranoia about southern Europe, including the Balkan approaches to the Romanian oilfields and refining facilities at Ploesti. Oil supplies and movements were elements of the contrasting strategic depth and operational vulnerability of the two alliance systems. Exposure to Allied aircraft from Italy, as the Allies advanced doggedly through the peninsula, was to place yet more of a burden on German air defences. In addition, German-occupied industrial centres in Italy were heavily bombed, notably Milan and Turin.

Germany's allies were also affected by the air assault. Thus, heavy bombing attacks on Bulgaria began on 19 November 1943. The raids, especially that of 30 March 1944 on the capital Sofia, indicated clearly the shift in fortunes and encouraged a decline in enthusiasm for continued support for Germany. Bulgaria abandoned Germany in September, although the Soviet advance into the Balkans was a far more significant factor in this than the bombing.

Distrust of the West characterised the Soviet response to Operation Frantic, which entailed American bombers using Soviet bases to refuel and rearm, having bombed the Axis zone. This operation

NEWSMAP

FOR THE ARMED FORCES

243rd Week of the War — 125th Week of U. S. Participation

Digging a shelter against enemy shells is the first order of business after Lt. George Scheoneck, Little Rock, Ark., and Lt. Gerald Priebe, St. Paul, Minn., landed their reconnaissance plane on the Anzio beachhead.

A Yank examines the rear compartment, housing the control wire, of a German remote-controlled miniature tank. The explosives-carrier, called the "doodlebug" by the Allies, was knocked out by machine gun fire in Italy.

No records are provided as to which of the several treatments offered is here being administered. Barber is Pvt. George H. Olin, Coudersport, Pa. Customer is Pvt. Clarence T. McFall, Richardson Park, Del., who is with the MP's. The emporium is somewhere in Italy.

THE WAR FRONTS

AIR OFFENSIVE: A list of the targets covered by the Allied air forces in the past month would sound like a guide-book index to Nazi-occupied Europe. They ranged as far north as Oslo in Norway and as far east as Ploesti in Rumania. Emphasis was swinging to such tactical targets as rail yards, airfields and various defensive installations which the enemy has prepared to oppose the coming land assault. So successful was this campaign that the British Ministry of Economic Warfare felt justified in announcing that enemy rail lines no longer could carry the load required of them.

Still a high priority objective was the enemy's aircraft industry. Heavy blows at Friedrichshafen, site of the Dornier works, and Oslo, where the Kjeller airframes factory repairs Luftwaffe fighters and troop carrying planes, were among the production centers which were hit during the past week.

Luftwaffe fighter attacks were dropping off due to losses, decreased production and the attempt to maintain reserve forces to combat the land invasion, but American heavy bombers encountered unusually large numbers of the enemy when they revisited Berlin. About 1000 Liberators and Fortresses, escorted by as many fighters, gave the German capital a going over that compared with any delivered previously. Anti-aircraft was extremely heavy in addition to the German fighter support, and 63 heavy bombers and 14 U. S. and RAF fighters failed to return. The enemy lost 88 planes in the same fight.

Allied Mediterranean-based planes hit rail centers, viaducts and port areas in northern Italy and southern France. Genoa and Toulon were among the targets listed.

In the month of April about 97,000 tons of bombs had been dropped on enemy territory by Allied planes from Britain and Italy, and in the last four days of the month bombs rained on Europe at the rate of six and a half tons a minute right around the clock.

Reports in London indicated that the operations of the 8th USAAF against the German aircraft industry on Feb. 20, 22, 24, and 25 constituted one of the most striking victories of the war. Four of sixteen primary factories attacked were knocked out of productions, seven were heavily or severely damaged, two suffered considerable damage and three suffered minor damage. During the attacks 381 enemy planes were destroyed and 548 damaged in the air and on the ground. Air battles fought later have shown that the damage limited the efforts of German front-line fighter squadrons.

Lieut. Gen. Doolittle, Commander of the 8th USAAF, said that 1300 German fighters were knocked out in the air and on the ground by American airmen during April. This figure is substantially more than the German aircraft production for the month. U. S. losses for the same period were 359 bombers and 144 fighters.

ITALY: Small scale offensive action by American troops on the right of the Anzio beachhead placed them within two miles of the Appian Way. This was the principal deviation from the routine ground patrols continuously carried out by both sides at Anzio and Cassino. Air activity varied with the weather. Some days there was practically no flying. On the other hand, one day saw 2000 sorties carried out.

Anti-aircraft gunners with the Fifth Army have rolled up an appreciable score of enemy aircraft to their credit. From the start of the Italian campaign until April 22 they knocked down 426 enemy planes and an additional 281 were marked as probables. The gunners have been getting about 30 per cent of all enemy planes that come over.

CENTRAL PACIFIC: A convincing illustration of the freedom with which our Navy can move through the enemy island area of the Central Pacific came last week in the shape of a powerful U. S. task force bombardment of Japanese strongholds in the Carolines.

The tremendous fleet assaulted Truk, Satawan and Ponape after coming directly from the Hollandia-Humboldt Bay action at New Guinea where it had supported the landing there of Southwest Pacific forces.

Carrier-based planes dropped a record weight of 800 tons of bombs in the Truk area, and cruisers and battleships moved inshore to bombard the big bases supporting islands at Ponape and Nomoi. One hundred and twenty-six enemy planes were destroyed at Truk.

The planes hit Truk April 29 and 30, Satawan, in the Nomoi Islands, April 30 and Ponape on May 1.

It was the third major fleet action within a month and ranked in effectiveness with that at Palau. None of the warships suffered any damage and 30 American flight personnel were reported missing.

Guam, in the Marianas Islands, was bombed for the first time by land-based bombers ranging 1100 miles beyond the Marshalls. Another notable long-range bomber action was a strike at Wake.

SOUTHWEST PACIFIC: An important reason for the weakened Jap position in the Southwest Pacific was revealed last week in the announcement that 1727 Japanese coastal vessels, barges and schooners have been sunk and 3548 damaged since April, 1942. This report from Gen. MacArthur's headquarters stated that the destruction has had the effect of "paralyzing enemy efforts to supply, reinforce or evacuate" the Japanese 17th and 18th Armies which are cut off and surrounded in New Guinea, New Britain, New Ireland and the Solomons. The headquarters estimated the 1727 small boats destroyed could move a 50,000-man army. In addition to the loss of the vessels several thousand Japs were killed, drowned or wounded in the attacks that wrecked the ships.

American infrantrymen moving in from two sides of Hollandia consolidated the victory in six days and took the three airfields. At the eastern end of the Jap position Allied troops captured Alexishafen, northwest of newly-taken Madang. As a result of the Hollandia victory it was estimated that about 60,000 Japs were now cut off and left to die of hunger and jungle diseases.

SOUTHEAST ASIA: The battles in both India and Burma were going in favor of the Allies, and while hard fighting continued in the peaks surrounding Kohima, in India, the town itself had been relieved and strong units of Fourteenth Army reinforcements were moving down the road from Dimapur.

Below Kohima, British and Indian troops were clearing Jap road blocks from the two-lane highway that connected that town with Imphal. Heavy fighting was reported southeast of Imphal.

More of the airborne troops known as "Chindits" landed in Burma and struck a new blow at Jap communications by knocking out 40 miles of the Mandalay-Myitkyina railway, in the vicinity of Indaw. The first "Chindit" swoop in this area started a month ago and the present one, London reported, gives promise of still better results. The action is in support of Lieut. Gen. Stilwell's Chinese units moving southward toward Mogaung and Myitkyina. Stilwell's troops were reported within 20 miles of Mogaung and about 35 miles from Myitkyina.

EASTERN FRONT: Air activity, in which the Red Air Force joined with the Soviet Navy to sink scores of barges, ferries and assorted small craft trying to maintain contact with Sevastopol, was the outstanding action on the Eastern front. Improved weather on the Central front also enabled the Red airmen to strike at German rail concentrations at Lwow and Brest-Litovsk.

German communiques repeatedly stated that the Soviets were getting a new offensive under way in southeastern Poland and reported heavy fighting between the Carpathians and the Upper Dniester River. On the other hand, the Soviets continued to state "there was nothing important to report."

TASK FORCE SITS FOR ITS PORTRAIT

Few pictures have given the feeling of size and power of one of our task forces as does this one. The ships shown here calmly anchored in one of the Marshall lagoons are only part of the mighty force that hammered the islands into submission. In the picture are nine aircraft carriers, a dozen batteships and cruisers, destroyers and supply ships stretching far into the background.

When a flash flood in the South Pacific jungles washes out an existing bridge the immediate call is for the Army

Mopping up on Bougainville: A Jap pillbox has been located and Sgt. Charles H. Wolverton of the 37th Division bites his tongue as he begins to heave a grenade.

NEWSMAP Prepared and distributed by ARMY INFORMATION BRANCH ARMY SERVICE FORCES 205 E. 42nd Street, NEW YORK 17, N. Y.

MONDAY, MAY 8, 1944 • WEEK OF APRIL 27 TO MAY 4 • Volume III No. 3F

began on 2 June 1944 after over 200 American bombers that had attacked Debrecen in Hungary landed in Poltava, Ukraine. The Soviets were opposed to the possible impact of such an American presence in Ukraine. As a result, there were to be very few missions, and Stalin refused to accept that the aircraft could provide help during the Warsaw Rising against German control.

The conflict in New Guinea faded from attention after 1942, and has never attracted much British coverage, but it remained a difficult struggle, waged in arduous fighting conditions, until the end of the war: the jungle, the mountainous terrain, the heat, the malaria and the rain combined to cause heavy Allied casualties. Allied fighting quality in New Guinea rose markedly, and successive Japanese positions were taken, in part thanks to amphibious assaults – for example, Finschhafen in October 1943, Saidor in January 1944 and Hollandia and Aitape that April, landings that successively outflanked the Japanese. Many positions, such as Hollandia, were important due to their airfields, and these successes were seen as an important adjunct to the advance on the Philippines.

On the back of the double-sided newsmap Harrison reproduced his spherical topographical view to draw Eurasia and North America together and outline four strategic choices he called 'Four Approaches to Japan': 'From Alaska', 'From Manchuria', 'From China-Burma' and 'From the S.W. Pacific' (see pages 182–183). The maps carry no explanatory text, but in a best-selling atlas in 1944 he introduced his readers explicitly to his themes:

> *'This map, together with the three that follow, show in perspective the various approaches to Japan. The first of the series, 'Japan from Alaska,' shows how the direct northern route cuts into the heart of the Japanese Empire. 'Japan from the Solomons,' on pages 44 and 45, is a reminder of the vast distances in the southern Pacific and of the importance of the Japanese stronghold of Truk. 'Japan from China,' pages 46 and 47 shows the huge continental mass that Japan is trying to subdue; it indicates the close geographical relationship that can be put to work in Allied offensive action; at the same time it demonstrates the difficulties of supply – all U.S. material must now pass into China over the worst succession of mountain ranges in the world. The fourth map, 'Japan from Siberia,' pages 48 and 49, shows not only how close Siberia is to Japan, but also how vulnerable Vladivostok and the Maritime Provinces are to attack from Manchuria.'*

While arresting, the maps were in practice deeply flawed as guides to strategy, and for physical as well as political reasons. Alaska might offer a route to the Kurile Islands, Hokkaido and the remainder of Japan, but as a military prospect the route was greatly limited due to the problems for America of building up supplies in Alaska and, more generally, the issues posed by cold, ice and the dark winter. In cartographic terms, Alaska might be closer to Japan than was California, but the logistical possibilities of the latter were far greater for operations in the Pacific. Politically, the route was not viable unless the Soviet Union co-operated, but it had a non-aggression treaty with Japan, which was not broken until August 1945. Nevertheless, Japanese concern about a possible American threat to Hokkaido led to the stationing of significant forces there. Conversely, that Japanese presence failed to prevent the development of the route from Alaska to the Soviet Far East as the most significant source of Western materiel for the Soviet military.

A WORLD IN FLUX

Esso Standard Oil Company war map, 'Fortress Europe', 1944.

Produced exclusively by General Drafting Company in New York, the makers of the then well-known Esso-branded road maps, this is the 'Invasion Edition' of 'Esso War Map II that features Fortress Europe [and] The World Island' (1944), with its cover artwork of a North American B25 Mitchell bomber successfully attacking an unidentified location.

The first Esso war map had appeared in 1942, since when (as the map explains) the development of a global war with 'the Americans in the thick of the fighting everywhere' had necessitated a different type of maps to allow the public to follow 'the strategy of the Allies as it develops from day to day throughout the world'. Esso's mapmakers stress the role of oil in the conflict ('oil is ammunition use it wisely'), which for war manufacture was 'indispensable'. Not only its availability was important but also its ease of movement. The Allies were stronger in both respects.

The concept of the 'World Island' referred to in the title, was based, as the accompanying map text notes, on the work of Halford Mackinder, a British geopolitician and specifically his *Democratic Ideals and Reality* (1919), of which an American edition appeared in 1942. He argued that it was necessary for commentators to distinguish between reality, in the shape of geopolitical realities, and democratic ideals, and claimed that it was important to understand the fundamental geographical inequalities of nations.

In his 1904 paper 'The Geographical Pivot of History' Mackinder had focused on a Eurasian 'heartland', control over which could lead to the dominance of Eurasia as a whole. In his 1943 essay 'The Round World and the Winning of the Peace', Mackinder presented his original account as motivated by concern about Russia and also arising from the growth of German power, and looking ahead focused on the forthcoming strength of the Soviet Union and on the need to prevent any resurgence of Germany as an aggressive power. The latter was to be achieved by obliging Germany to face the certainty of war on two fronts: with the Soviet Union in the heartland and with sea power based on the United States, Britain as a forward stronghold and France as the defensible bridgehead.

Mackinder presented the Soviet Union and the Western alliance as friendly, but as the inset map and section 'Peace Conference Problems' noted, by July 1944 the nature of the subsequent peace was a matter dividing the Allies.

One major difference was contrasting views over Poland by Britain and the Soviet Union. Looking back to rivalry from the beginning of the Soviet Union, Stalin informed Churchill that the Poles were 'incorrigible' – adding, in March 1944, that there could not be 'normal relations' between the Soviet Union and the London-based Polish government-in-exile, and that the Polish view of the frontier was unacceptable. Churchill was willing to be helpful about the frontier, but offered a view on power that was very different to that of Stalin:

> *'Force can achieve much but force supported by the good will of the world can achieve more. I earnestly hope that you will not close the door finally to a working agreement with the Poles which will help the common cause during the war and give you all you require at the peace. If nothing can be arranged and you are unable to have any relations with the Polish Government, which we shall continue to recognise as the government of the ally for whom we declared war upon Hitler, I should be very sorry indeed. The War Cabinet ask me to say that they would share this regret....'*

In August–September 1944, Soviet forces were not willing to continue their westward offensive, when the Germans were under pressure from the Poles who had risen in Warsaw, mainly because Stalin did not wish to see Polish nationalists in control of their capital instead of his own protégés. Other reasons were largely excuses; Stalin did not want political options, then or on other occasions, that were determined by logistics, or indeed by military problems.

The competing strategies for the post-war world also affected discussion of wartime strategy in the shape of the direction and speed of Allied advances. For all powers, planning for the post-war era in part involved a determination to insert forces into areas judged to be of particular significance. This strategic dimension drew on a wide range of concerns, including the prospect of post-war conflict.

Henry Morgenthau, the highly influential US Secretary of the Treasury, proposed in 1944 that Germany should be divided into two and deindustrialised. As with ideas of partition, notably from the French, at the end of the First World War, this approach was not pursued in full: Germany again was to lose territory to neighbouring powers, but with the exception of the reversal of the 1938 union with Austria it was not to be partitioned into separate German states. East and West Germany emerged essentially because of differences between the occupation powers, and not as a war goal.

Aside from the risk of German *revanche*, the British Chiefs of Staff were, from 1944, actively considering post-war threats to the British world. In a United States that had thrown off its inter-war isolationism, the feeling was growing that a continent dominated by the Soviet Union was just as undesirable as one controlled by Germany. The Esso map summarised the state of flux as follows: 'The face of the world is changing. Countries have come—and gone. Boundaries have changed. Place names are different. New cities have sprung up. New trade routes have been laid out. Distances and natural barriers do not mean what they used to. As the world shrinks, our personal knowledge of it must be expanded.'

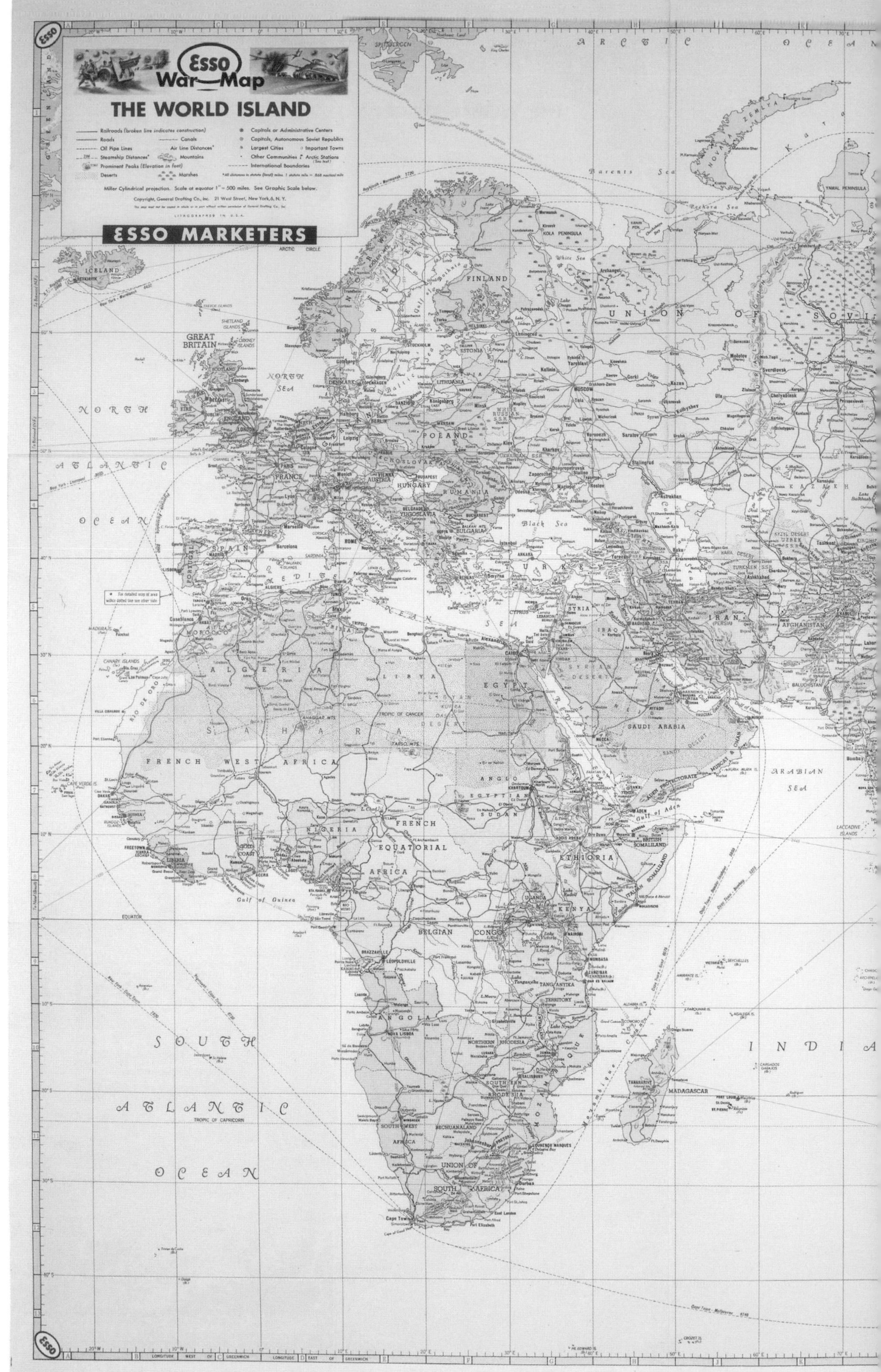

Esso
Esso War Map
THE WORLD ISLAND
Railroads (broken line indicates construction)
Roads
Canals
Oil Pipe Lines
Air Line Distances*
Steamship Distances*
Mountains
Prominent Peaks (Elevation in feet)
Deserts
Marshes
Capitals or Administrative Centers
Capitals, Autonomous Soviet Republics
Largest Cities
Important Towns
Other Communities
Arctic Stations (See text)
International Boundaries
Miller Cylindrical projection. Scale at equator 1" = 500 miles. See Graphic Scale below.
Copyright, General Drafting Co., Inc. 21 West Street, New York, 6, N. Y.
LITHOGRAPHED IN U.S.A.
ESSO MARKETERS
ARCTIC OCEAN
ARCTIC CIRCLE
ICELAND
GREAT BRITAIN
NORTH SEA
NORTH ATLANTIC OCEAN
FINLAND
UNION OF SOVI
POLAND
FRANCE
SPAIN
HUNGARY
RUMANIA
Black Sea
TURKEY
IRAN
AFGHANISTAN
MOROCCO
ALGERIA
LIBYA
EGYPT
SAUDI ARABIA
ARABIAN SEA
SAHARA
FRENCH WEST AFRICA
ANGLO EGYPTIAN SUDAN
NIGERIA
FRENCH EQUATORIAL AFRICA
ETHIOPIA
Gulf of Guinea
BELGIAN CONGO
KENYA
ANGOLA
NORTHERN RHODESIA
SOUTHERN RHODESIA
MADAGASCAR
SOUTH WEST AFRICA
BECHUANALAND
UNION OF SOUTH AFRICA
SOUTH ATLANTIC OCEAN
TROPIC OF CAPRICORN
INDIA

Esso
A R C T I C O C E A N
Laptev Sea
East Siberian Sea
TAIMYR PENINSULA
NOVO-SIBIRSKIE IS.
DE LONG ISLANDS
Chukchi Sea
ARCTIC CIRCLE
VERKHOYANSK MTS.
CHUKOTSKI PENINSULA
Gulf of Anadyr
A L A S K A
S O C I A L I S T R E P U B L I C S
Bering Sea
Sea of Okhotsk
KOMANDORSKIE IS.
A L E U T I A N I S L A N D S
TANNU TUVA
M O N G O L I A
MONGOLIAN PEOPLE'S REPUBLIC
GOBI (SHAMO) DESERT
MANCHURIA
SAKHALIN
HOKKAIDO
KURILE ISLANDS
Sea of Japan
KOREA
HONSHU
J A P A N
TOKYO
C H I N E S E R E P U B L I C
Yellow Sea
East China Sea
SHANGHAI
CHUNGKING
RYUKYU ISLANDS
FORMOSA
HONG KONG
P A C I F I C O C E A N
TROPIC OF CANCER
HAWAIIAN ISLANDS
INTERNATIONAL DATE LINE
BURMA
THAILAND
FRENCH INDO CHINA
HAINAN
Gulf of Tonkin
South China Sea
LUZON
MANILA
PHILIPPINE ISLANDS
MARIANA ISLANDS
BAY OF BENGAL
ANDAMAN ISLANDS
NICOBAR ISLANDS
CEYLON
Sulu Sea
MINDANAO
C A R O L I N E I S L A N D S
MARSHALL ISLANDS
MALAY STATES
SINGAPORE
SUMATRA
BORNEO
Celebes Sea
CELEBES
EQUATOR
GILBERT ISLANDS
BISMARCK ARCHIPELAGO
NEW GUINEA
SOLOMON ISLANDS
JAVA
BATAVIA
N E T H E R L A N D S E A S T I N D I E S
PHOENIX ISLANDS
ELLICE ISLANDS
TOKELAU OR UNION ISLANDS
SAMOA ISLANDS
Arafura Sea
Timor Sea
Gulf of Carpentaria
CORAL SEA
SANTA CRUZ IS.
NEW HEBRIDES
FIJI ISLANDS
TONGA OR FRIENDLY ISLANDS
COOK ISLANDS
NOUMEA
TROPIC OF CAPRICORN
O C E A N
GREAT SANDY DESERT
A U S T R A L I A
GREAT VICTORIA DESERT
Great Australian Bight
S O U T H P A C I F I C O C E A N
Brisbane
Sydney
CANBERRA
Tasman Sea
TASMANIA
NORTH ISLAND
NEW ZEALAND
WELLINGTON
SOUTH ISLAND
Graphic Scale for Esso World Island Map
Statute (Land) Miles
100
200
300
400
500
500 Miles
FIGURING MILEAGES ON THIS MAP
For distances running approximately east and west the accompanying scale can be used. Since the map scale increases north or south from the equator, be careful to use the line on the scale corresponding to the average latitude of the line between the desired points. For example, the mid-point of the line Dutch Harbor—Tokyo is about 45° north. Hence use scale for 45°.
Distances approximately north and south can be figured by finding the number of degrees of latitude between the points by reference to the marks along left and right edges of the map. Each degree equals about 69 statute miles. Note that while the number of miles to a degree is constant, the actual length of graduations increases north or south from the equator, due to the nature of the projection used on this map.
OIL IS AMMUNITION
USE IT WISELY !

OBJECTIVE PARIS

George Horace Davis, 'D-Day', 17 June 1944.

Published in the *Illustrated London News* just 11 days after the event, this aerial view provides an arresting and unconventional perspective, with Paris on the horizon, offering a strategic point to the invasion. It was accompanied by the caption, 'A panoramic map of the coastal regions from Cape Gris Nez to Cherbourg Peninsula'. The panoramic view reflected the growing influence of aerial photography and also made it much easier for the general public to understand than a standard map.

The scale of the assault is ably captured, both in terms of the invasion armada and with reference to Allied air control, although, somewhat confusingly, many of the aircraft are shown as coming from the area of German control. On 6 June itself, 156,115 men landed, and by 11 June there were 326,547. The Allied losses on 6 June were fewer than had been feared, in part due to the success of Allied deception and also thanks to complete sea and air dominance and to the poor implementation of what was anyway a mistaken German strategy. However, the subsequent breakout from the invasion zone proved more difficult than had been anticipated – a theme that, understandably, was not captured in this map. This included the difficulty of representing the *bocage* (the local terrain of mixed hedged pasture and woodland) at this scale.

The Allies had assumed that the Germans, having been unable to hold their coastal fortifications, would fall back in order to defend a line, probably that of the River Seine. Instead, the Germans chose to fight hard, for both the coast and the rest of the territory. This obliged the Allies to fight in the difficult terrain of the *bocage*, with its thick hedgerows and sunken lanes. Although the Caen plain is flat, Normandy was not good tank country. The landscape both greatly affected cross-country performance and assisted the German defence. Allied armour, doctrine and tactics were not well suited to the *bocage* and the opportunities it offered to defenders.

The Germans had ample experience from the Eastern Front of mounting a defence, as well as good equipment in the form of anti-tank guns and heavy tanks. Resting on the defence, the Germans enjoyed the advantage of firing first, and at close range, and from a stable position. Entering open ground exposed Allied tanks to serious risk, which led to more indirect support using dead ground. This situation put a renewed emphasis for the attackers on infantry–armour co-operation, but that is easier in doctrine than in practice, and the *bocage* made coordination particularly difficult.

With their individual units often lacking adequate training, experience, quality equipment, command and doctrine, the Allies both faced a hard battle and fell behind the anticipated phase lines for their advance. Allied casualty rates in the breakout were far higher than in the initial landings. Despite air attacks, especially on bridges, the Germans were able to reinforce their units in Normandy, although the delays forced on them both ensured that the Allies gained time to deepen their beachheads and obliged the Germans to respond in an ad hoc fashion to Allied advances, using their tanks as a defence force rather than destroying the beachheads. When the German armour was eventually employed in bulk, on 29–30 June, it was stopped by Allied air attacks. Another armour counter-attack was defeated on 11 July.

In the Battle of Normandy, the Germans learned how to adapt in the face of concentrated firepower and air attack, and they also reacted well to defending the *bocage*, whereas the Allies, notably the British and the Canadians, found it difficult to break through and restore manoeuvre. The Allied tanks failed to achieve what the Americans and British (and Germans) had expected from tanks.

Normandy's numerous hedges and sunken lanes provided excellent cover for opposing tanks and for anti-tank guns, and the *bocage* obstructed observation, movement and lines of fire, each of which individually, and even more in combination, greatly affected the capabilities of the attacking armour. Helped by the cover afforded by the *bocage*, the defending Germans also used sticky bombs against tanks. In Normandy small unit actions became the norm, and these tested the ability to develop new tactics. Both sides did so, while having to confront a lack of the necessary experience.

There were also serious operational limitations for warfare dominated by tanks, not least their vulnerability to anti-tank guns, which had become much more powerful since 1940, and the problems of communications between tanks and infantry. Radios were installed in tanks to aid coordination with supporting air power, while telephone sets were placed on the backs so that infantry could communicate with the tanks. German anti-tank weapons included the highly effective 88mm dual-purpose (anti-aircraft and anti-tank) gun, the PaK 38 and PaK 40 anti-tank guns, and the Panzerfaust and Panzerschreck anti-tank weapons, but the Germans suffered from a relative shortage of ammunition. Moreover, the Allies had far more anti-tank guns.

ENGLISH CHANNEL
CAPE GRIS NEZ
BOULOGNE
GOURNAY
PARIS
MANTES
DIEPPE
ROUEN
R. SEINE
R. EURE
LOUVIERS
EVREUX
BACQUEVILLE
ELBEUF
VERNEUIL
VEULETTES
YVETOT
LILLEBONNE
R. RISLE
BRIONNE
FECAMP
BONBAC
BERNAY
ETRERAT
HARFLEUR
HONFLEUR
LISIEUX
LE HAVRE
CAPE DE LA HEVE
MOUTH OF THE SEINE
GACÉ
TROUVILLE
DEUVILLE
VIMOUTIER
HOULGATE
AMMEVILLE
DIVES
DOZULÉ
GREVECOUR
ARGENTAN
AERODROME
FALAISE
MOUTH OF RIVER ORNE
OUISTREHAM
ARGENCES
CANAL
COLOMBELLES
ROCQUANCOURT
LION SUR MER
R. ORNE
CAEN
BRETTEVILLE
RISING GROUND
ILE DE LA GRUNE (ROCKS)
SOUTHAMPTON TO CAEN - 120 MILES
I DE BERNIERS (ROCKS)
ST. AUBUN SUR MER
CARPIQUET
EVERCY
COURSEULLES
FLERS
ASNELLES
BAY of THE SEINE
UNDULATING CULTIVATED GROUND RISING INLAND
AR.ROMANCHES
BAYEUX
TILLY
TREVIERES
BALLEROY
VIRE
LITTRY
FOREST OF CERISY
CAUMONT
CAPE PERCÉE
GRAND CAMP
LISON
AIREL
ISIGNY
R. VIRE
ST. LO
ST. MARIE
CAPE BARFLEUR
BARFLEUR
ST. VAAST
CARENTAN
ST. MERE EGLISE
SAINTENY
ST. PIERRE
QUETICHOU
ST. JARES
MONTEBOURG
PERIERS
CHERBOURG PENINSULA
CAPE LEVY
ROCKY COAST
VALOGNES
TOURLAVILLE
SAUVEUR LA VICOMTE
PLAIN OF AY
COUTANCES
PORBAIL
AY SANDS
ILE PELÉE
BREAKWATER
OCTEVILLE
CARTARET
TO SOUTHAMPTON 83 MILES
CHERBOURG
LES PIEUX
BARNEVILLE
CLIFFS AND ROCKS
G H DAVIS 1944

"D" DAY

MUKHTUYA
OLEKMINSK
UST MAYA
VITIM
TOMMOT
MAGADAN
OKHOTSK
OLA
UNION OF SOVIET SOCIALIST REPUBLICS
TIGIL
AYAN
SEA OF OKHOTSK
TRANS-SIBERIAN RAILROAD
NIZHNE KAMCHATSK
KOMANDORSKIE (U. S. S. R.)
LAKE BAIKAL
UDSKOE
BARGUZIN
IRKUTSK
NIKOLAEVSK
PETROPAVLOVSK
ULAN UDE
CHITA
SRETENSK
NERCHINSK
UST BOLSHERETSK
KUIBYSHEVKA
BLAGOVESHCHENSK
KOMSOMOLSK
SAKHALIN
ALTAN BULAK
AIGUN
PARAMUSHIRO
MANCHOULI
HAILAR
NUNKIANG
SHIKUKA
CHALANTUN
STALINSK
KHABAROVSK
ULAN BATOR
TSITSIHAR
KURILE ISLANDS (JAPAN)
MONGOLIA
ODOMARI
HARBIN
MANCHURIA
WAKKANAI
HSINKING
KIRIN
OTARU
SAPPORO
TOLUN
VLADIVOSTOK
MUKDEN
HAKODATE
SEISHIN
CHENGTEH
AOMORI
PEIPING
ANTUNG
SEA OF JAPAN
KANKO
TIENTSIN
PORT ARTHUR
HEIJO
GENSAN
NINGSIA
TAIYUAN
KOREA
NIIGATA
SENDAI
NORTH
SEOUL
LANCHOW
TSINAN
TSINGTAO
NAGOYA
TOKYO
JAPAN
TSINING
OSAKA
KAIFENG
FUSAN
YOKOHAMA
MOPPO
CHANGAN
KURE
SASEBO
NAGASAKI
KOCHI
300 MILES
NANKING
CHINKIANG
CHINA
HANKOW
SHANGHAI
KAGOSHIMA
WUCHANG
HANGCHOW
CHUNGKING
NANCHANG
AMAMI
600 MILES
CHANGSHA
WENCHOW
NANSEI ISLANDS (JAPAN)
KWEIYANG
FOOCHOW
CHICHI
BONIN ISLANDS (JAPAN)
KWEILIN
TAIHOKU
KUNMING
VOLCANO IS.
AMOY
FORMOSA
900 MILES
CANTON
PESCADORES
HONG KONG
HANOI
HAIPHONG
KIUNGSHAN
HAINAN
1200 MILES
WAKE
LAOAG
APARRI
MARIANAS ISLANDS (JAPAN)
HUE
LINGAYEN
FRENCH INDO-CHINA
SOUTH CHINA SEA
MANILA
1500 MILES
SAIPAN
ROTA
PHILIPPINES
GUAM
MARSHALL ISLANDS (JAPAN)
ENIWETOK
CAMRANH BAY
SAIGON
ILOILO
1800 MILES
YAP
WOTJE
BUTUAN
KWAJALEIN
PALAU IS.
ZAMBOANGA
DAVAO
TRUK
PONAPE
2100 MILES
JALUIT
LABUAN
SANDAKAN
BRUNEI
NORTH BORNEO
CAROLINE ISLANDS (JAPAN)
MALAY STATES
SARAWAK
2400 MILES
GILBERT ISLANDS (GT. BR.)
KUCHING
MANADO
BORNEO
TERNATE
PONTIANAK
BALIKPAPAN
SCHOUTEN IS.
CELEBES
2700 MILES
NEW IRELAND
BANDJERMASIN
AMBOINA
N. E. NEW GUINEA
RABAUL
MADANG
BATAVIA
SEMARANG
MAKASSAR
NEW GUINEA
SOLOMON ISLANDS (GT. BR.)
NETHERLANDS INDIES
BOUGAINVILLE
CHOISEUL
3000 MILES
LAE
NEW BRITAIN
SANTA ISABEL
JAVA
SOERABAJA
PAPUA
DILI
(PORTUGAL)
PORT MORESBY
GUADALCANAL
KOEPANG
TIMOR
SANTA CRUZ ISLANDS (GT. BR.)
DARWIN
PINE CREEK

THE THREATENER IS THREATENED

Edwin L. Sundberg, 'Quadruple Threat to Japan', 3 September 1944.

This map by the *New York Sunday News* staff artist Edwin L. Sundberg provides a sense of impending doom for Japan: 'Japan, which was the threatener a couple of years ago is now Japan, the threatened.' The map is organised in terms of 'bombing distances' to Tokyo, marked at 300-mile (480-kilometre) intervals, and indicates the significance of the American gain of Saipan Island, which had been captured that summer. Arrows (see pages 194–195) highlight 'the four chief worries of the Son of Heaven' (a reference to the Japanese emperor). Although that was not its intention, the map made it apparent that the conquest of the Philippines would be a diversion from the key axis of aerial assault that was offered by Admiral Chester W. Nimitz's seizure of island bases. The bombing of Japan from Saipan began in November.

Most of the large quantities of supplies the US Tenth Air Force had been flying into China were for the American air assault on the Japanese, and not munitions for the Chinese armies. However, from the summer of 1943 the Americans had also been planning for a Chinese advance on Guangzhou (Canton)–Hong Kong to link up with an American advance, via Taiwan, to the coast of China. These advances were to be followed by joint operations to clear northern China and establish bases from which Japan itself could be attacked. Confident that an air assault on Japan could make a major difference to the war, the Americans sent the first Boeing B29 Superfortresses to become operational to China. From there, they bombed targets in Japanese-occupied China, Korea and southern Japan, but most of Japan was beyond the range of raids mounted from the Chinese bases. In practice, however, the strategy had weak fundamentals, not least because it relied on the coincidence of too many variables. In addition, Chinese military and logistical deficiencies were serious.

By 1944, there was serious tension, as Jiang Jieshi (Chiang Kai-shek of the Republic of China) resisted pressure for Joseph Stilwell, an American, to become commander of all Chinese forces, and also American demands that Jiang seek a coalition with the communists in order to put pressure on the Japanese. In practice, Jiang could not mount a defensive front and could not switch to an offensive mode due to a lack of equipment, a weak senior officer corps (which for political and cultural reasons he could not significantly alter) and his heavy losses of professionals from 1937. The Americans were insufficiently attentive to this. The irascible Stilwell, moreover, could not get on with anyone.

The map did not capture the success of Operation Ichi-Go, which had been launched by Japan in China in April 1944, leading to the capture of American air bases. The Japanese pressed on to overrun much of southern China, although that only ensured that a disproportionate share of Japanese military resources were devoted to China. Moreover, Japanese successes in China could not be translated into a broader strategic achievement, or even into knocking China out of the war. Indeed, having more territory to control proved a strategic burden and, faced with mounting casualties, the Japanese ran out of impetus in early 1945. Nevertheless, the American hope of launching an air assault on Japan from China was redundant by the time this map was produced. Japan had demonstrated it could manage, notwithstanding American naval success, by establishing an overland route between central China and Indo-China, but did so without making clear the limited strategic, military and economic value of this alternative in the face of American capabilities. In the event, there was no equivalent in China, for either Japan or the Americans, to the strategic possibilities increasingly available to the Americans in the Pacific. Moreover, these possibilities permitted the Americans to use their clear superiority in naval strength, amphibious capability and air power.

Politics was a key part of the process, because Japan had maintained, against Hitler's wishes, its non-aggression agreement of 13 April 1941 with the Soviet Union. As a result, the Soviet Union would not allow its territory to be used for American air attacks on Japan. The Soviet Union's refusal to breach its neutrality lessened the strategic significance of the North Pacific, including the Aleutians, and left the Americans with the more difficult alternatives of bases in China and/or the western Pacific. The former were difficult to supply and protect whereas the latter were hard to conquer. Correspondingly, despite a request by Hitler, Japan did not attempt to block American supplies to the Soviet Far East, and the route was an important one that was enhanced by the building of the Alaska Highway, the proposal for which was approved by the US Army on 6 February 1942 and the US Congress five days later. In practice, most supplies were sent by sea from American west coast ports, including the large number of trucks/lorries that were to prove important to Soviet mobility in 1944.

QUADRUPLE THREAT to JAPAN

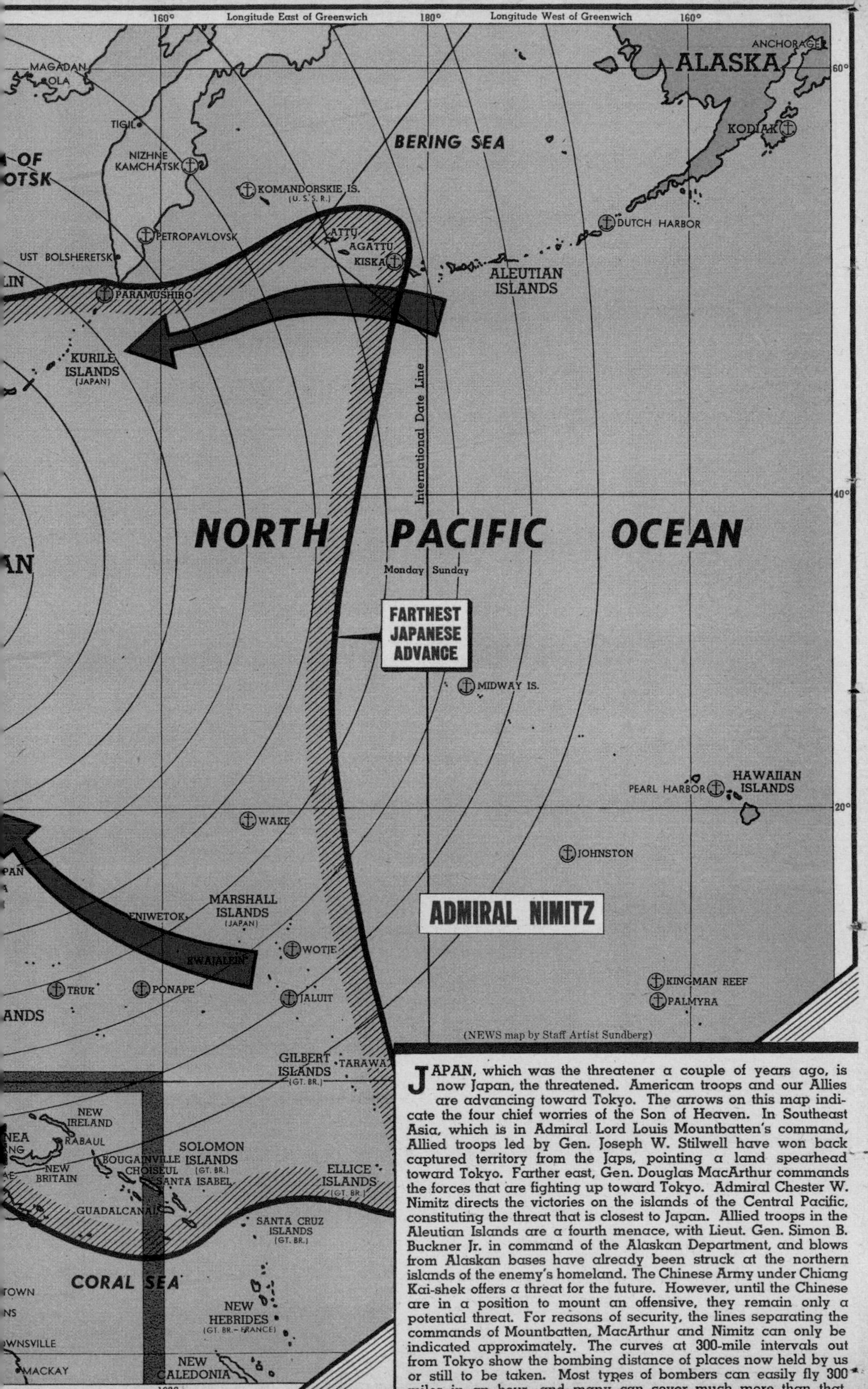

JAPAN, which was the threatener a couple of years ago, is now Japan, the threatened. American troops and our Allies are advancing toward Tokyo. The arrows on this map indicate the four chief worries of the Son of Heaven. In Southeast Asia, which is in Admiral Lord Louis Mountbatten's command, Allied troops led by Gen. Joseph W. Stilwell have won back captured territory from the Japs, pointing a land spearhead toward Tokyo. Farther east, Gen. Douglas MacArthur commands the forces that are fighting up toward Tokyo. Admiral Chester W. Nimitz directs the victories on the islands of the Central Pacific, constituting the threat that is closest to Japan. Allied troops in the Aleutian Islands are a fourth menace, with Lieut. Gen. Simon B. Buckner Jr. in command of the Alaskan Department, and blows from Alaskan bases have already been struck at the northern islands of the enemy's homeland. The Chinese Army under Chiang Kai-shek offers a threat for the future. However, until the Chinese are in a position to mount an offensive, they remain only a potential threat. For reasons of security, the lines separating the commands of Mountbatten, MacArthur and Nimitz can only be indicated approximately. The curves at 300-mile intervals out from Tokyo show the bombing distance of places now held by us or still to be taken. Most types of bombers can easily fly 300 miles in an hour, and many can cover much more than that.

LEYTE GULF

R.M. Chapin, *Time*, 'I Have Returned', 30 October 1944.

The title of this Robert Chapin Jr map from *Time* refers to a speech made in 1944 by General Douglas MacArthur, who epitomised the US desire and determination to recapture the Philippines to reverse his flight from them in 1942, when he had vowed: 'I came through and I shall return.'

MacArthur insisted that the Americans had obligations to their Filipino supporters, but there was also concern about breaking the link between Japan and Japanese-controlled Southeast Asia and the naval forces stationed there. Moreover, as yet, there was no proof of the success of B29 bomber raids on Japan. The invasion of the Philippines began on 20 October 1944 when the island of Leyte was invaded by four divisions. Resistance was initially light with the main threat that of a Japanese naval attack on the poorly protected landing force. There was growing pessimism in Japan: the head of the Naval Operations section asked on 18 October that the fleet be afforded 'the chance to bloom as flowers of death'.

The Battle of Leyte Gulf (23–26 October) was the largest naval battle of the war. In Operation Sho-Go (Victory), the Japanese sought to lure the American carrier fleet away, employing their own carriers as bait, and then using two naval striking forces, under Vice Admirals Takeo Kurita and Shoji Nishimura, respectively, to attack the vulnerable landing fleet. This overly complex scheme posed problems both for the ability of American admirals to read the battle and control its tempo, and, as at Midway in 1942, for their Japanese counterparts in following the plan. In a crisis for the American operation, one of the strike forces was able to approach the landing area and was superior to the American warships there. However, instead of persisting, the strike force retired. The net effect of the battle, which overall was dominated by American naval air power, was the loss of four Japanese carriers, three battleships, 10 cruisers, other warships and many aircraft. This defeat of the surface fleet followed that of the carrier air force at the Battle of the Philippines Sea on 19–20 June.

However, the struggle on Leyte Island became more difficult because the Japanese sent 50,000 reinforcements. As a result, a supporting invasion was launched on the west coast on 7 December. Luzon was not invaded until the Americans landed in Lingayen Gulf on 9 January 1945, picking the same site chosen by the Japanese in December 1941, 120 miles (190 kilometres) north of well-defended Manila.

For report documenting above map see TIME, OCTOBER 30, 1944 issue

DEUTSCH SÜD-AMERICA
DESIGNED BY
SCHIKLGRUBER & Co
IDIOT'S
ELIGHT
DEUTSCH EUROPA
DEUTSCH AMER
DEUTSCH AFRIC
DEUTSCH AUSTRALIE
ARTHUR SZYK

Propaganda

6.

'MODERN GERMANY, shown red, with Hitler's Mein Kampf dream of further expansion indicated by red lines.'

Legend on the 1939 version of Ernest Dudley Chase's 'Europe: A Pictorial Map' (originally 1938).

'The Map Maker', a caricature by Arthur Szyk, a Polish Jewish émigré in the United States, in the March 1942 issue of *Esquire* magazine, depicts a scene at 'Schiklgruber & Company Fancy Maps'. Welcoming General Hideki Tōjō, the Japanese Prime Minister, to the Axis, Goebbels declares, 'Now that you've joined us, the Führer will make a special map for you!' Hitler meanwhile paints a map of 'Deutsch Süd-America'.

Propaganda was designed to sustain domestic support for the war and to affect opinion abroad. For all combatants, there were major concerns about morale and resilience, anxieties that prompted attempts to gather intelligence and to influence opinion. This situation was certainly true of democracies, and reflected a broader process of establishing consensual democratisation there. In 1937, John Buchan, the Governor-General of Canada, who had played an important role as Director of Information in the successful First World War British propaganda effort, declared in a speech to the Canadian Institute of International Affairs:

> *'The day has gone when foreign policy can be the preserve of a group of officials at the Foreign Office, or a small social class, or a narrow clique of statesmen from whom the rest of the nation obediently takes its cue. The foreign policy of a democracy must be the cumulative views of individual citizens, and if these views are to be sound they must in turn be the consequence of a widely diffused knowledge.'*

Yet, totalitarian societies, including Germany and the Soviet Union, also worried about public opinion, and had to get their mechanisms of authority and power to work to that effect, which was a difficult task. The Gestapo responded with great anxiety to the terrified public mood after the devastating British bombing of Hamburg in July 1943, when maybe 35,000 were killed in raids that proved the Luftwaffe could not protect German civilians.

As part of an increasingly visual age, maps had a major role to play in propaganda, and were used in a variety of mediums including film. *Why We Fight* is a series of seven motivational films, made from 1942 to 1945 and commissioned by the US government, most of which were directed by Frank Capra. The series included *Prelude to War* (1942), which depicted a hemisphere of light and another of dark dictatorship, with the New World being surrounded and then conquered, while the maps of Germany, Italy and Japan were transformed into menacing symbols.

Propaganda maps in films could be animated and thus more readily grasp the imagination of the public. In the German film *Sieg im Westen* (*Victory in the West*, 1941), animated maps create an impression of inevitable military success, with the Germans being those who had the initiative and over 30 such maps displayed Germany's rapid success on the Western Front in 1940. This film in turn became an inspiration to other work.

Target audiences

Maps were primarily produced for nationals of the state in question. Families at home were then able to know where their fathers, husbands and sons were, and propaganda maps encouraged the public to feel that success was both news and a prospect. Maps exaggerated the contribution made by the forces of the state producing the map. For example, the German map of the invasion of Poland in September 1939 published in the book *Die Soldaten des Führers im Felde* (1940) ignored resistance and, understandably, made German advances far more prominent than those of their Soviet counterparts. Italian maps in 1940–42 showed the expansion of Axis control.

In turn, British newspapers produced maps of British successes, although not only those. At a time when American and Australian forces were driving back the Japanese, the *Daily Express* on 16 July 1943 showed them rolling back the rays of the Rising Sun, Japan's key symbol, the narrative located on a map of the Pacific. More definitively, the issue of 2 April 1945 has Allied soldiers from both east and west rolling up the map of Europe, with the defeated Hitler looking on.

Maps were also designed to influence opinion among allies, neutrals and people in conquered areas. In particular, the struggle for American public opinion in 1939–41 had an effect on cartographic accuracy and purpose. The German Library of Information in New York produced propaganda maps that were felt to be effective, notably *The War in Maps* (1941), and in 1940 it had justified the invasion of France, Belgium and the Netherlands by publishing *Allied Intrigue in the Low Countries White Book No. 5*, a German Foreign Office work on the Anglo-French policy of extending the war. A map of a proposed advance into Belgium

ARTHUR SZYK,
'The Map Maker', March 1942.

SCHIKLGRUBER & COMPANY
FANCY MAPS
DEUTSCH EUROPA
DEUTSCH AMERICA
THE MAP MAKER
by ARTHUR SZYK
"Now that you've joined us, the Führer will make a special map for you!"

ARTHUR SZYK, 'THE MAP MAKER', MARCH 1942.

On the premises of 'Schiklgruber & Company Fancy Maps' Hitler presents 'special' fantasy maps for his allies, Mussolini, Franco and the Mufti of Jerusalem, a group to which Japan's General Hideki Tōjō is being welcomed. In practice, they are all dupes because Germany will dominate everything (Mussolini's map is 'German Dominion of Italy'). At Hitler's feet lies a yellow book titled *Idiot's Delight*. He is shown with Goebbels and Göring, while Himmler lurks behind the map. The list of customers next to the book includes Petain (France), Pavelić (Croatia), Antonescu (Romania), King Boris (Bulgaria), Horthy (Hungary), Mannerheim (Finland), Quisling (Norway) and Hácha (Bohemia and Moravia).

However, as Mussolini appears to have discovered, the reality offered was not that of, for example, the 'German-Spanish World Empire Map' that Franco is holding, but one of allies being duped. The likely character of the sequence of maps already done – 'Deutsch Australien', Deutsch Africa', 'Deutsch America' and 'Deutsch Europa' – is shown by the 'Deutsch Süd-America' on which Hitler is working, his brush dripping blood red. The threat of his plan to the United States is made more explicit because Mexico and the West Indies are shown as part of the new Reich.

South America had been a sphere of interest to Germany and Italy, and the Germans had made diplomatic, espionage and economic attempts to build up support there. In part, these attempts reflected the desire to exploit opportunities, not least those presented by local German settler populations and by sympathetic authoritarian governments, such as the Peron dictatorship in Argentina. In part, the attempts were a product of the global aspirations of key elements in the German government. As in the case of Mexico in the First World War, there was also a desire to weaken the United States by causing problems in its backyard. Thus, although it had many flaws, there was a strategic intention underlying Germany's Latin American policy. But it provoked and had a central problem with implementation: the inability of Germany to give teeth to its hopes.

This inability was a reflection of British naval strength, which underlined the maritime background to any Latin American strategy, as well as Germany's focus on operations in Europe. Yet, despite the flaws of Germany's Latin American policy, there was the potential to cause trouble. This potential was one of the victims of Hitler's decision to declare war on the United States. Instead, the Rio Conference in January 1942 saw the creation of the Inter-American Defense Board, which was designed to coordinate military matters throughout the Western hemisphere. In effect, the United States assumed responsibility for the protection of the region, and it, not Germany, benefitted from Britain's declining role, which was accelerated with the sale of British assets in order to fund the war.

In practice, despite being portrayed as dupes, Nazi Germany's allies followed their own priorities and managed their commitments accordingly. Thus, Bulgaria sought to restrict its activity as an ally of Germany to Balkan expansion. Seeking to remain neutral, Bulgaria did not declare war on Britain until 13 December 1941 and never declared war on the Soviet Union. No troops, not even volunteers, were sent to fight the Soviets. The Finns stopped attacking the Soviets once they had regained the territory lost in 1941.

Tensions between Germany's allies were not eased by the character and content of German alliance politics. The allies were kept in the dark, there were no summits equivalent to those of the Western Allies, Intelligence was not much shared, other than between Germany and Italy, and there was a general failure to sustain co-operation. All of this was in keeping with the more general character in Hitler's concept of Europe, namely his preference for a racial hierarchy over an acknowledgement of the legitimate political aspirations of others. Irrespective of this, the Germans suffered from a lack of plans for collaboration and from the absence of any relevant organisation of staff.

Arthur Szyk (1894–1951) was a Polish-Jewish artist who had settled in the United States in 1940 with the support of the British government and the Polish government-in-exile. He was an accomplished caricaturist who drew on medieval traditions of illuminated manuscripts. In 1942, he depicted Hitler as the Anti-Christ.

between Brussels and the River Sambre that accompanied this supposedly illustrated the Anglo-French plans.

The effectiveness of German mapping led to criticism in the United States, for example by the sociologist and propaganda expert Hans Speier, a refugee from Nazi Germany who wrote the article 'Magic Geography' published in *Social Research* (1941), which was followed by Louis Quam's 'The Use of Maps in Propaganda' in *Journal of Geography* (1943). There were also propagandist maps produced by American newspapers, such as that in the isolationist *Chicago Herald and Examiner* entitled 'If We Enter a World War—and Lose!', which warned about the danger of partition among the Axis victors, demonstrating that with a map and adding a picture of New York becoming a 'No Man's Land' under air attack.

Influencing the viewer

A prominent and dramatic instance of attempting to influence opinion in conquered areas was one aimed at a French audience by the German propaganda department based in Paris, the Propaganda-Abteilung Frankreich. Churchill was depicted as an octopus reaching out to attack the French Empire, as at Dakar (1940) and Syria (1941), with the attacks being bloodily repelled (see page 214). The idea of Churchill spreading his tentacles around the world had been used before the war, when he was portrayed anti-Semitically as a Jewish octopus. From the beginning of Operation Barbarossa in June 1941, a key theme of German propaganda was that of a crusade against communism, which was intended to suggest a united Europe under German leadership.

In turn, the Allies sought to influence opinion in occupied Europe and elsewhere. Thus, in 1943, a poster by the Scottish artist Frederick Donald Blake, who produced work for the Ministry of Information (as well as maps for the *Daily Express*), was printed in French, Dutch, Spanish, Greek and Arabic. The French-language one was entitled 'Plus de 1,000,000 de tonnes de l'Axe coulees dans la Mer du Nord and dans La Manche' (translating as 'More than 1,000,000 tons of Axis shipping sunk in the North Sea and the Channel'). The vivid impression was of seas filling up with sunk or damaged German shipping.

Reportage and propaganda very much overlapped in the case of maps for the military. Army newspapers contained maps – for example, *Yank*, an American weekly with a circulation of over three million, contained NGS maps. Maps were used by the Americans for the 'Newsmap' series, which were similar to the 'ABCA Map Reviews' published by Britain's Army Bureau of Current Affairs (ABCA), employed for instructional purposes and also to maintain morale. The newsmap for the week of 23–30 August 1944 included a map of the Celebes (Sulawesi), another of southern France looking north from the Provencal landing beaches, and ones of France and the Balkans showing Allied advances. Individual units also produced maps: in October 1944 the US XIX Corps' newspaper *Tomahawk* published, while the unit was fighting in Germany, 'Deutschland Unter Allies' with a text that promised success. German units also created maps – such as the 5th Infantry Division as it advanced on the Eastern Front in 1941; its map provided an exemplary account of the operation (see pages 216–217).

Some maps were educational. 'This is Ann ... she drinks blood!' (November 1943) was provided to the American military to show the areas where malaria was a risk. A US Army captain Theodor Geisel, who served with Frank Capra in the Signal Corps, drew Ann, the anopheles mosquito (see pages 222–2213. Geisel is better known by his peacetime pen name Dr. Seuss.

In addition, morale-shredding maps were published and distributed, sometimes dropped from aircraft. They were intended for civilians or for opposing troops – such as with the German leaflets aimed at Allied forces in Italy, where progress was slow and casualties high. 'It's a Long Way to Rome' (1944) was impressive, but in the more striking 'Speaking of Timetables' (1944), a death's head uses calipers to demonstrate the slow rate of Allied progress up the map of Italy and suggests an arrival in Berlin of about 1952 (see pages 226–227). So also with the American leaflet 'The Hour Is Drawing Near!', which presents the islands captured by the Americans and argues that the Japanese government knows that defeat is inevitable, and the 1944 leaflet for Japanese soldiers that translates as 'The South Pacific is the South Pacific. Japan is Japan.' (see page 229), which shows the Japanese in the Dutch East Indies separated from Japan by the American position in the Philippines as well as American warships.

More specifically aimed at civilians were leaflets emphasising the devastation stemming from bombing. Illustrated with a map in the shape of an aerial photograph, 'Das War Hamburg' ('That Was Hamburg') (see pages 218–219) was dropped after the deadly British raid on the city in July 1943 and it urged the German people to overthrow Hitler. In turn, 'Shadow over England' (September 1944) showed Britain in the shadow of a V1 rocket (see page 228).

The German reading of the First World War, and in particular of Germany's failure in 1918, led to a great emphasis being placed by the Nazi regime on propaganda in Germany because it was argued that civilian society was more vulnerable than the military to war-weariness, and that the Home Front was therefore very fragile. Obedience was insufficient; there had to be positive mobilisation for the war effort. Hitler also argued that propaganda should be aimed at the feelings and that if it was pitched at an intellectual level it should be geared to those of limited intelligence. Intimidation became more pronounced in Germany as the war went badly and rumour flourished.

Alliance politics

The search for allies was part of the propaganda effort. In 1937, the prominent Ukrainian geographer Volodymyr Kubiyovych prepared and edited an atlas, *Ukrayiny i sumizhnyk krayin* (*Atlas of the Ukraine and Adjoining Countries*), in Ukrainian and English. It was published in what was then Lvov in Poland and represented an assertion of Ukrainian identity against Soviet rule. In 1941, at a time when Kubiyovych, as head of the Ukrainian Central Committee in Cracow, was co-operating with the Germans, he prepared an atlas in German based on his 1937 work, but it was not published.

More generally, the many problems of alliance politics were a subject kept well from the eyes of contemporaries. Instead, these politics were especially prone to propaganda that was designed to make alliances appear natural and strong. This was particularly so of Anglo-American propaganda about the Soviet Union during 1941–45, and of German propaganda in 1940–43 about the value of the alliance with Italy.

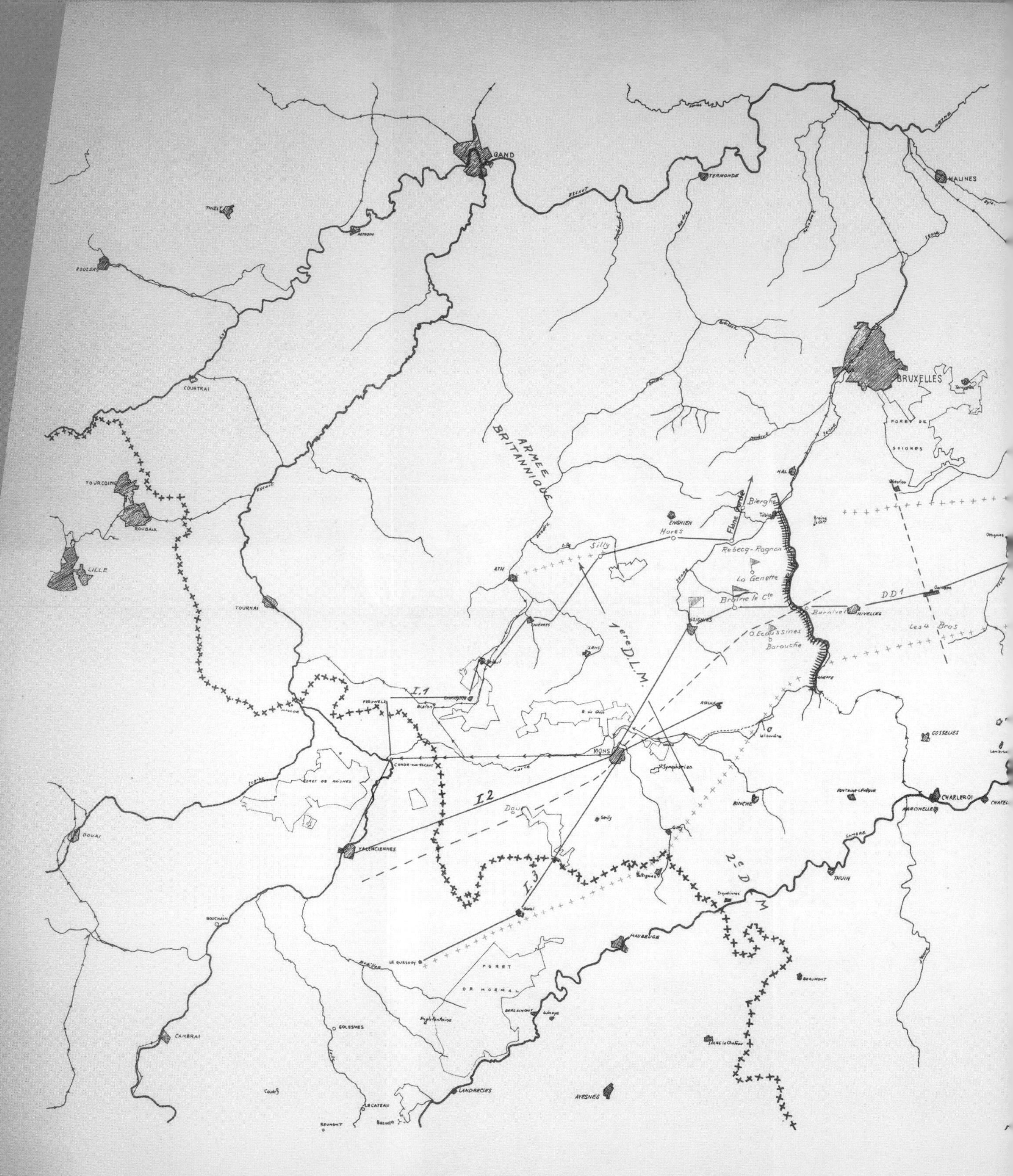
GAND
ARMEE BRITANNIQUE
BRUXELLES
LILLE
1ere D.L.M.
2e D.L.M.
DD1
I.1
I.2
I.3
Silly
Hoves
Bierghes
Rebecq-Rognon
La Genette
Braine le Cte
Ecaussines
Les 4 Bras
MONS

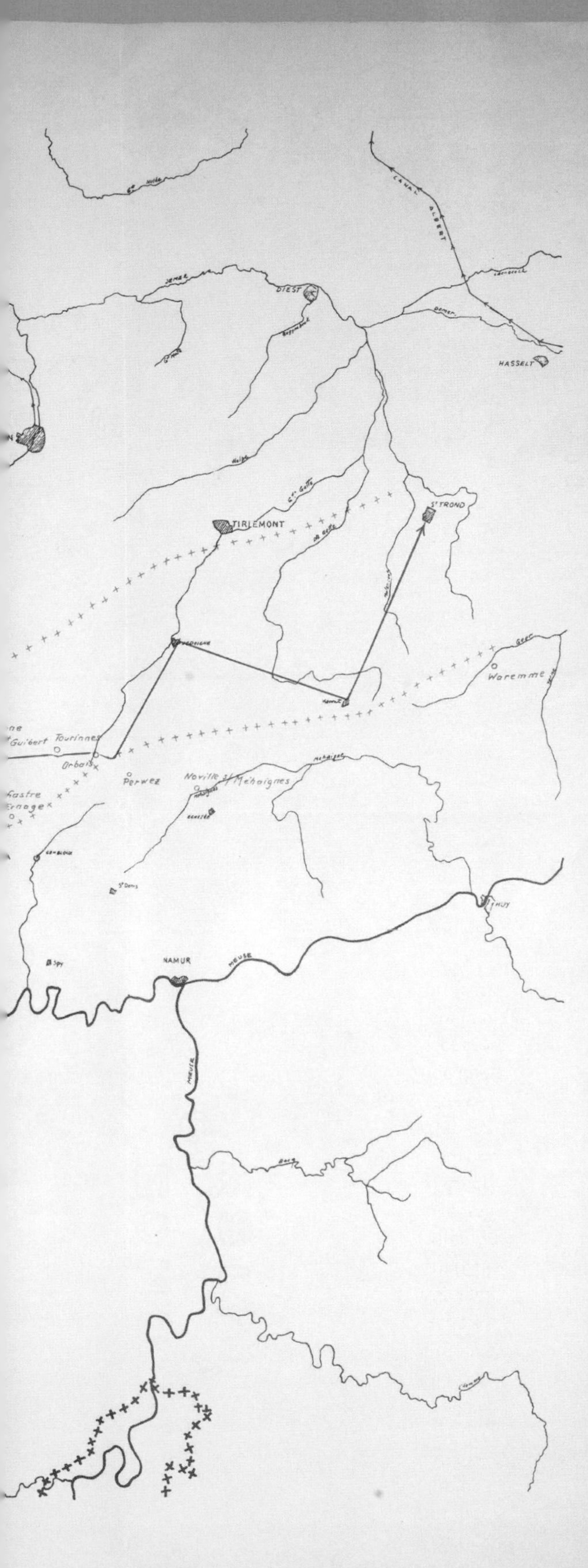

<u>ALLEGED ALLIED MILITARY PLANS, 1940.</u> In 1940–41, the German Foreign Office published White Books intended to discredit the Allies. The White Books contained Allied documents that the Germans claimed to have seized from places like the Polish Foreign Office archives in Warsaw, although the Poles claimed not to have left any incriminating papers. In German hands any real documents were easily misrepresented and disseminated as Nazi propaganda; for example, about Allied interest in 1939–40 in extending the war to Scandinavia and the Balkans, and also material on British plans to bomb the Soviet oilfields at Baku, which were used to supply Germany during the 1939–41 Nazi–Soviet Pact.

This map was published as evidence of an alleged Anglo-French plan to invade the Ruhr with the co-operation of Belgium and the Netherlands, and thus was justification for the German attack on the West in May 1940. Published by the German Library of Information in New York, the map did not prove the spurious claims. In fact, the Anglo-French plans for moving forward were rather different: they focused on the problems posed by the Belgians and the Dutch being neutral and refusing to jeopardise their neutrality by joint defence preparations. Instead, the British and the French would be able to respond to a German invasion of either country only after it had begun. To have abandoned Belgium would have been unacceptable politically, as well as risking the loss of part of the Allied order of battle, not least by also leaving the Dutch exposed.

As a result, the French and British moved their mobile reserve into Belgium before they were aware of the main direction of the German attack. Assuming that the Germans would advance north of the Meuse and were committed to a methodical battle, the French and British therefore planned to engage them in the same area, only for the actual German attack to come through the Ardennes, which was not really tank country and where the French had deployed relatively few troops, tanks and reserves.

A JAPANESE WORLD

Japanese map, 'The World Moving Toward a New World Order', July 1940.

Propaganda maps took many forms, including posters and jigsaws. In Japan propaganda maps even appeared on kimonos, but this example is believed to be from *Schoolmates-Students on the Front Lines*, a magazine published to mobilise young people for the war effort. From the mid-1930s militaristic and nationalistic material circulated widely in Japan, much of it containing maps, photographs of Japan's leaders and *gunka*, which were military songs children would sing during collective group activities.

The map exaggerates Japan's impact in part by presenting the central and northern Pacific as being under its sway, and by doing likewise as far as the Bay of Bengal and the waters of the east Pacific. In practice, although the range of Japanese operations was extraordinarily impressive, the impact in the waters off western and eastern Australia, the Indian Ocean and the central Pacific was far more limited than the map suggests.

The curved arrow moving toward Japan from the US west coast reads 'Goal is to invade the Far East'. A line of warships arrayed east of Japan reads 'Japan is invincible'. The strategic importance of and Japanese interest in Singapore is clearly illustrated, and further to the west popular opposition to British rule in India is depicted.

In practice, the Battle of the Coral Sea on 7–8 May 1942 was to end the possibility of a Japanese threat to Australia, while the American victory in the Battle of Midway on 4–5 June marked a key failure for the Japanese in the central Pacific. These battles also saw the loss of crucial Japanese naval assets. Prior to these battles, however, Japanese hopes were mirrored by Allied fears. Rumour, even panic, gripped areas fearing attack by Japan, including California and Australia. In response, coastal positions were hastily fortified, including the approaches to ports, notably Auckland, Sydney, Melbourne and Perth. Japanese-Americans were detained and expelled from coastal areas.

Conquests of Japan's Axis ally Germany were also shown in *Schoolmates-Students on the Front Lines*, such as a conveyor belt in the Atlantic illustrating American supplies to Britain, but being challenged by German submarines in the Atlantic. A chart showed German air power as far stronger than that of Britain and another depicted Italian pressure on the Allies in Egypt.

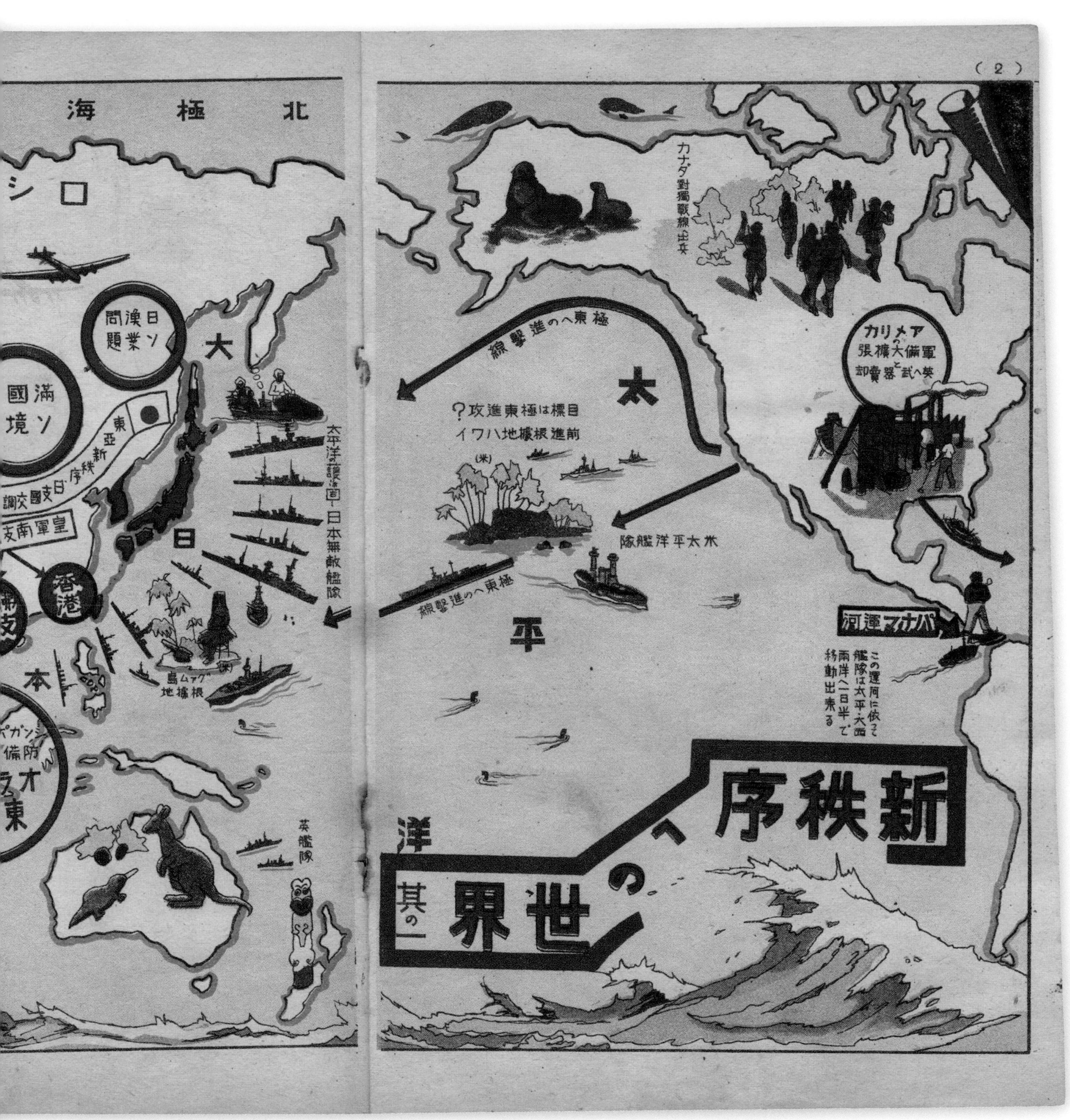

(2)
北極海
日ソ漁業問題
満ソ國境
香港
大
日
本
極東への進擊線
目標は極東進攻?
前進根據地ハワイ
米太平洋艦隊
極東への進擊線
グアム島根據地
(米)
英艦隊
カナダ對獨戰線出兵
パナマ運河
この運河に依って艦隊は太平・大西兩洋へ一日半で移動出來る
太
平
洋
新秩序への世界
其の一

BRITAIN'S VAST AIR OFFENSIVE—OVER 700 R.A.F. ATTACKS ON GERMANY.

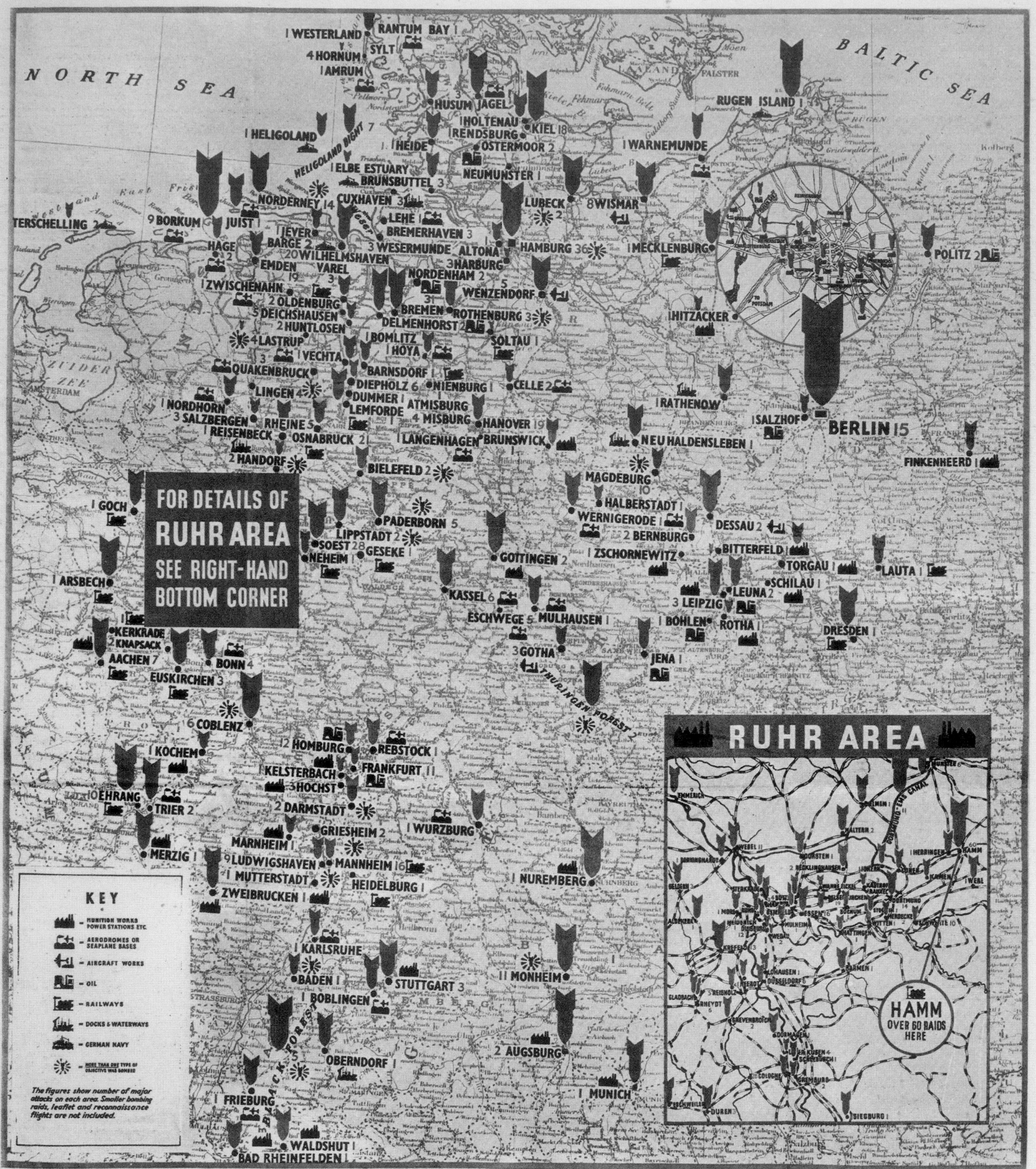

THE R.A.F. ONSLAUGHT ON GERMANY: A COMPREHENSIVE MAP SHOWING THE VAST EXTENT OF BRITAIN'S BOMBING OFFENSIVE SINCE THE OUTBREAK OF WAR, EMBRACING ARMAMENT WORKS, AERODROMES, DOCKS AND NAVAL BASES, OIL DEPOTS AND REFINERIES, GOODS YARDS, RAILWAY JUNCTIONS, BARGES AND SHIPPING.

BOMBING ONSLAUGHT

Ministry of Information, 'Britain's Vast Air Offensive—Over 700 R.A.F. Attacks on Germany', 12 October 1940.

At the time of the heavy German air assault on Britain in the Blitz, the British Air Ministry and the Ministry of Information provided information, including maps. This one was published in *The Illustrated London News* of 12 October and drew attention to British air attacks on Germany: Bomber Command had mounted 3,239 sorties in September. 'Britain's Vast Air Offensive' emphasised the range of economic and military targets covered and how much of Germany was being bombed, but no information was presented about the damage caused. The Ruhr, Germany's industrial heartland, was given a particular focus in the map alongside the claim that every town of importance there had been bombarded. The north Rhine city of Hamm, the location of huge railway marshalling yards for the Ruhr and Cologne areas, was marked as a particularly heavy target and had suffered more than 60 raids between May and December 1940, but the damage was not lasting.

As a result of heavy aircraft casualties in daylight bombing in 1939, the RAF had bombed unescorted and at night, but that meant that precision was not possible. As a result, it shifted in December 1940 to area bombing. The bombers used – the Whitley, the Hampden and the more impressive Wellington – did not match the specifications of the Lancaster. The latter became operational in March 1942 and extended the range and increased the intensity of bombing. Earlier, the small bombers, such as the Hampden I and the Blenheim, were not fast enough and had poor defensive capabilities. Indeed, the concept of the small bomber was not fully thought out. These aircraft were withdrawn from bombing in 1942 and 1943 respectively.

The RAF attacks served a strategy of long-term economic warfare. They also served to assure British public opinion, then suffering from the Blitz, that the country was striking back. Bombing also assumed more importance because British forces were unable to compete with the Germans in continental Europe. The response of Britain, at that stage the operationally weaker power, was to seek strategic advantage from indirect attack, in the shape of bombing, blockade and subversion, each of which was designed to hit Germany's economy and national morale. Demonstrating German vulnerability was a currency that had meaning in terms of British public opinion and pre-war doctrine about air power contributed greatly to the strategy. The belief was that serious damage could be inflicted on the German war economy, which would affect German resilience. Moreover, this was a strategy that made sense of Britain's power and position: there was an existing bomber force, so that strategy could be fitted to existing force structure, and the country was within bombing distance of major German ports and industrial sites, notably the Ruhr. In a blow to German prestige, a British bombing raid meant that Vyacheslav Molotov, the Soviet Foreign Minister, had to take shelter on 13 November 1940 when on a visit to Berlin.

The British strategy was pursued despite the existence of many problems, including the limited accuracy of bombing, the deficiencies of the bombers initially available, the lack of long-range fighter escorts and the extent to which the use of aircraft and crew for bombing ensured that fewer of both were available for anti-submarine patrols and ground support for the army. A British strategic review of August 1941 noted:

> *'Bombing on a vast scale is the weapon upon which we principally depend for the destruction of German economic life and morale. To achieve its object the bombing offensive must be on the heaviest possible scale and we aim at a force limited in size only by operational difficulties in the UK. After meeting the needs of our own security we give to the heavy bomber first priority in production. Our policy at present is to concentrate upon targets which affect both the German transportation system and morale.'*

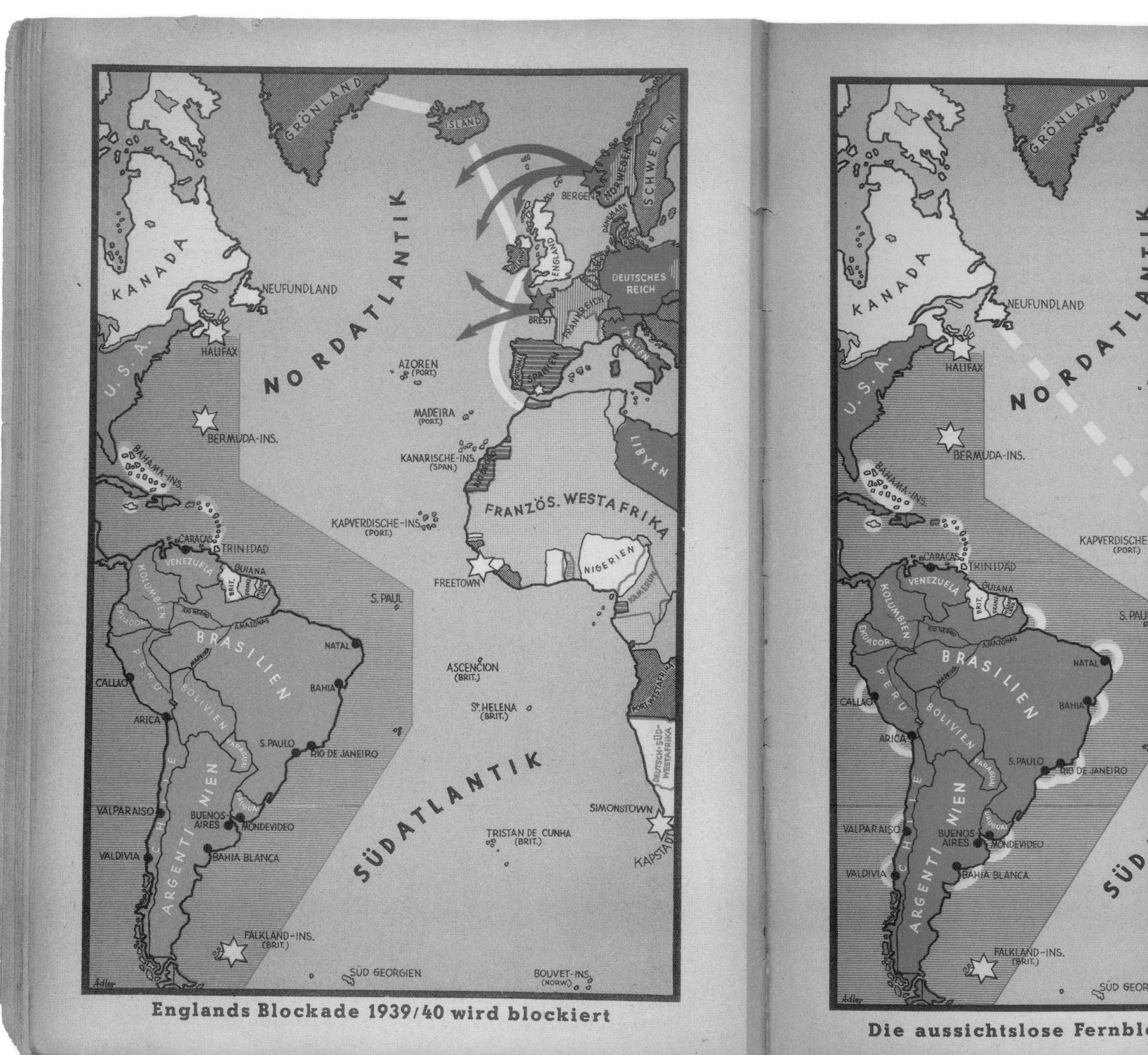
Englands Blockade 1939/40 wird blockiert
Die aussichtslose Fernblo

GLOBAL GERMANY

Ernst Adler, maps of the increased German presence in the Atlantic, from *Der Krieg 1939/40 in Karten*, 1940.

Blockade, according to *Der Krieg 1939/41 in Karten*, was a major aspect of British strategy: 'Inglorious hunger blockades, and cowardly assaults on weaker opponents, are the traditional war policy of Britain.' At the outset, the British government was sceptical about Germany's capability to sustain a long war and was confident that, as in the First World War, the Allied forces in France would be able to resist attack, a confidence that was a necessary corollary of any alliance with France that required the dispatch there of a British army. Hoping to intimidate Hitler by a limited war, especially with the damage that would result from a blockade, British strategy relied on forcing him to negotiate, or on exerting pressure that would lead to his overthrow.

The theme of the first map, 'The Blocking of the English Blockade', is the greater range of German naval power (rather than the British blockade) now that Atlantic waters can be shown to be vulnerable to a Kriegsmarine profiting from the new naval bases of Bergen in Norway and Brest in France, both of which are defended against British air attack. The map also reveals the strategic significance of the Portuguese and Spanish islands in the Atlantic.

In the second map, 'The Hopeless Long-Distance Blockade of Europe', the Germans express a belief that they have broken Britain's blockade of Europe and will soon occupy Britain itself, whereupon the belief that the war and a long-distance blockade can be continued from Canada will be exposed as a fantasy: 'The British bases in the Atlantic have been used thus far exclusively for smaller ships [than battleships]. They would be insufficient even if a floating dock or two were established in South Africa and Canada.'

Unwilling to focus on submarine warfare, the Kriegsmarine, building on the attitudes and policies developed prior to the First World War, wanted Germany to become a power with a global reach provided by a strong surface navy. As Britain had become a more obvious opponent, so plans for the size of the Kriegsmarine became more ambitious, with plans in 1938 and 1940 replacing the more limited one of 1934, which had been directed essentially against France. The Kriegsmarine also sought Atlantic bases from which it would be possible to threaten the convoy routes that brought Britain crucial supplies, as well as to increase German influence in South America and to challenge US power in the continent.

Hitler, who on 26 September 1940 declared his interest in the Canary Islands, certainly hankered after global domination, and wanted Germany to regain the African colonies it had lost to Britain, France and their allies. Those on the Atlantic – Togo, 'Kamerun' and 'Deutsch Süd Westafrika' (Namibia) – were marked on the map within red lines, serving as a reminder of Germany's imperial past.

This was an aspect of Hitler's general determination to reverse the verdict of that world war, but regaining these colonies was very much tangential to his central concern with reversing the Versailles peace settlement as far as Europe was concerned and, moreover, with creating a new Europe. Separately, the Kriegsmarine was not central to his concerns, while its interest in the establishment of bases around the Atlantic and in the Indian Ocean was scarcely credible.

In practice, the sinking of the *Bismarck* on 27 May 1941 was a demonstration of German weakness in surface shipping, not least because the British deployment included two battleships, two carriers and 15 cruisers. Hitler then ordered Germany's other surface raiders to be concentrated in Norwegian waters. The withdrawal, on 12 February 1942, of major warships from Brest, the leading Atlantic base in German hands, lessened the threat posed by these warships to the Allied position in the Atlantic, and thus their potential danger as a fleet-in-being. The following month, a British attack wrecked the dry dock at St Nazaire, the only dry dock on France's Atlantic coast big enough to accommodate the *Tirpitz*. The establishment of American bases in Greenland and Iceland was significant, not least in terms of providing air cover from the latter against German submarines.

SPONTANEOUS HELP IN BRITAIN'S
AEROPLANE SYMBOLS INDICATE CASH GIFTS THUS:–
=£500,000 =£50,000 =£10,000
Empire countries themselv
aircraft and/or operating
Schemes are shewn thus
British Communitie
CANADA
U.S.A. ★
NEWFOUNDLAND
EUROPE ★
MADEIRA
GIBRALTAR
MALTA
EGYPT ★
CYPRUS PALESTINE & TRANSJORDAN
MIDDLE EAST ★
INDIA, BURMA, HONGKONG, CHINA ★ THAILAND ★
CUBA & HAITI ★
BERMUDA
Br. HONDURAS
CURAÇAO AND ARUBA ★
BAHAMAS, JAMAICA WINDWARD & LEEWARD Is. TRINIDAD
GAMBIA, SIERRA LEONE
NIGERIA
SUDAN
ADEN
KOWEIT BAHREIN OMAN
CEYLON
DUTCH IN SINGAPORE
GOLD COAST
KENYA, UGANDA, TANGANYIKA, NYASALAND, N. RHODESIA, ZANZIBAR, ETC.
Br. GUIANA
BELGIAN EMPIRE
S. RHODESIA
BRITISH MALAYA
CENTRAL & S. AMERICA ★
OTHER PARTS OF AFRICA
BASUTOLAND, BECHUANALAND, SWAZILAND
MAURITIUS
UNION OF S. AFRICA
AU
FALKLAND Is.
DIRECTOR OF PUBLICITY, PRIME MINISTER'S DEPARTMENT, WELLINGTON, NEW ZEALAND.
DECEMBER Is
PRINTED FOR H.M.STATIONERY OFFICE BY CHROMOWORKS LTD. WILLESDEN, LONDON. 51-2213.

COMMONWEALTH CAMPAIGN

Chromoworks, Ltd., 'Spontaneous Help in Britain's Air War', 1 December 1941.

This 1941 map shows the financial support ('cash gifts' using aircraft symbols) for Britain's air war from around the British Empire and Commonwealth, as well as that from well-wishers and expatriates, notably in South America and the United States, and from overseas Dutch and Belgian support. The image is of global effort and power converging on Europe. At that point, 1 December 1941, Japan had not joined in the war against Britain. Australia, Canada, New Zealand, India, South Africa and Southern Rhodesia (now Zimbabwe) are all identified (by using the RAF roundel) as 'Empire countries themselves manufacturing aircraft and/or operating Air Training Schemes' (the scheme in Canada being particularly important). There was also a tremendous contribution of manpower that is not shown in this map: for example, the Royal Canadian Air Force sent 94,000 men abroad, many of whom served in the RAF. In 1941, 23,676 aircrew trained under the British Empire Air Training Scheme, 16,653 of them in Canada, 3,344 in Australia, 1,417 in Southern Rhodesia, 1,292 in New Zealand and 970 in South Africa.

The map underplayed the developing tension within the Commonwealth over the growing threat from Japan, a threat that led Australia and New Zealand in particular to press hard for the movement of their military units back from the Middle East, a measure that was resisted by Churchill so that some units remained to fight at El Alamein. Nevertheless, the theme of the Commonwealth rallying round was correct, and ensured that subsequent statements that 'Britain Fought Alone' need to understand Britain as meaning the British Empire and Commonwealth. This was true not only in the air, but also at sea and on land. Australian and New Zealand warships played a significant role against both Germany and Japan, but the most important imperial ally at sea was Canada, which by the end of the war had the world's third largest navy. With Halifax, Nova Scotia, as the key base, Canadian warships played a fundamental role in anti-submarine operations in the North Atlantic.

Canadian units were also important to the defence of Britain against threatened invasion in 1940. The Canadians also occupied Iceland in 1940, attacked Dieppe in 1942, took one of the five D-Day beaches in Normandy and played a major role in the subsequent advance through France and the Low Countries into Germany.

Australian, New Zealand, Indian and African troops played a central role in the conquest of Italian East Africa and North Africa, of Vichy Syria, Lebanon, Madagascar and Iraq, and were crucial in the struggles with Japan and Italy. The contribution by India was the most significant numerically and made a substantial difference, especially to the war on land. There was no conscription in India, but about two million men served in what became the largest volunteer army of the war. This turnout was despite the anti-British 'Quit India' campaign of many nationalists in 1942, which resulted in around 40,000 nationalists led by Chandra Bose fighting for Japan in the Indian National Army. Loyal Indian forces not only contained disaffection on the Home Front but also provided effective forces able to fight in a variety of terrains. In their recruitment in India, the British showed considerable flexibility, both promoting many Indians within the army and also recruiting from what had been regarded as non-martial races.

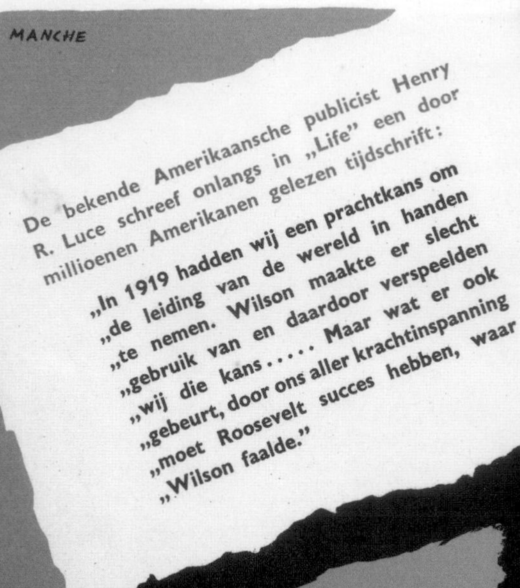

De Dollarp

ALASKA
1876
1898
1848
CALIFORNIË
Nw. MEXICO
TEXAS
HAWAII
CUBA
PORTORICO
1898
1941
1942
1898
PHILIPPIJNEN
1941

Voor de Amerikanen is het Volkenrecht het recht van de wildernis. Amerika voor de Amerikanen beteekent, dat ieder land waar de Yankees hun vlag willen planten, plotseling Amerika is en onder de Monroeleer valt. De geschiedenis van Amerika is één imperialistische rooftocht, waarin met dollars inplaats van kogels werd geschoten! De traditioneele politiek van Amerika is, door middel van het uitbuiten der Europeesche oorlogen, zich te verrijken met de Europeesche belangen in den Atlantischen en in den Grooten Oceaan. Zoo dacht ook Roosevelt dezen oorlog uit te buiten ten bate van de Amerikaansche wereldheerschappij! Reeds heeft het stervende Engeland zijn offer gebracht.

Maar voor het eerst heeft Amerika verkeerd gereken
omsingelen thans Amerika in de twee Oceanen. De P
bij Pearl Harbour. De vloten van den Atlantische
koopvaardijschepen vallen ten offer aan den doodelijk
De vangarmen van de Dollarpoliep worden afgesned
nationale plutocratie, geconcentreerd in fort Knox,
door de onweerstaanbare legers der jonge volken, door

Goedgekeurd door de Afd. Propaganda van het Departement van Volksvoorlichting

AN ENEMY EVERYWHERE

Louis Mache, 'The Dollar Octopus', 1942; SPK, Vichy poster, c.1942; and Pat Keely, 'The Indies Must Be Free', 1944.

The image of the octopus was used for propaganda purposes by both Allied and Axis forces. The motif was a longstanding one – it had been directed against then expansionist Russia in Fred Rose's 'Serio-Comic War map for the year 1877' as well as in a French propaganda map of 1917 'La Guerre est L'Industrio Nazionale de la Prusse'.

The propaganda poster 'Confiance: Ses Amputations Se Poursivent Méthodiquement' ('Be Assured: The Amputations are Proceeding Methodically'; see overleaf, page 214) is signed by the artist 'SPK' who created other Vichy collaborationist propaganda but about whom nothing is known. If not created for Vichy it was the work of the Propaganda-Abteilung Frankreich, the German propaganda unit in occupied France. It plays upon old Anglo-French colonial rivalries and fears. Winston Churchill is an octopus and his tentacles represent British military interventions in 1940–41 in the Middle East and Africa, including a totally unsuccessful attempt to seize Dakar, the major port in French West Africa, and the attack on French warships at Mers El-Kébir in Algeria that left many Vichy servicemen dead. However, the invasion of the French colonies of Lebanon and Syria was a complete success. The poster's cut tentacles misleadingly suggest that the British have been thwarted everywhere. In East Africa, another 'amputation' – the Italian occupation of British Somaliland in August 1940 – was followed, in February 1941 by the conquest of Italian Somaliland and then, in March, by the reconquest of British Somaliland. Vichy had a powerful fleet and was essentially a Mediterranean empire, controlling southern France as well as French North Africa, Lebanon and Syria – and, further afield, colonies in West Africa, Southeast Asia and Madagascar. Attacks on Vichy forces in 1940 made it easier for Vichy to co-operate with Germany and compromised support for its opponent, Charles de Gaulle, the leader of the Free French.

'De Dollarpoliep' ('The Dollar Octopus'), left, is by the Dutch Nazi painter Louis Emile Manche and was published by the German propaganda agency in occupied Holland, 1942. His brother died fighting the Soviet forces. Accused by Manche of seeking world domination, the USA has 11 tentacles (referring to years when the USA made a territorial acquisition) and is presented as encircled – under threat from German U-boats in the Atlantic and Japanese aircraft and submarines in the Pacific – with its tentacles beginning to be severed. The occupation of the Dutch colonies of Curaçao and Surinam – important sources of oil and bauxite for industry – was a response to the German conquest of the Netherlands: with Dutch consent, British troops moved in in 1940 and were replaced by the Americans in February 1942. Bases were established in Bermuda and Newfoundland by agreement with the British. British forces had occupied Iceland, a Danish colony, after the German conquest of Denmark in 1940, and the Americans subsequently replaced the British forces. Manche's panel at the left includes the accusation:

> *'The history of America is of imperialist plunder, where dollars were fired instead of bullets! It is the traditional policy of America, through the exploitation of European wars, to enrich themselves with the European possessions in the Atlantic and in the Pacific Ocean. Roosevelt also plans to use this war to benefit American world domination! [America] is seizing what the dying England has sacrificed!'*

Patrick Cokayne Keely was an English illustrator who created posters for the Ministry of Information and also produced several for the London-based Dutch government-in-exile, including this 1944 lithograph (overleaf, page 215) with a menacing Japanese octopus: 'Indie Moet Vrij! Werkt en Vecht Ervoor' ('The Indies Must Be Free! Work and fight for it!'). With a clear and bold use of imagery, colour and wording, the call to drive Japan from the Dutch colony was ably presented but it had little impact in the Dutch East Indies where, alongside the brutality of Japanese occupation, nationalist independence sentiment was growing.

CONFIANCE...

NORVEGE

ALLEMAGNE

MERS EL-KEBIR

LYBIE-EGYPTE

SYRIE

DAKAR

SOMALIE

SES AMPUTATIONS SE POURSUIVENT MÉTHODIQUEMENT

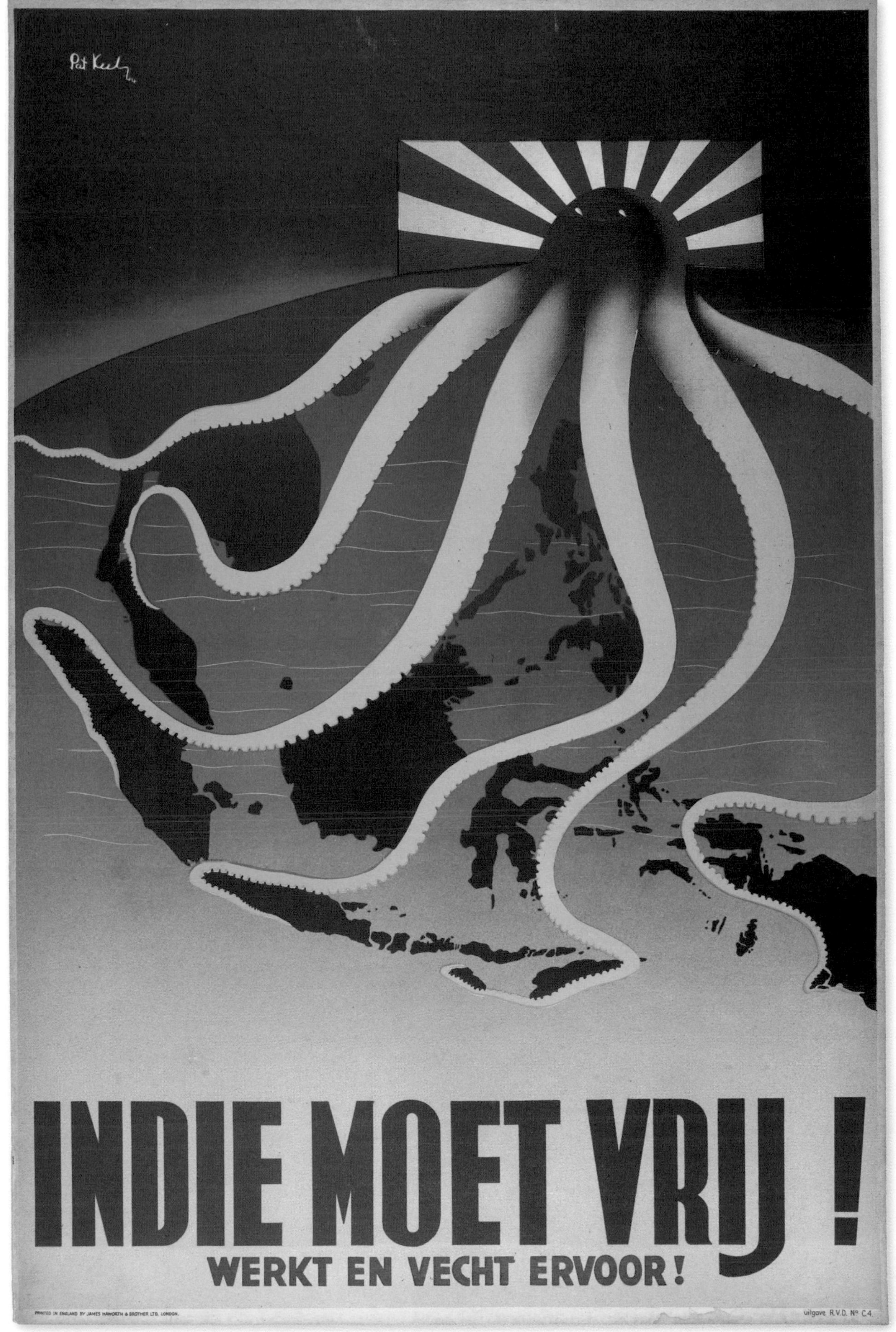
Pat Keely
INDIE MOET VRIJ !
WERKT EN VECHT ERVOOR!
PRINTED IN ENGLAND BY JAMES HAWORTH & BROTHER LTD. LONDON.
uitgave R.V.D. Nº C.4.

ULM DIVISION IN THE EAST

Fold-out pictorial map illustrating the war at German divisional level, 1942.

This large map is from Mit *Ums im Osten: Eine Bildfolge vom Einsatz der Ulmer Infanterie-Division* ('With Us in the East: A Sequence of Images from the Ulm Infantry Division'), a book, published in Stuttgart in 1942, about the successful advance toward Moscow of the 5th Infantry Division in 1941. The book was edited by Rittmeister (a cavalry rank) Wolfgang Köstlin and is intended to be a morale-boosting depiction of inexorable German strength and success (the tanks and planes are all German, except for the wreckage of a crashed Soviet aircraft west of Minsk), with Soviet units surrounded and defeated. Advancing German units look organised; the Soviets appear chaotic.

Wehrmachtpropaganda, the Wehrmacht's Propaganda Division, approved the map, and the compass rose is decorated with the division's symbol of Ulm Cathedral (pointing north) atop a seagull's wings. The introduction reveals that the book, which chronicles the battlefield as well as the people encountered in the east, is to serve as a permanent reminder of the sacrifices made by the men of the division.

After initial gains, advancing several hundred kilometres, the Germans were driven back in the Battle of Moscow in a major Soviet counter-attack launched on 5–6 December 1941. Although the presentation in the map of Soviet prisoners being marched south is misleading, because many were harshly treated from the outset, the part played by horse transport in the German advance is more accurate – a theme otherwise underplayed in propaganda at the time, which focused on the role of mechanisation in German warmaking.

The map is an illustrated timeline of the division's actions in German Army Group Centre: fighting along the Bialystok–Minsk front in June 1941, then in the battle for Smolensk in July and August. At both Bialystok and Smolensk pockets of Soviet units with hundreds of thousands of men were encircled, with the infantry support of the armour provided by the likes of the Ulm Infantry Division being what helped to make the encirclement work. A counter-attack mounted by the tanks of the Soviet Red Army's Western Special Military District proved to be a total failure.

Subsequently, thanks to the convergence of tanks as shown in the map, advancing to the northeast from the southwest with the existing move eastward of Army Group Centre, as part of Operation Typhoon (the drive to capture Moscow, launched on 30 September), another pocket was created at

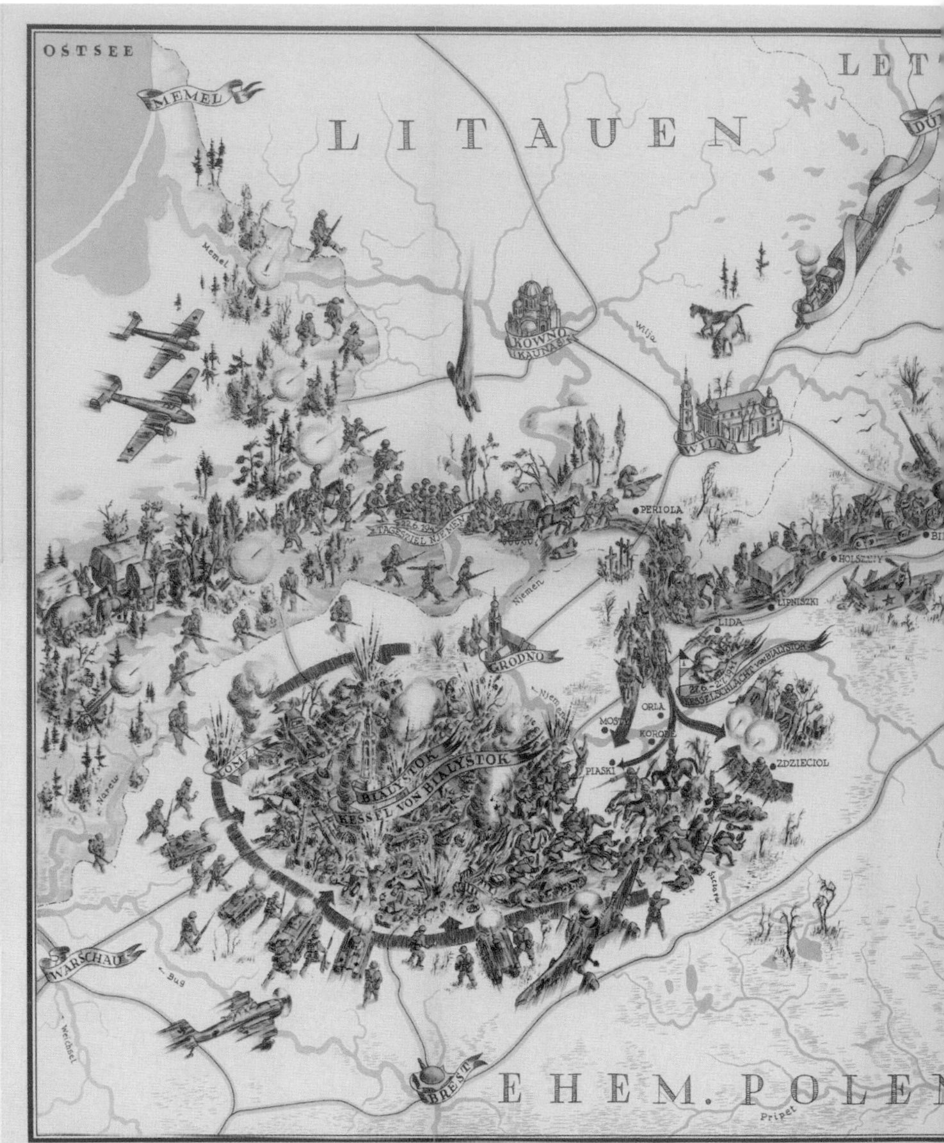

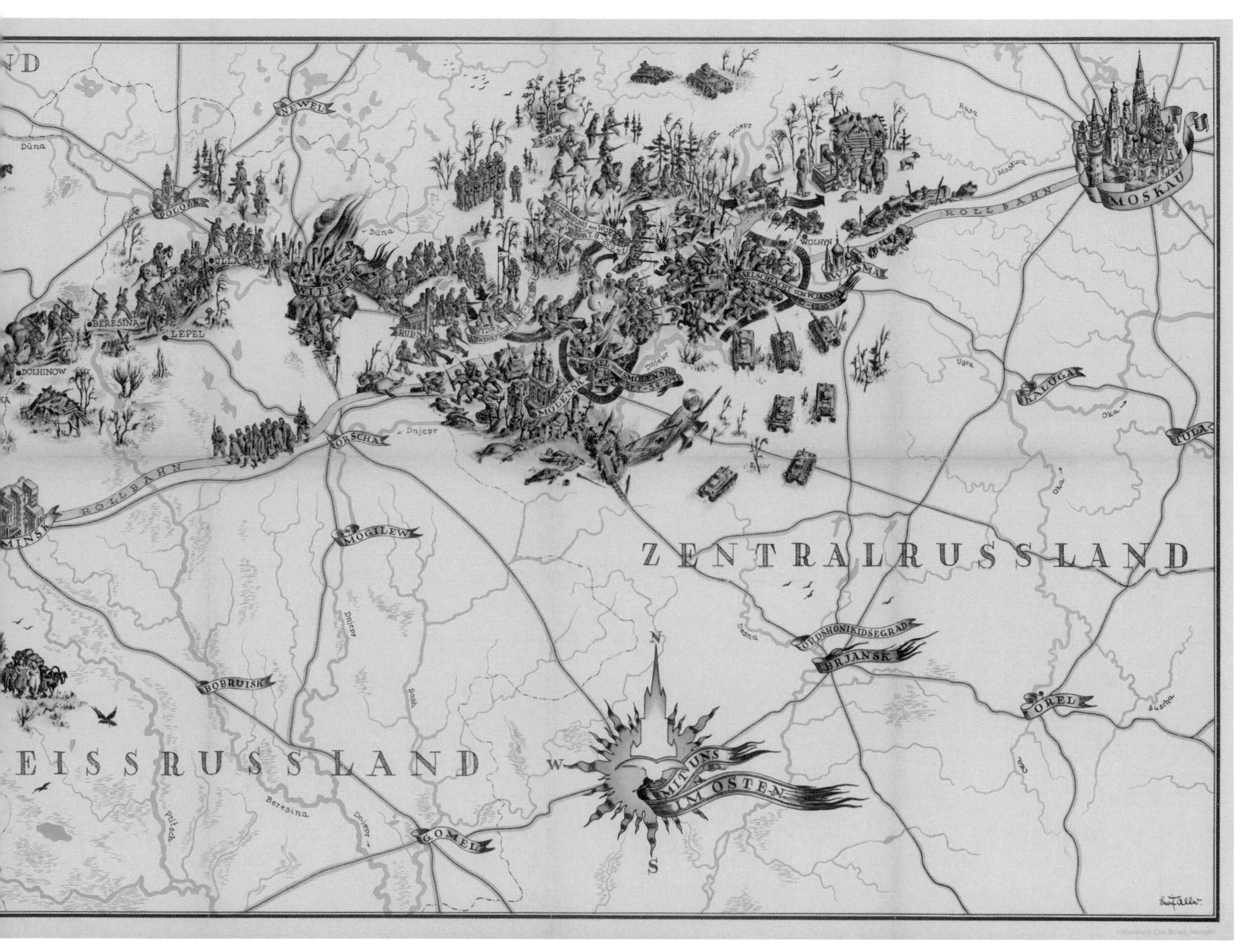

Viazma ('Wjasma' in the map). The map shows the Ulm Infantry Division engaged in the final drive for Viazma (2–10 December 1941, the last date provided), when at about the same time some panzer units had reached within 12 miles (19 kilometres) of Moscow. These successive victories represented the high water mark of the division's activities, and essentially the high point of the German invasion. The invasion had also put a heavy burden on the advancing German forces. Although the Soviet units were suffering greater losses, German units were wearing out – experiencing reductions of 30–40 per cent of their original strength – and winter was closing in, with the rain, the mud, the cold and poor roads. Troops transported in from the Soviet Far East in November alone had grown the Soviet reserve by 1.8 million men, most of whom were committed to the defence of Moscow. By the time this map was being given official approval, the grim reality of Operation Typhoon and the prospect of a sustained struggle would have been known.

ANNIHILATION INTENSIFIED

Ministry of Information, 'Das War Hamburg' ('That Was Hamburg'), July 1943.

This propaganda leaflet (what the Air Ministry referred to as a 'white bomb') was dropped by the RAF following the British air assault on the city over four days in 1943, and it reads: 'That Was Hamburg' – 'Part of an aerial view after the RAF attacks' – 'The red-framed image section is shown enlarged on the reverse'. The other side of the leaflet has a close up of the area highlighted on the front, next to a description of the destruction: 'The buildings are either completely destroyed or burned so that nothing but empty walls remain...' The text on the back declares that the war is lost, the Luftwaffe is helpless, every industrial town is threatened with the fate of Hamburg and that Hitler 'fights to gain time – time for the destruction of Germany'. On the left of the front of the leaflet are paraphrased extracts from a speech that Winston Churchill gave on 14 July 1941 to honour the British civil defence forces' response nationwide to the Blitz, a speech that has become known as 'You Do Your Worst and We Will Do Our Best'. The leaflet reads: 'We will now bomb Germany on an ever larger scale, month after month, year after year, until the Nazi regime has either been exterminated by us, or – better still – until the German people themselves have put an end to it.' In the actual speech Churchill had declared:

> *'We ask no favours of the enemy. We seek from them no compunction. On the contrary, if tonight the people of London were asked to cast their vote whether a convention should be entered into to stop the bombing of all cities, the overwhelming majority would cry, "No, we will mete out to the*

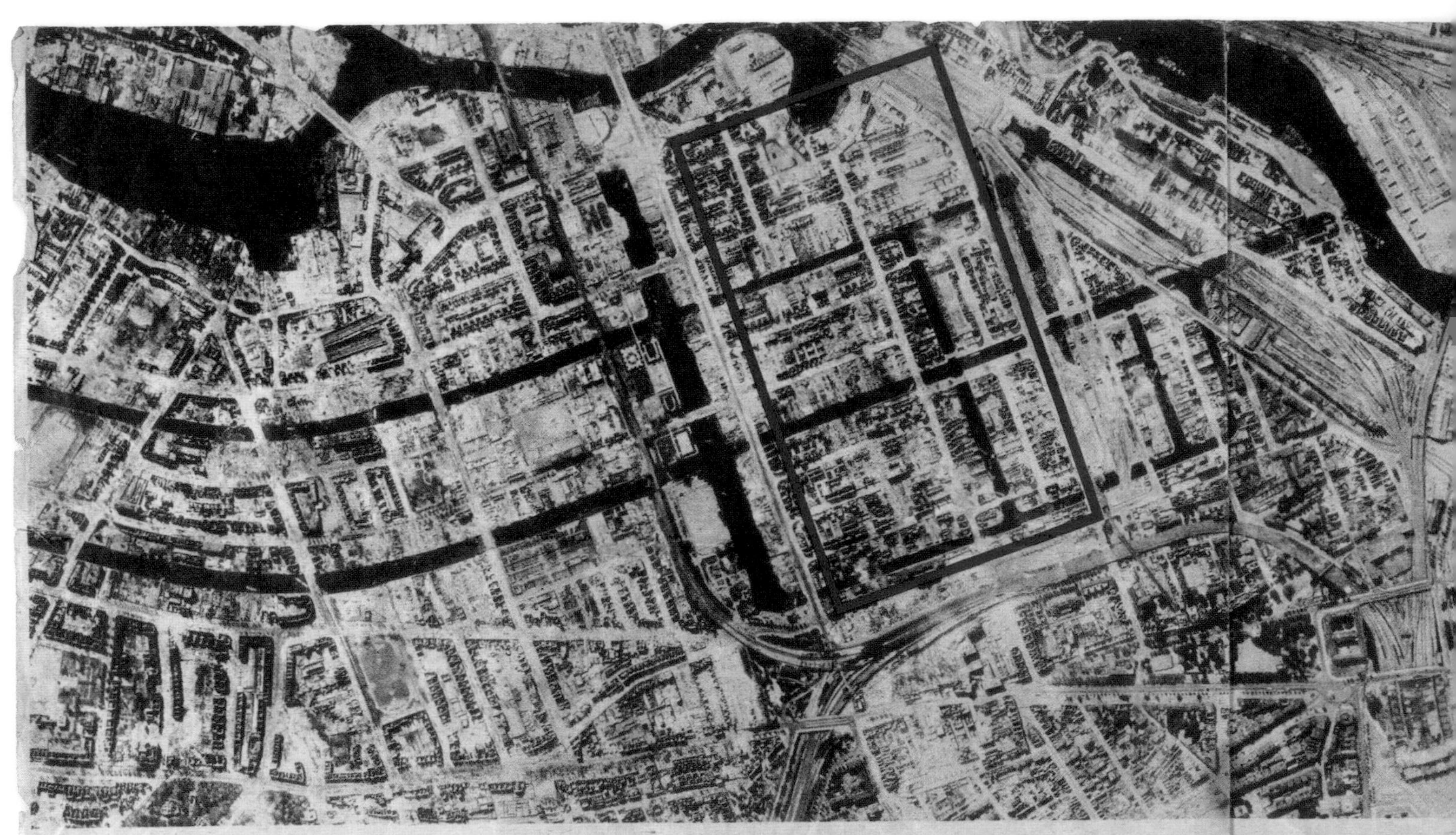

„Wir werden von nun an Deutschland in immer grösserem Masstab mit Bomben belegen, Monat für Monat, Jahr für Jahr, bis das Naziregime entweder von uns ausgerottet ist, oder — besser noch — bis ihm das deutsche Volk selbst den Garaus macht."
Churchill, 14. Juli 1941.
DAS
HAM

Germans the measure, and more than the measure, that they have meted out to us." The people of London with one voice would say to Hitler: "You have committed every crime under the sun. Where you have been the least resisted there you have been the most brutal. It was you who began the indiscriminate bombing. We remember Warsaw in the very first few days of the war. We remember Rotterdam." ... Perhaps it may be our turn soon; perhaps it may be our turn now.' ...

'We have now intensified for a month past our systematic, scientific, methodical bombing on a large scale of the German cities, seaports, industries, and other military objectives. We believe it to be in our power to keep this process going, on a steadily rising tide, month after month, year after year, until the Nazi regime is either extirpated by us or, better still, torn to pieces by the German people themselves.'

The 1943 raids on Hamburg killed up to 40,000 people, notably on 28 July as the result of a firestorm created by a combination of incendiary and high explosive bombs. The impact was horrifying: those killed were either suffocated or burned to death. The raid, however, badly affected German morale; it led to the partial evacuation of cities, including Hamburg and Berlin, and helped to bring about a marked increase in criticism of the Nazi regime.

Goebbels recorded his alarm in his diary and the regime's concern about civilian morale helped encourage an emphasis on home defence. The needs of air defence ensured that around 60 per cent of German military production in 1943 was devoted to aircraft and anti-aircraft guns, while Luftwaffe strength was increasingly concentrated in Germany to oppose the Anglo-American Combined Bomber Offensive, which lessened the availability of aircraft for ground support of German forces on the front lines. Much of the Luftwaffe continued to be employed on the Eastern Front, and aircraft were available to contest Anglo-American forces in Italy, but fewer than would otherwise have been the case.

In response to the Hamburg raid the Germans developed new night-fighting methods and technology (including adapting bombers to act as night fighters) that caused British losses to mount from the late summer.

AR
RG
Teil einer Luftaufnahme nach den Angriffen der RAF Der rotumrandete Bildausschnitt ist umseitig in Vergrösserung wiedergegeben.

BRITAIN-SPEARHEAD OF AT
SUPPLIES TO RUSSIA
TO LOFOTEN AND VAAGSÖ
FOOD
SHIPS
GUNS
SHIPS
SHIPS
ELECTRIC POWER
TANKS
AIRCRAFT
COAL
LONDON
FOOD
SUPPLIES AND TROOPS TO AFRICA, MIDDLE EAST, INDIA, PERSIA, AUSTRALIA, ETC., ETC.
BREST
LORIENT
ST. NAZAIRE
BRUNEVAL
Gasworks
Troops
Airfields
Railways
Railways
DIEPPE
Troops
BOULOGNE
Power Stn.
Factories
Airfields
Bridges
Renault Wks
Matford Wks
PARIS
Schneider-Creusot Wks.
TURIN
MILAN
GENOA
EMDEN
BREMEN
HAMBURG
KIEL
LÜBECK
ROSTOCK
OSNABRÜCK
MÜNSTER
HAMM
GELSENKIRCHEN
DORTMUND
ESSEN
DUISBURG
KREFELD
DÜSSELDORF
COLOGNE
HANOVER
KASSEL
LUDWIGSHAFEN
MANNHEIM
SAARBRÜCKEN
KARLSRUHE
BOMBER COMMAND
FIGHTER COMMAND
Night and day the bombers of the R.A.F. strike deep into the enemy's war machine . . . Almost daily R.A.F. fighters sweep over enemy occupied territory to strafe communications and strategic targets . . . Coastal Command aircraft strike at U-boats and
F. DONALD BLAKE

PRODUCTION AND DESTRUCTION

Frederick Donald Blake, 'Britain – Spearhead of Attack', 1943.

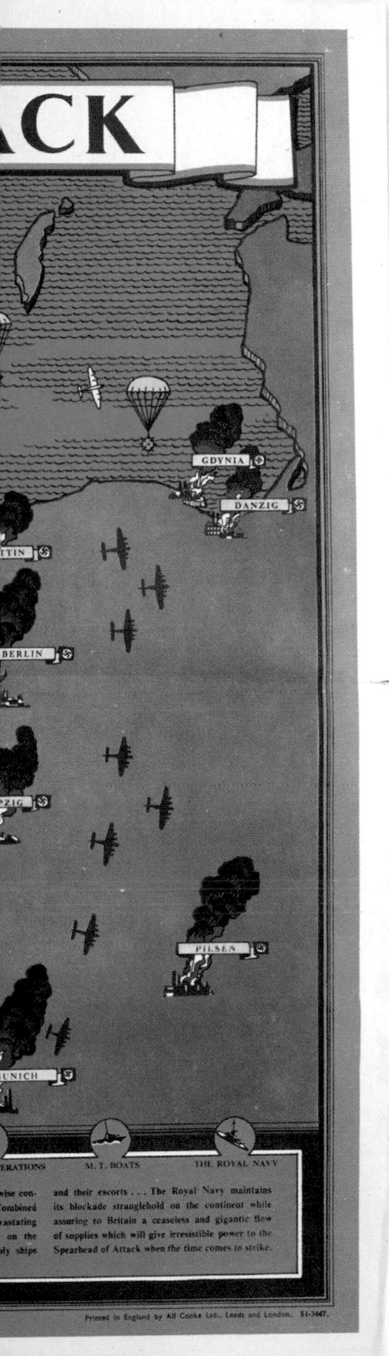

This poster, by the British artist Frederick Donald Blake, appeared in French, Arabic, Dutch, Greek and Portuguese language versions, just as his 'The Battle of the Atlantic' had. A similar poster, in the same five languages, covered the Mediterranean. Scotsman Blake produced work for the Ministry of Information as well as war maps for the *Daily Express*, and he served as an air raid fire officer.

The contrast in the poster is stark: whereas Britain has ample food, fuel, power and weaponry, continental Europe is a scene of destruction and Germany's industrial Ruhr is ablaze as 'Night and day the bombers of the RAF strike deep into the enemy's war machine'. Described in the legend as 'Combined Operations', the boats in the North Sea going to the Lofoten and Vaagsö islands refer to commando raids, in March and December 1941, that helped lead the Germans to station more troops in Norway than was necessary in strategic terms, which was a success for the British. Designed for specific purposes, these attacks also expressed British concern that the conquest of Norway gave Germany naval access to the Atlantic. Air-dropped mines are also shown descending by parachute, and ships head north carrying 'supplies to Russia'.

This propaganda was part of a wider campaign designed to demonstrate that Britain was active despite the fact that it had not launched any 'Second Front' invasion of Western Europe. American policymakers were strongly opposed to what they saw and decried as the Mediterranean obsession of British policy, notably the invasion of Italy. The Soviet Union and its supporters were even more suspicious, and Churchill had photos sent to Stalin of the bomb damage inflicted on the Krupp armaments works at Essen, which was accurately presented as reducing the pressure on the Soviet Union because the availability of more German air power would have been operationally useful to Germany on the Eastern Front, rather than in defending Germany against bombing. There was also concern about what impact a failure to mount a Second Front would have on domestic opinion in occupied Europe. Indeed, the Germans were pressing on with their fortifications and were inflicting heavy casualties on the European resistance movements.

THIS IS Ann.. ...she drinks blood!

Her full name is Anopheles Mosquito and she's dying to meet you!

Her trade is dishing out MALARIA! If you'll take a look at the map below you can see where she hangs out.

She can knock you flat so you're no good to your country, your outfit or yourself. You've got the dope, the nets and stuff to lick her if you will USE IT.

Use a little horse sense and you c lick Ann. Get sloppy a careless about her and she'll bat you down just as surely as a bom a bullet or a shell.

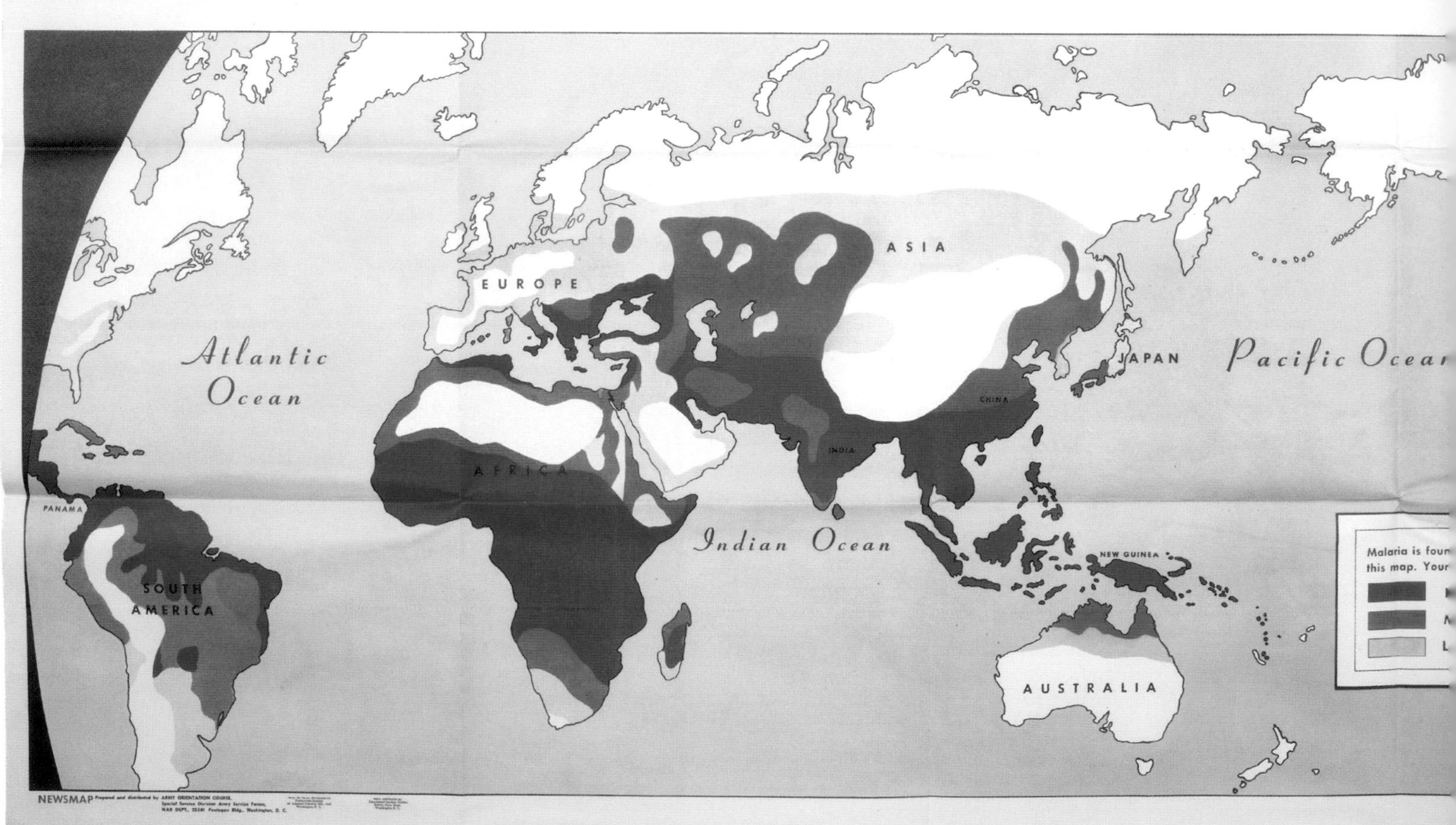

PERILS OF THE TROPICS

Theodor Geisel, 'This is Ann ... she drinks blood!', 8 November 1943.

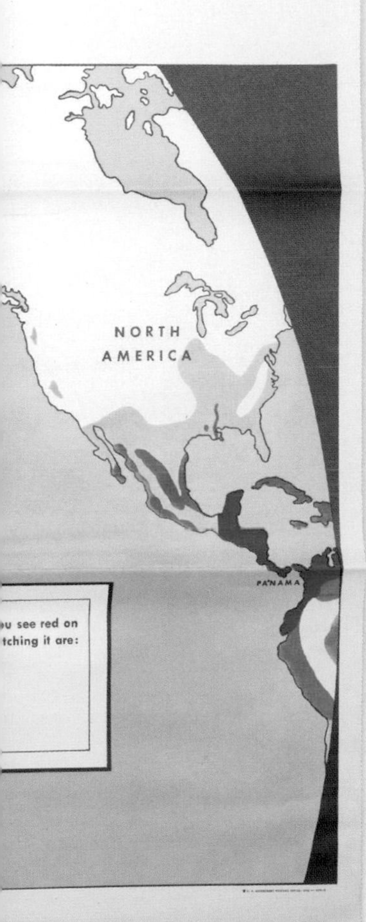

This illustration, from a newsmap, is about the threat from malaria, the primary carrier of which was the *Anopheles minimus flavirostris* species of mosquito, nicknamed 'Ann' by service personnel. The artist, Theodor Geisel, also wrote and illustrated dozens of children's books under the pen name Dr. Seuss. The artwork was designed for US forces and prepared and distributed by the Army Orientation Course Special Division Army Service Forces, a propaganda arm of the War Department based in the Pentagon, and was considered an important part of preparation for troops confronting environments that were totally new to them.

The overwhelming majority of Americans came from areas where the chance of catching malaria was non-existent or remote, and most US citizens had never been abroad. The 1898 Spanish–American War had driven home to the US Army the problems of operating in the tropics, in that case Cuba, Puerto Rico and the Philippines, and these had been underlined in the 1910s, 1920s and 1930s by the dispatch of US Marine Corps units to Haiti, the Dominican Republic and Nicaragua. The US Army had developed a considerable proficiency in preventative disease control (including carefully selecting base sites and using protective nets), and this had been seen in the effort to combat malaria and yellow fever in the building of the Panama Canal, finished in 1914.

In the Second World War, the situation was accentuated by conscription, which brought in recruits who were not necessarily fit and healthy enough for the tropics, or well enough informed. The poster was designed to help with that education campaign, with the emphasis placed on using the preventative measures and the kit provided.

In 1943, American forces were operating not only in areas where disease was readily understood as an issue, notably the Solomon Islands and New Guinea, but also in the Mediterranean where the risk was less well understood. This was true in particular of Tunisia and Sicily and then likewise of southern Italy. In 1942, there had been a singular failure in response, notably with the defence of the Philippines (when it was estimated 24,000 out of 75,000 defending troops were infected). In the first stage of the New Guinea campaign, from July 1942 to January 1943, the Americans and Australians suffered heavier losses to malaria than to combat with the Japanese, and the US 32nd Division was so badly affected that it was out of action for nearly a year.

From 1943 anti-malaria discipline improved, and as a result sickness rates fell. So also for the British and Australian forces fighting the Japanese. A report by General Stanley Savige on the operations of the 3rd Australian Division in the Salamanca area of New Guinea in late 1943 noted: '...the rules of hygiene and sanitation are anti-malarial precautions must be strictly observed at all times, no matter how hard the fighting, or how weary the troops.'

Similar health campaigns were waged in support of improving sexual hygiene, notably against venereal disease, which was also debilitating and again affected many troops.

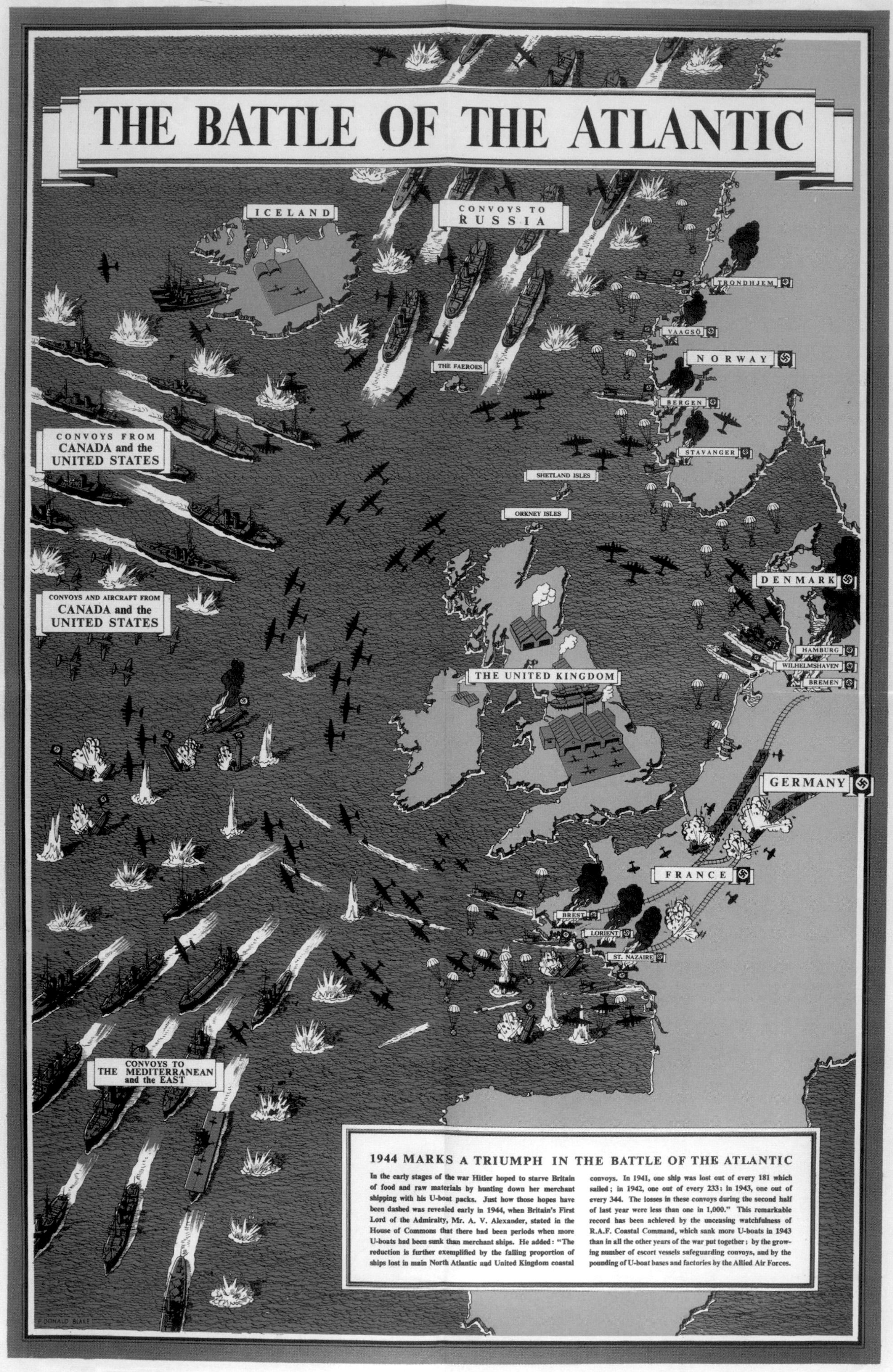

THE BATTLE OF THE ATLANTIC

ICELAND

CONVOYS TO
RUSSIA

TRONDHJEM

VAAGSÖ

THE FAEROES

NORWAY

BERGEN

STAVANGER

CONVOYS FROM
CANADA and the
UNITED STATES

SHETLAND ISLES

ORKNEY ISLES

CONVOYS AND AIRCRAFT FROM
CANADA and the
UNITED STATES

DENMARK

HAMBURG

WILHELMSHAVEN

BREMEN

THE UNITED KINGDOM

GERMANY

FRANCE

BREST

LORIENT

ST. NAZAIRE

CONVOYS TO
THE MEDITERRANEAN
and the EAST

1944 MARKS A TRIUMPH IN THE BATTLE OF THE ATLANTIC

In the early stages of the war Hitler hoped to starve Britain of food and raw materials by hunting down her merchant shipping with his U-boat packs. Just how those hopes have been dashed was revealed early in 1944, when Britain's First Lord of the Admiralty, Mr. A. V. Alexander, stated in the House of Commons that there had been periods when more U-boats had been sunk than merchant ships. He added: "The reduction is further exemplified by the falling proportion of ships lost in main North Atlantic and United Kingdom coastal convoys. In 1941, one ship was lost out of every 181 which sailed; in 1942, one out of every 233; in 1943, one out of every 344. The losses in these convoys during the second half of last year were less than one in 1,000." This remarkable record has been achieved by the unceasing watchfulness of R.A.F. Coastal Command, which sank more U-boats in 1943 than in all the other years of the war put together; by the growing number of escort vessels safeguarding convoys, and by the pounding of U-boat bases and factories by the Allied Air Forces.

CONVOYS SAFEGUARDED

Frederick Donald Blake, 'The Battle of the Atlantic', 1944.

British artist Frederick Donald Blake (1908–97) has here updated a 1943 version of this map poster, which was published for distribution in multiple languages. Blake presents industrial Britain, with its shipyards and factories, as an unsinkable island runway from which warplanes and servicemen radiate out both to protect transport and to search for the enemy, while warships and merchant ships continue to criss-cross the surrounding oceans. In the mid-Atlantic the U-boats are depicted in disarray. Meanwhile, infrastructure and military bases in continental Europe are ablaze or being bombarded.

The celebration of success in the Battle of the Atlantic reflected the ability of the Allies both to inflict more damage on the Germans and provide more shipping. The percentage of Allied ships sunk had fallen considerably in 1943, and continued to fall, while the ratio of ships sunk to U-boats destroyed also fell. That was the crucial year because the ratio was 2:1, compared to 14:1 in 1940–42. The number of submarines sunk per year was much greater than in the First World War, while the percentage of Allied shipping lost was lower. The Germans faced a strategic issue posed by Allied shipbuilding as well as tactical and operational challenges imposed by better Allied proficiency in anti-submarine warfare.

The success achieved by stronger convoy defences was combined in a more effective overall anti-submarine strategy, while the Germans proved less adept than the Allies in changing tactics and technology. The use of air bases in the Portuguese-controlled Azores closed the mid-Atlantic 'Air Gap' against the U-boats, notably what the Germans termed the 'Black Pit' west of the archipelago. In October 1943, the British were allowed by the Portuguese to establish the Lajes air base and in November 1944 the Americans followed with the Santa Maria base. This reflected both Allied pressure and the astute calculation of the shift in relative advantage by neutral Portugal.

Moreover, the availability of more Allied aircraft meant that submarines from their French bases travelled across the Bay of Biscay under the surface and thus used up some of their supply of electricity because the recharging of batteries had to be done on the surface. There was also an important shift in the relative quality of personnel; U-boat losses also cost the Germans experienced crew. The Allies had to prevail in the Battle of the Atlantic not only because it was crucial to the provision of imports to feed and fuel Britain but also because of the need to build-up military resources there. This was the background to the invasion of France, and it underlined the strategic quandary that was faced by Germany, which had an intractable conflict on the Eastern Front that was likely to be joined by fresh commitments in France.

ATTRITION IN ITALY

German propaganda leaflet and poster, 'Speaking of time-tables' and 'It's a Long Way to Rome', April 1944.

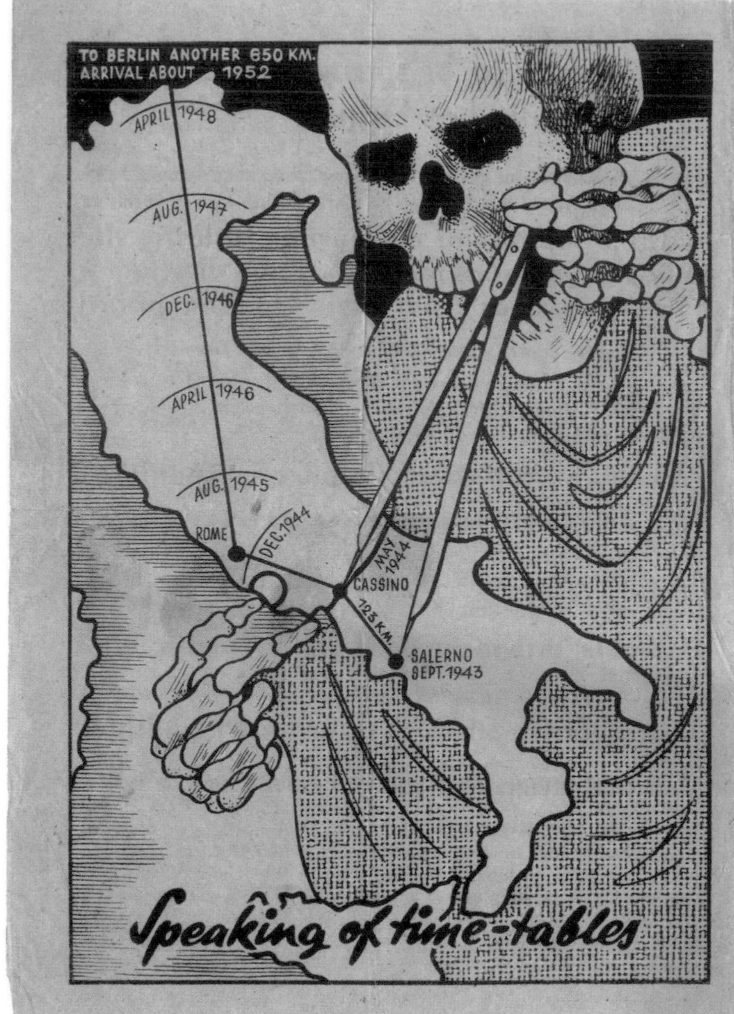

The invasion of mainland Europe long expected by the Germans came in early September 1943 when the Allied forces landed in southern Italy from Sicily. German propagandists seized upon the ability of the Germans to slow the Allied advance northward in Italy and inflict heavy casualties. This leaflet and poster from 1944 were aimed both at the population under German occupation elsewhere in Europe (the poster is known to have been put up in the streets of Paris and the style is recognisably French) and advancing Allied troops in Italy, which included French, Polish and Indian units. The poster claims that a garden snail travelling at 0.8 metres a minute for 191 days would have advanced 320 kilometres (200 miles) by 1 April 1944, whereas the Allies had achieved only 180 kilometres (112 miles). In fact, a snail could actually have covered 220 kilometres (138 miles), but the essential point was nevertheless correct. As slow as the advance up the peninsula had been to that point (hampered too by the weather), after Monte Cassino had finally fallen in mid-May 1944 the Allies had moved rapidly to liberate Rome by 5 June 1944 (well before December 1944 as forecast by the German 'time-tables' leaflet).

The Germans had proved able to build up their forces faster than the Allies, helped by rail and road systems which were not yet seriously disrupted by bombing. Indeed, their response to the overextended Salerno landing proved more determined than had been anticipated, not least in preventing the Allies from using Montecorvino airfield, which had been captured on landing day. It proved impossible to use the ports of Salerno and Vietri once they had been captured because the shallow beachhead was only consolidated and expanded with considerable difficulty. On 12 September, a German counter-attack tested the Allies hard, exhausting their reserves and leading both to the dropping of US airborne forces into the beachhead and to a heavy reliance on support from warships. Blocked, the Germans began to withdraw to north of Naples on 16 September. The Allies were left in control of southern Italy, but the Germans had established a strong line across the peninsula, making plentiful use of the mountainous terrain to create effective defensive positions, from which they were to fight ably and cause heavy casualties for the attacking Allies. Monte Cassino, a key feature in the well-defended Gustav Line and marked on the leaflet, proved the most difficult obstacle.

The pace of advance in Italy was slower than in France or Eastern Europe because the defensive density was higher, the front was narrow and the terrain more difficult. Allied forces took appreciable casualties without either strategic results or an attritional wearing down of Axis strength; and the German propaganda inevitably referred to these factors. However, Anzio led to the commitment of German reserves (including aircraft from Germany, France, Greece and northern Italy) that might otherwise have been sent elsewhere. Moreover, Allied bases in Italy helped to support partisan resistance activity in Yugoslavia, and were used to mount bombing missions against targets in southern Europe. In each case, the cost-benefit was better than that which would have arisen from an invasion. Italy-based American air attacks in April–August 1944 destroyed most of Romania's refineries, which in 1943 had produced 2,406,000 tons of petroleum products for Germany.

To Churchill, the Balkans presented an opportunity, not only to harry the Germans but

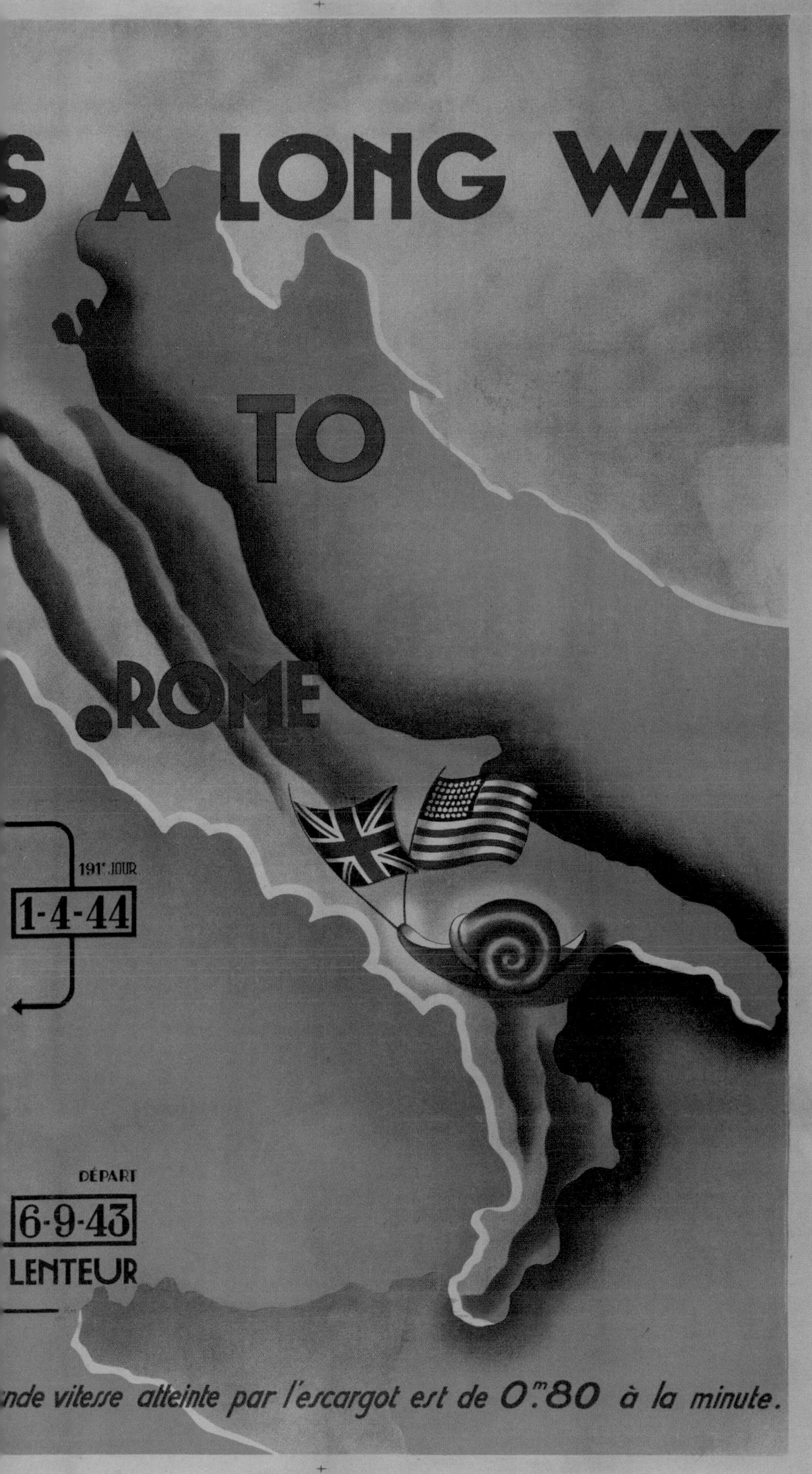

also to pre-empt Soviet advances, notably by advancing through the Ljubljana Gap from Trieste via Slovenia toward Vienna. In practice, this route would have provided the Germans with significant defensive opportunities as well as exposing the British to very grave problems with logistics and the weather. To the Americans, operations into the Balkans were a distraction from defeating the Germans in France, although Roosevelt appreciated that with no invasion of France due until 1944, it was necessary to remain engaged in conflict with the Germans in Italy after the fall of Sicily.

The Allies assigned substantial forces to the campaign in Italy. At Cassino alone 25 divisions, 2,000 tanks and over 3,000 aircraft were committed in the eventually successful fourth battle of Monte Cassino in May 1944. The Allies suffered heavy losses, as in the Liri Valley where the Germans had built tank obstacles, laid minefields and cleared lines of fire as killing zones. The static defences included Panther tank turrets concreted into the ground. More Allied tanks were knocked out in taking this position on 23 May than on any other day of the Italian campaign.

RETALIATION WEAPONS

German leaflet, 'Shadow over England', September 1944.

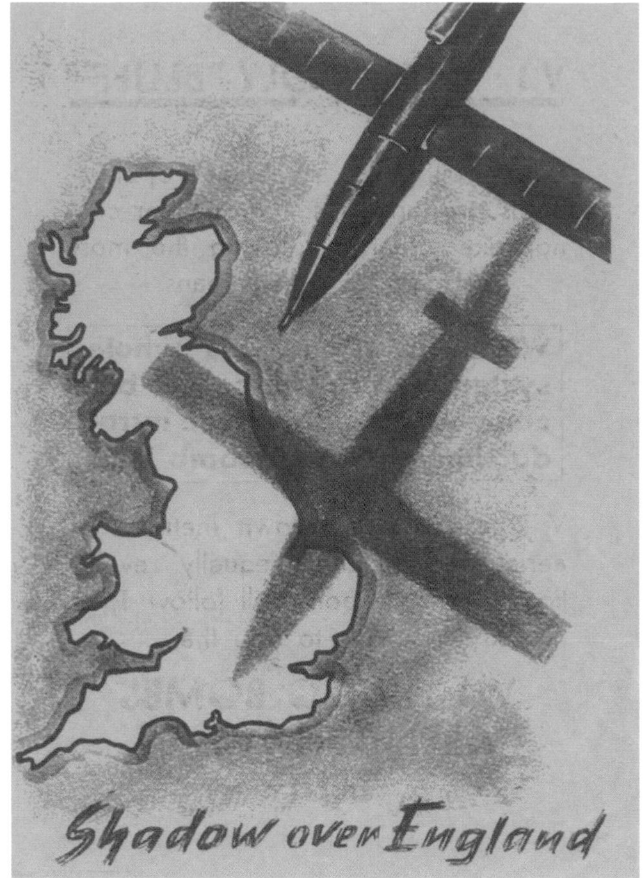

This German leaflet threatening rocket attack was intended to intimidate Londoners. Ground-to-ground V1s, the rocket here illustrated ('V' being for Vergeltungswaffen, or 'revenge/retaliation weapon'), were launched at Britain from 13 June – some 9,521 of them in total – although they could be destroyed by anti-aircraft fire thanks to the use of the proximity fuse. London was hit by 2,419 V1s, which people referred to as 'doodlebugs' or 'buzz bombs' because of a distinctive noise they made. The rockets were launched from sites on the French coast and their use reflected the inability of the Germans to sustain conventional air attacks on Britain, as well as Hitler's fascination with new technology and the idea that it could bring a paradigm leap forward in military capability and satisfy the prospect of retaliation for British bombing. (The same or similar leaflets, part of a 'Shadow over England' series, were dropped on Allied troops advancing through Italy from August 1944 onwards, produced by German propaganda officers there – Abschnitts Offizer Italien. Text on the reverse of the leaflets reproduced reports from UK and Allied newspapers about the 'devastation' happening on the British home front.)

There had been growing interest in rocketry in the 1920s and 1930s, notably in Germany and the Soviet Union where they combined modernity, futurism and the promise of leapfrogging the major powers.

Britain as a whole was presented in this leaflet as a victim of V1 attacks, with the map dramatising the nature of the threat, but that was misleading, not least due to inaccuracy and unreliability. The distance-control of modern missile systems was absent, while it also lacked a mapping system in terms of terrain navigation technology. Many of the V1s fell short of their designated target, London, and instead landed elsewhere in southeast England.

From 8 September, V2s were launched at London. Travelling at up to 3,000mph (4,800kmh), they could be fired from a considerable distance, and could not be destroyed by anti-aircraft fire or sent off course by a fast fighter aircraft tipping its wing, as the V1s could be. London was hit by 517 V2s, with over 2,700 Londoners killed and many more injured. The V1 bombardment of London resumed on 3 March 1945 when long-range V1s were fired from the Netherlands. However, the lack of an effective guidance system was but one of the many serious design flaws that indicate the questionable nature of assessments that see the V2 as the basis of a viable space programme. Indeed, after the war both the USA and Soviet Union focused instead on longer-range bombers.

The Germans were short of rocket fuel for the V2s, while the payload was constrained because the V2 had to re-enter the Earth's atmosphere and the nose cone heated up due to friction. Built using slave labour, the rush to force the V2 into service led to a high margin of error. The rockets were seen as a way to reassure the German public that the Allied bombing campaign was being met with reprisals.

The Germans hoped to develop the capacity to attack the United States by means of multi-stage rockets, space bombers and submarine-launched missiles. This technology appeared to offer an alternative to developing German naval power, which Allied naval strength precluded.

Air and missile attacks in the war killed 29,890 Londoners and seriously injured 50,507. Within Greater London 116,000 houses were destroyed and 288,000 badly damaged. The casualties and ruin inflicted by rockets led to RAF efforts to target production facilities, as well as to the capture of launching pads during the advance from Normandy.

RELIANT ON IMPORTED RESOURCES

US Sixth Army, 'The South Pacific is the South Pacific. Japan is Japan', October 1944.

This propaganda leaflet was published by the Psychological Warfare Branch of the US Sixth Army for distribution to Japanese soldiers to challenge their morale. The title is at the top left and the map shows the US Navy and the Americans in the Philippines (centre), where they landed on the island of Leyte from 20 October 1944, the 'Netherlands East Indies' (lower left) and 'Japan' (top right). The aircraft carriers, warships and fighter aircraft of the US Navy are preventing the oil of the East Indies (produced in the oilfields of Sumatra and Borneo) from reaching Japanese industry. The text on the verso ('Supply and War Potential') quotes Army Minister Sugiyama telling the Japanese Diet in September that supplies were a major problem, and then points out that the landing in the Philippines will only make the situation much worse: 'What is the possibility that these critical supplies desperately needed by the Japanese munitions industry will really reach Japan?... Even if you are trying to fight with all your might, if supplies are not sufficient is it really possible to do your job sufficiently?' The US Army established an interest in 'special warfare' in 1941 and by 1942 the capability to wage psychological warfare (defined as destroying the will of the enemy to achieve victory and depriving the enemy of the support, assistance or sympathy of neutrals) through propaganda had been absorbed into the Office of War Information. Most US Army operational work took place at theatre level, and much unconventional warfare experience was gained in the Philippines in 1941–45.

The Japanese, of course, were aware of the crucial significance of resources and feared an attack on the oilfields, which were essential for naval operations. Pre-war Japan imported around 90 per cent of its oil, mostly from the USA, and had stockpiled some in preparation for war, when it intended to seize the oilfields of Southeast Asia. In the event, in 1945, Australian amphibious forces were launched against the oil production areas in Borneo, while British carrier-based aircraft attacked the well-protected refining facilities on Sumatra. The last oil tanker from Southeast Asia reached Japan in March 1945.

The map focused on American surface shipping, but in reality submarines proved more significant as a strategic force, greatly lessening the economic value Japan derived from its conquests. Island nation Japan was heavily dependent on the import of rubber, tin and rice, whereas the wartime German empire was a land one with few maritime routes.

The Japanese did not inflict enough casualties to cause a deterioration in American submarine leadership or in the quality and motivation of the crew. Indeed, the United States became the most successful practitioner of submarine warfare in history, helped by the ability to decipher Japanese signals. The Japanese in the Pacific proved less effective than the Allies in the Atlantic at convoy protection and anti-submarine warfare and devoted far fewer resources to them, notably by not providing adequate air support for convoys.

However, judged by developments in the war, the attack on Japanese morale failed. Indeed, from 25 October 1944, the Japanese increasingly turned to *kamikaze* suicide attacks, flying aircraft into ships, to counter American naval superiority. Moreover, their troops' willingness to fight on even in hopeless circumstances (often to the death) ensured that there were few prisoners – and also led the Americans to seek to take few. Although there was some killing of Japanese troops trying to surrender, this was not common; there was no equivalent to the systemic killing of SS troops by the Soviets, let alone the brutal treatment of Soviet prisoners by the Germans and of Allied prisoners by the Japanese.

GREENLAND
Baffin I.
Hudson Bay
Davis Strait
Reykjavik
ICELAND
NORWAY
North Sea
BRITISH ISLES
London
Paris
FRANCE
Bay of Biscay
SPAIN
Madrid
PORTUGAL
Lisbon
Gibraltar
Algiers
TUNIS
DOMINION OF CANADA
Churchill
Port Nelson
Edmonton
Vancouver
Regina
Winnipeg
Victoria
Seattle
Portland
Montreal
Ottawa
Quebec
NEWFOUNDLAND
St. Johns
Halifax
Boston
New York
Philadelphia
Washington
Chicago
UNITED STATES
San Francisco
Los Angeles
Savannah
New Orleans
Galveston
Gulf of Mexico
Tropic of Cancer
MEXICO
Veracruz
Mexico
Bermudas
Bahama Islands
CUBA
JAMAICA
WEST INDIES
BRIT. HOND.
HONDURAS
GUATEMALA
SALVADOR
NICARAGUA
COSTA RICA
PANAMA
Colon
Caribbean Sea
Caracas
TRINIDAD
VENEZUELA
Georgetown
GUIANAS
COLOMBIA
Bogota
ECUADOR
Quito
Equator
PERU
Callao
Lima
BOLIVIA
La Paz
Sucre
BRAZIL
Belem (Para)
Recife (Pernambuco)
São Salvador (Bahia)
Rio de Janeiro
PARAGUAY
YERBA MATE
Asuncion
CHILE
Valparaiso
Santiago
URUGUAY
Montevideo
Buenos Aires
ARGENTINA
Falkland Is.
Magellan's Str.
Cape Horn
SOUTH AMERICA
PACIFIC OCEAN
ATLANTIC OCEAN
Tropic of Capricorn
Azores
Madeira
Canary Is.
C. Verde Is.
MOROCCO
Marrakesh
ALGERIA
Dates
RIO DE ORO
FRENCH WEST AFRICA
Dakar
Bathurst
Timbuktu
GAMBIA
PORT. GUINEA
SIERRA LEONE
Freetown
LIBERIA
Accra
Lagos
NIGERIA
GOLD COAST
Gulf of Guinea
AFRICA
Ascension
St. Helena
ANGOLA
Benguella
SOUTH WEST AFRICA
Walvis Bay
Tristan da Cunha
Churchill to Liverpool 2936 (Summer)
Montreal to Glasgow 2698 to L'pool 2760 (Summer)
N. York to L'pool 3045, to Southampton 3091
LUMBER SHIP
PASSENGER & MAIL LINERS
OIL TANKER
FRUIT SHIP
COLLIER
Colon to Liverpool 4530
New York to Rio de Janeiro 4748
Rio to Southampton 5084, to Gibraltar 4192
Southampton to Cape Town 5947
Panama to Wellington 6488
They shall beat their swords into plowshares, and their spears into pruninghooks: nation shall not lift up sword against nation, neither shall they learn war any more.
ISAIAH B.C. 760–698
The only excuse for war is that we may live in peace unharmed
CICERO B.C. 106–43
SYMBOLS
80
40
0

Retrospective

7·

'The experience of war was neither clear-cut nor decisive for many peoples and nations, especially those chosen by the major powers as targets for ideological policies.'

'The Polish War' in The Times *Atlas of the Second World War* (1989), p.206.

The geopolitics of what kind of world would follow the war remained an issue while hostilities continued. There were non-specific projections of the future such as MacDonald Gill's 'The Time and Tide Map of the Atlantic Charter' (1942), which was a benign account of a peaceful and commercial world. There was also much discussion of specifics, notably the future political settlement for Eastern Europe and Germany, and consideration of how this settlement might affect wider international relations. In the United States, Henry Morgenthau, the Secretary of the Treasury, proposed in 1944 that Germany should be divided into two and deindustrialised, a proposal that was not pursued. Robert Chapin Jr's map 'German Jigsaw' (*Time*, 21 February 1944) similarly considered a division: the south was to be a separate state, while Pomerania, Silesia and East Prussia went to Poland and Denmark also made gains; the future of the Saar and the Ruhr were left unclear. Division was also a theme in Arnold Brecht's book *Federalism and Regionalism in Germany: The Division of Prussia* (1945), in which maps were used to present, on the basis of the voting trends of Weimar Germany, democratic and nationalist regions. Separately, the eastern border of Poland became a matter of vexed negotiation between the Western Allies, notably Britain, and the Soviet Union, during which maps were deployed. Richard Edes Harrison's 'Land of the Setting Sun' (1943) outlined that Japan would be stripped of its gains from past territorial expansion.

From 1944, the British Chiefs of Staff were actively considering post-war threats to the British world. This included concern about the Soviet Union and Chinese Nationalist pressure on British India. For example, the Post-Hostilities Planning Staff produced a map in 1944 about projected Soviet lines of advance against India, a map that looked back to nineteenth-century British anxieties. This map was consistent with the view in British India that Baluchistan (and the Herat–Kandahar–Khojak–Bolan route) formed India's 'front porch', as opposed to the side entrance via Kabul and the Khyber Pass. To the British in 1944, western Afghanistan also formed a kind of pivot on which a Soviet force might turn toward southern Iran and, more particularly, the Straits of Hormuz bottleneck to the entrance of the Persian Gulf, an area that remains of geopolitical significance to this day.

Concern about post-war confrontation ensured a determination to seize Axis maps and the plates used to print them, as the occupying Americans did in Japan in 1945.

Toward a new global order

Western geopoliticians assessed the prospects of a new global order and, in particular, the future roles of the Soviet Union and Germany. Halford Mackinder, Britain's leading geopolitician, wrote an essay, 'The Round World and the Winning of the Peace', which appeared in the American journal *Foreign Affairs* in 1943. In this, the Soviet Union was seen as an ally against any German resurgence, which, on the basis of 1919–39, was a major concern.

In contrast, Nicholas Spykman, a Dutch immigrant who was head of the Institute for International Studies at Yale, argued that a Europe dominated by the Soviet Union would be as dangerous as one run by Germany. Instead, rebutting the powerful American isolationism of the inter-war years, he pressed for interventionism by the United States, including to further a Europe in which it would play a role, because the continent was seen as a key interest. In his *The Geography of the Peace* (1944), Spykman developed a 'rimland thesis' as a form of forward protection for the United States, an approach that looked toward the post-war geopolitics of 'containment' of the Soviet Union and, from 1949, of membership of the North Atlantic Treaty Organization (NATO).

The war in retrospect

While the war years had included mapping for the subsequent peace, including the 1944 map of planning for the reconstruction of the City of London, peacetime saw the mapping of the recent war, in part in order to assist military education and also to underline particular political lessons. The former was the case with staff rides and wargaming as 'lessons' were learned.

The role of the defeated could be underplayed. German atlases long ignored, or seriously downgraded, the Holocaust; in contrast, the *Jewish History Atlas* (1969) by Martin Gilbert provided maps of

LESLIE MACDONALD GILL, 'The "TIME & TIDE" Map of THE ATLANTIC CHARTER', 1942.

EASTERN EUROPE'S WAR IN MAPS

Political 'lessons' were particularly promoted in the historical atlases produced, under tight state control, in communist countries. Lenin had pointed out that visual aids, including maps, could be important for political education.

Alongside historical atlases published after the war, for example the *Atlas Istorii SSSR dlia srednei shkoly* (*Historical Atlas of the USSR for High Schools*, 1949–54), came dynamic and informative wall charts of the conflict and its operations. Even in 1990 the *Atlas Istorii SSSR* (1990) for fifth-year students understandably employed heroic symbols for the Soviets in retreat, but not for their German counterparts.

So also with the Soviet client states. In Hungary's historical atlas for secondary schools, *Történelmi atlasz* (1961), the map of the 'liberation of Hungary in 1944–45' (see pages 240–241) used the borders of Hungary after 1945. Budapest's War History Institute issued four wall maps in 1968–70, showing first the 1848–49 war of independence, second the 'patriotic war' of the Hungarian Soviet Republic in 1919, third Hungarians against Fascism, with an inset map of resistance operations in Budapest in 1941–45, and fourth a map of the 'liberation' of Hungary in 1944–45.

Atlases often stressed positive relations with Russia/the Soviet Union and minimised or ignored problems. The role of Soviet forces in driving the Germans from Eastern Europe was presented as part of a continuum looking back to Russia driving back the Turks from the Balkans in the nineteenth century, as in Romania's *Atlas Istoric* (1971), which mapped the insurrection of 1944 (see pages 242–243) and the operations of the Romanian army in helping to liberate Transylvania, Hungary and Czechoslovakia. The *Atlas Po Bulgarska Istoriia* (1963) mapped resistance in 1941–44, liberation (Soviet conquest) in 1944 and subsequent operations in Yugoslavia and Hungary. A separate atlas of the Bulgarian resistance in 1941–44 had already been published in 1958.

Polish historical atlases did not emphasise Polish resistance to Russia's role in the Partitions of Poland of 1772–95 and, while the *Atlas Historyczny Polski* (1973) showed the 1919–20 Polish-Russian war, it ignored the Russian role in its four maps of the German conquest of 1939. The *Atlas Historyczny Polski* (1967) had a map of territorial changes to the Polish state that compared the frontiers of 1018, 1634, 1939 and post-1945 to show that the first and last corresponded most closely, so as to suggest that Poland's eastward expansion in the meanwhile, one reversed by Russian conquests, had been an aberration. Three of the 24 maps in *Nasza Ojczyzna* (*Our Fatherland*, 1981–82) were devoted to the war. The *Historical Atlas of Poland* (1981), a state-approved work presenting Polish history to the wider world, mentioned the large number of Poles killed during the war, including in Auschwitz, without noting that many of them were Jews.

In *Historický Atlas: Revolučního Hnutí* (1956), Czech readers were given maps that emphasised the positive role of the Soviet Union. In the map of the war with Japan, a prominent role was accorded to the war in China, which reflected credit on the Chinese communists, and to the Soviet invasion of Manchuria in 1945; whereas the Americans and British play a comparatively insignificant part.

For East Germany, the first volume (1975) of the atlas on world history depicted resistance to the Nazi regime, thus providing an anti-Fascist pedigree for the post-war East German regime. The *Istorijski atlas oslobodilačkog rata naroda Jugoslavije 1941–1945* (*Historical Atlas of the Liberation War of the People of Yugoslavia 1941–1945*), published in 1952 by the Military History Institute of the Yugoslav Army, in the aftermath of the 1948 break with the Soviet Union, did not emphasise Soviet actions. From a more limited baseline in mapping capability, Chinese mapping preferred to discuss the civil war with the Nationalists in 1946–49, rather than the wartime conflict with the Japanese, much of which was borne by the Nationalists.

With time, there were changes in communist historical cartography. In Poland, *Nasza Ojczyzna* (*Our Fatherland*, 1981–82) showed for 1939 both the German and the Soviet invasions. The map of Poles fighting the Germans in the war depicted the activities of Poles who were fighting alongside the Western Allies, while the map of the liberation of Poland stressed the role of the Poles rather than the Soviets. In Hungary, the 1985 edition of *Történelmi Atlasz* depicted the wartime aggrandisement of Hungary when allied to Germany.

Anti-communist atlases had very different content. In Iwo Pogonowski's *Poland: A Historical Atlas*, published in New York in 1987, Anglo-French inaction at the time of the 1939 invasions of Poland by Germany and the Soviet Union was severely criticised; although there was little practical that either Allied country could have done. Pogonowski depicted the 1942 plan by General Wladyslaw Sikorski, head of the Polish government-in-exile, for a pro-Western federation along Russia's western border, a scheme that he claimed had been betrayed by the United States. A map on 1944–47 recorded 'Civil War and Gigantic Deportations'.

A different mapping of the war followed the collapse of the communist regimes of 1989–91. The 1991 edition of *Történelmi Atlasz* retitled the liberation map as 'Military Operations in Hungary 1944–1945'. The *Történelmi Világaatlasz* emphasised the losses after the First World War to explain Hungarian expansionism in 1938–41. *Atlas Historyczny Swiata* (1992) contained a map marking Polish prisoner-of-war camps in the Soviet Union after the Nazi–Soviet Pact, including the site of the Katyn massacre, and also maps and plans that made clear the role of Free Polish Forces, while *Atlas Historyczny dla klasy VII–VIII* mapped episodes of Russian aggression, including in 1939, and provided a map dealing with the military role of expatriate Poles including a picture of Sikorski. The Lithuanian secondary school historical atlas, *Mokyklinis Visuotines Istorijos Atlases* (1993), placed Soviet and German gains in the same image in the map of Europe in 1939–41, which suggested equivalence. With its troubled wartime legacy, Croatia in the early 1990s did not produce an atlas dealing with the war.

Jewish partisan activities and also maps of revolts in the ghettoes and concentration camps of Nazi-occupied Europe.

British and American atlases had an understandable tendency to concentrate on the role of their forces, and this was notably so of American publications. Thus, the focus on the war with Japan was on the Americans, and not on the conflict in China or that in Burma or, eventually, on the Soviet invasion of Manchuria in 1945. For a long time, the war on the Eastern Front received insufficient attention in Anglo-American atlases, and particularly so after the Battle of Kursk in 1943. This meant that the significant Soviet offensives in late 1943 and 1944 were ignored, or at least underplayed.

In The Times *Atlas of the Second World War* (1989), a work that focused on operational maps, different projections and perspectives, such that north was not always at the top, were used to good effect. The atlas also offered a number of very different global battlefield perspectives – for example, 'The American War' or 'The Polish War', which ably made the point in its caption that 'the experience of war was neither clear-cut nor decisive for many peoples and nations, especially those chosen by the major powers as targets for ideological policies'. Maps of a front as a whole were employed to anchor more detailed treatments, as with 'The Eastern Front November 1942–May 1943', which was supported by five specialist maps, notably on Field Marshal Eric von Manstein's successful counter-offensive in February–March 1943. There were problems, however, in representing scale, let alone fighting quality. For example, in the maps of the Eastern Front the greater number of army groups on the Soviet side contributed to a sense of German weakness, but in practice the size of such groups varied greatly.

Maps have also featured in films about the war. Thus, in *Darkest Hour* (2017), an account of the crisis of 1940, Churchill was shown looking at a map that depicted the British units in Calais and Dunkirk, a map designed to indicate the importance of the defence of Calais as a means to protect the forces at Dunkirk. The scene captured Churchill's interest in maps, but did not itself add much to the story.

The retrospective use of maps to depict the war is part of the process by which our post facto understanding of the war was 'constructed' in terms of the narrative of choice and the relative significance of events, and therefore of the role of particular participants. Controversies over these points continue, and there is no reason to anticipate any changes in this respect. Indeed, if anything there has recently been an accentuation in the differences, notably with the argument that the focus on Soviet operations on land has led to a downplaying of the significance of the Western Allies' success in winning the war with Germany in the air and at sea. The United States and Britain also conquered Italy, and the United States was far more responsible for the defeat of Japan than the Soviet Union, although China played an important role in keeping much of the Japanese Army engaged.

In the future, it is likely that the relative rise of Asian powers, and notably of China and India, will ensure that a very different reading of the war is offered. In particular there will probably be a focus on China's war with Japan, and thus on a global conflict that actually began in 1937. For India, there is the issue of collaboration by some with Japan and, very differently, of the All-India Congress's 1942 'Quit India' campaign, which was directed against the British. As a result, a more complex reading may be offered, or the war may be underplayed. India reaching out to Japan for support against China, a trend in recent years, is also likely to be an issue affecting the treatment of the war. At all events, the future mapping of the war will reflect, at least to a degree, current politics, just as mapping at the time was a response to the pressing needs of contemporary campaigning and political mobilisation.

TRADE AND FREEDOM

Leslie MacDonald Gill, 'The "TIME & TIDE" Map of THE ATLANTIC CHARTER', 1942.

This map, designed by British artist Leslie MacDonald ('Max') Gill, first appeared in 1942 in the British weekly *Time and Tide*, which had commissioned it to celebrate the August 1941 joint statement agreed between the United States and Britain (four months before Pearl Harbor) that set out the two countries' hopes for a just post-war world. Four months later the hopes were seen as war aims.

Drawn up by Churchill and Roosevelt during the Placentia Bay conference off Newfoundland, the eight-point Anglo-American plan was quickly dubbed the 'Atlantic Charter' by the *Daily Herald*. The men's signatures, obtained by Gill, are reproduced at the end. MacDonald Gill's map provides the text of the charter and imagery of a bountiful world in a design that suggests that the Nazi threat will be defeated, just as has been promised in the sixth article: '... after the final destruction of Nazi tyranny, they hope to see established a peace which will afford to all nations the means of dwelling in safety within their own boundaries, and which will afford assurance that all the men in all the lands may live out their lives in freedom from fear and want.' The map was published separately in 1943 as a poster by The London Geographical Institute, established by the cartographer George Philip & Son, Ltd., in Britain and Denoyer-Geppert in the United States.

This morale-raising map, as *Time and Tide* saw it for its British readers (although so too did colonial peoples around the world seeking self-determination), is a comment on the difficult days of 1942. Oceanic trade is a key means of this new world. The presentation of the world represents the British idea of the centrality of trade, and one reference draws on a biblical theme with the quotation from Isaiah, which begins, 'They shall beat their swords into plowshares, and their spears into pruning hooks', accompanied by an illustration in the bottom left corner. The rays of the sun emanating from the Atlantic Charter imply a divine purpose is joined to that of Churchill and Roosevelt. The rays illuminate the continent of Europe, despite most of it then being under Axis control. Germany basks in the sun as an equal with any other nation.

The depiction of a world of resources connected by maritime trade, and with no reference to aircraft or railways, very much reflects British assumptions rather than those of the major Continental powers. Moreover, the trade is shown as focused on Britain or its imperial centres, such as Perth, Singapore and Wellington. These emphases drew on themes of imperial development, such as Halford Mackinder's book, *Britain and the British Seas* (1902) and the bronze sculptural relief above the main entrance of the British Empire Building (620 Fifth Avenue), constructed in the 1930s as part of the Rockefeller Center in New York, which consists of nine male and female figures representing major industries of the British Empire: fish, coal and a seaman's anchor (maritime trade) from the British Isles; salt, tobacco and sugar from India; wool from Australia; grain from Canada; and cotton from Africa. In a similar vein the map uses pictorial symbols (explained in a central legend at the bottom) to highlight the resources available in the countries of the world, which injects strategic importance into the map – given the fact that at the time of publication a global war was raging for control of or access to areas where these things were being produced.

Multiple roundels throughout the map offer multicultural vignettes from the daily life of people around the world, including: a lumberjack in Canada, a camel herder in Arabia and a peasant in China. The map also has five display quotations, mostly on the subject of war and peace, as well as the passage from the Book of Isaiah there are elegant extracts from Aristotle, Cicero, Alexander Pope and Ralph Waldo Emerson. The interconnectivity of the world is endorsed in the statement's article eight, and in January 1942 a 'Joint Declaration of the United Nations' was signed by 26 nations.

Churchill and Roosevelt attended the Placentia Conference on warships. Churchill based himself on the new battleship the *Prince of Wales*, which had been laid down in 1937. Seriously damaged by the German battleship *Bismarck* on 23 May 1941, in the encounter in which the battlecruiser *Hood* was sunk, the *Prince of Wales* was to be sunk by land-based Japanese naval torpedo bombers off Malaya on 10 December 1941. At the time of her sinking, alongside the battlecruiser *Repulse*, she had the best radar suite of any operational warship in the world, including close-in radar for her anti-aircraft guns and radar for her main guns; as well as good compartmentalisation. None of these could save a ship that seemed part of an endangered order.

&TIDE" Map of THE ATLANTIC CHARTER

REVISITING THE BATTLEFIELDS

Michelin, 'Bataille de Normandie', 1947 edition.

This Michelin map for the Battle of Normandy sites, with text in English and French, was first produced in 1947 and has subsequently been republished. At a scale of 1:200,000, the map shows the main sites of the battle in summer 1944. The memorialisation of the battle from the outset concentrated on the invasion beaches. This reflected the impressive nature of the events of D-Day, the largest amphibious invasion ever mounted, and also the fact that many of the positions remained in situ. This is particularly the case on Omaha Beach, where what remained of the Atlantic Wall defences testifies not only to the difficulty of the task that faced the Americans but also to the destructive power of the Allied bombardment faced by the German defenders. There is less to be seen on the other beaches, although what was – in 1944 – a German position on Utah has had its remains preserved as a museum. It forms an incongruous contrast to the bathers on the sandy beach. Among the most impressive thing to see on any of the beaches are the remains of the Mulberry harbour at Arromanches and some German gun casements further inland at Longues-sur-Mer.

There is less left to see for the airborne attacks, notably those by the American 82nd and 101st Airborne Divisions behind Utah Beach. The disorganised nature of the airdrop ensured that there was no one site of key importance, although Sainte-Mère-Église is an iconic one, where John Steele's parachute became caught on the spire of the church in the first Normandy village to be liberated by the US Army. Pegasus Bridge on the other flank, near Sword Beach, provides a more appropriate focus for the commemoration of the British airborne assault, which was more focused because of the use of gliderborne troops rather than parachutists.

The graveyards of the respective combatants also became important commemorative sites, with the different styles of gravestone being instructive: there was least individuality in the German graveyards. The main map helps the tourist as much as possible by featuring special icons denoting battle or liberation dates, parachute drops, relics of war and seriously damaged places, as well as an inset showing the broader movements of the military forces.

The map also covered the breakout battles, including the Falaise Pocket and the advance into Brittany and toward the Seine, although there was far less on the ground to see than on the beaches, to which tourism was developing by 1947. In large part, this was because of the absence of fortifications. The

LE HAVRE
Arromanches
(Port Winston)
6 JUN 44
Gold Beach
6 JUN 44
Juno Beach
6 JUN 44
Sword Beach
Asnelles
Ver-Plage
Courseulles
Graye
Bernières
Luc
Lion
Hermanville-Plage
Colleville-Montgomery-Plage
Riva-Bella
Douvres
6-17 JUN
Pegasus Bridge
Bénouville
6 JUN
Ranville
5-6 JUN
Flooded Area
Zone inondée
Houlgate
Cabourg
Dives
Dozulé
Troarn
Villers
Blonville
DEAUVILLE
Trouville
Villerville
CAEN
9-19 JUL 44
Carpiquet
Vaucelles
Bourguébus
Bretteville
Cambremer
Mézidon
Pipe-line
Villers-Bocage
Aunay-s-Odon
Thury-Harcourt
Mt Pinçon
Esquay
Evrecy
Maltot
Grimbosq
Livarot
St-Pierre-s-Dives
Brit.
Cdn.
7 JUN 44

Gal EISENHOWER
Gal MONTGOMERY - 21 Ar. Gr.
Gal BRADLEY
Gal DEMPSEY
Gal HODGES
Gal CRERAR - 1 Ar.
Gal PATTON - 3 Ar.
12 Ar. Gr.
Gal LECLERC
CHERBOURG
COTENTIN
AVRANCHES
MORTAIN
FALAISE
CHAMBOIS
CAEN
ST-LÔ
PLUTO (PIPE-LINE)
CHERBOURG
St-Vaast-la-Hougue
Montebourg
Ste-Mère Eglise
Utah Beach
Plage Prést Roosevelt
la Madeleine
Pointe du Hoc
Omaha Beach
Arromanches (Port Winston)
Gold Beach
Ver-Plage
Asnelles
Port-en-Bessin
Colleville
St-Laurent
Vierville
U.S.A.
Isigny
Carentan
Flooded Area
BAYEUX
Barneville
St-Sauveur-le-V.
la Haye-du-Puits
Lessay
Périers
St-Jean-de-Daye
ST LÔ
Marigny
COUTANCES
Tilly-s-S.
Balleroy
Caumont-l'Éventé
Villers-Bocage
Aunay-s-Odon
Torigni
Tessy-s-Vire
Percy
Gavray
le Bény-Bocage
St-Charles-de-Percy
Granville
Villedieu-les-P.
VIRE
Condé-s-N.
Flers
AVRANCHES
BAIE DU MONT SAINT MICHEL
le Mont St Michel
Mortain
Domfront
Voie de la Liberté
Réimpression de l'édition 1947
Reprint of the 1947 edition
PNEU MICHELIN
46, Av. de Breteuil, 75324 PARIS CEDEX 07
Tél. (1) 45661234

Germans had instead used the dense hedgerows of the Normandy *bocage* country.

The Michelin company produced a number of maps for the growing tourism market, with the clear aim of satisfying American purchasers who in 1947 were both wealthy and able to travel, whereas British counterparts were affected by strict foreign exchange regulations. The sequel to the Normandy map was 'Voie de la Liberté' ('Road to Liberty'), which showed the route of the Allied forces from Normandy to Bastogne in Belgium, where a heroic American defence had blocked German forces in the Battle of the Bulge in December 1944. The map recorded the liberation dates for individual places, and American war cemeteries for both world wars, while the text described the key events. Understandably, the problems faced were not adequately covered, notably command flaws.

Several other maps also reflected the American contribution, notably the battle for Provence stemming from Operation Dragoon, the invasion in which American forces predominated, as well as the Battle of Alsace, covering operations there in November 1944 to March 1945. French forces had also played a major role in these two campaigns.

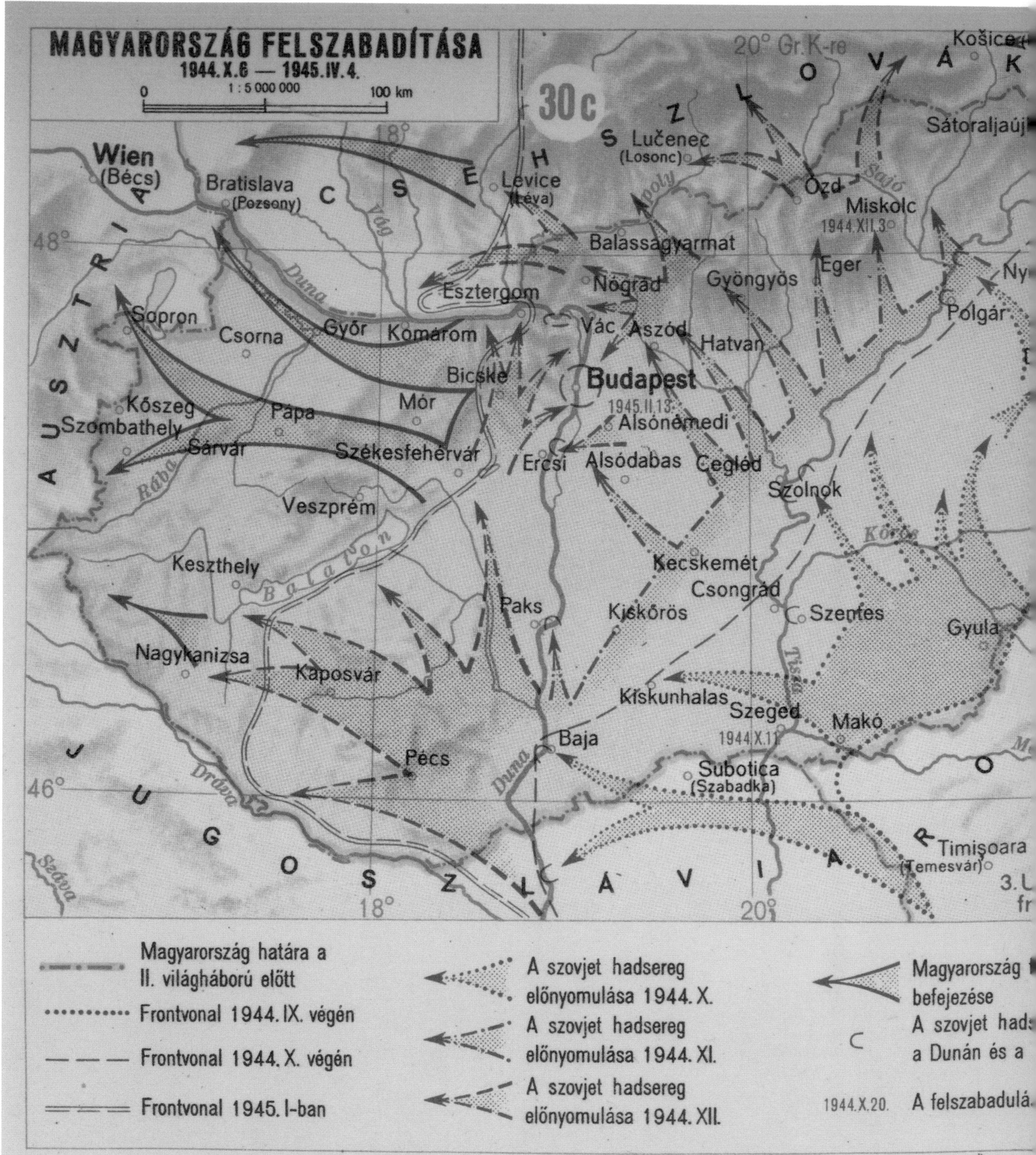
MAGYARORSZÁG FELSZABADÍTÁSA
1944. X. 6 — 1945. IV. 4.
0
1 : 5 000 000
100 km
30c
Wien
(Bécs)
Bratislava
(Pozsony)
Levice
(Léva)
Lučenec
(Losonc)
Ózd
Miskolc
1944. XII. 3.
Košice
Sátoraljaújhely
Balassagyarmat
Esztergom
Nógrád
Gyöngyös
Eger
Polgár
Sopron
Csorna
Győr
Komárom
Vác
Aszód
Hatvan
Bicske
Budapest
1945. II. 13.
Kőszeg
Szombathely
Pápa
Mór
Alsónémedi
Sárvár
Székesfehérvár
Ercsi
Alsódabas
Cegléd
Szolnok
Veszprém
Keszthely
Kecskemét
Csongrád
Paks
Kiskőrös
Szentes
Gyula
Nagykanizsa
Kaposvár
Kiskunhalas
Szeged
1944. X. 11.
Makó
Baja
Pécs
Subotica
(Szabadka)
Timişoara
(Temesvár)
AUSZTRIA
CSEHSZLOVÁKIA
JUGOSZLÁVIA
Duna
Vág
Rába
Balaton
Dráva
Tisza
Körös
Sajó
Ipoly
Száva
48°
46°
18°
20°
20° Gr. K-re
Magyarország határa a II. világháború előtt
Frontvonal 1944. IX. végén
Frontvonal 1944. X. végén
Frontvonal 1945. I-ban
A szovjet hadsereg előnyomulása 1944. X.
A szovjet hadsereg előnyomulása 1944. XI.
A szovjet hadsereg előnyomulása 1944. XII.
Magyarország
befejezése
A szovjet hads
a Dunán és a
1944. X. 20.
A felszabadulá

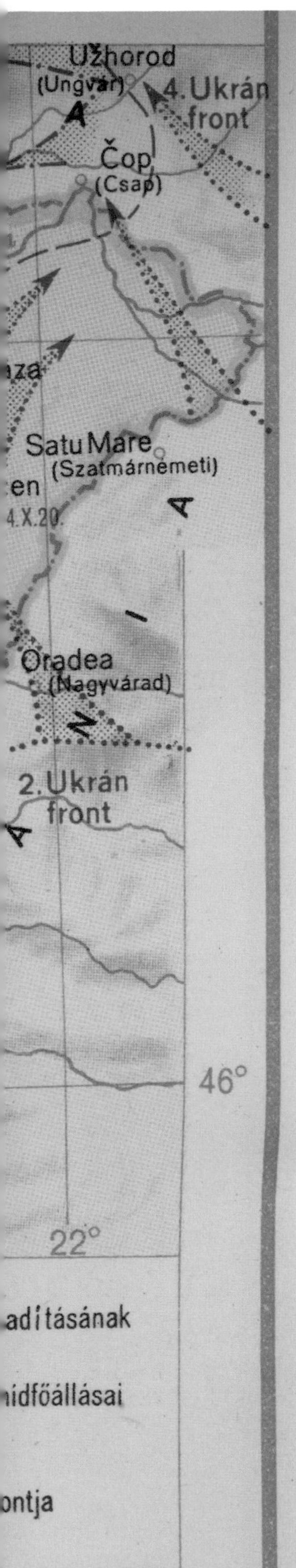

NATIONAL LIBERATION

Történelmi atlasz ('Historical Atlas'), 'Map of the liberation of Hungary in 1944–5', 1961.

With the Second World War having ushered in a communist regime in Hungary, the ruling communists emphasised the liberation role of the Soviet Red Army in ensuring victory against the Nazis and ignored the fact that Hungary had (like the Soviet Union itself) previously allied with Germany and benefitted territorially as a result in 1938–41. Post-war atlases only used Hungary's 1945 borders. Earlier gains made by Hungary, as a result of having allied with Hitler, in northern Transylvania and parts of Slovakia and Yugoslavia in 1938–41, were ignored because this territorial aggrandisement was a taboo subject during the years of communist rule.

The communists also underplayed the willingness of Hungary's wartime non-communist regime to abandon Germany. By 1943, the government under Miklóz Kállay was seriously considering how best to switch to the Allies without it leading to German occupation. After January 1943, Hungary did less fighting on the Eastern Front (having suffered badly at Stalingrad and Voronezh in 1942–43), and negotiated with Britain and the United States. This encouraged Churchill's interest in an advance into southeastern Europe via Yugoslavia.

In the event, thanks to effective German military intervention on 19 March 1944, Hungary's position in the Axis camp was actually strengthened, with the Hungarian army effectively forced by the Germans into taking a more active stance against the Soviets, while more systematic persecution of the Jews meant that large numbers of Jews were seized and sent to Auschwitz. By 1945, the Hungarian regime's lengthy collaboration with the Nazis had ensured that about half a million Hungarian Jews (two-thirds of the country's Jewish population) had perished in one of the largest deportations of the Holocaust.

In turn, Soviet battlefield success in reaching the Carpathians and, crucially, the defection of Romania, led the Hungarian head of state, Admiral Horthy, a conservative but not a Fascist, to stop the deportations (by then the only Jews remaining were confined to two ghettoes in Budapest), change the government on 29 August 1944, and to agree an armistice with the Soviets on 11 October.

In response, the Germans helped ensure a successful coup. Thanks to the installation of the puppet Arrow Cross government under Szálasi on 16 October, Hungary was unable to switch sides, but the morale of Hungarians fell, and in the bitter and bloody siege of Budapest against Soviet attack in 1944–45 – half a million men and 1,500 tanks – the morale of the German defenders proved higher than that of the Hungarians.

Hitler ordered his troops to hold the city at all costs. In part, the German motivation came from wanting to prevent a Soviet advance toward Austria and in part to retain the remaining area of oil production. However, with larger forces, the Soviets advanced into Hungary from both Romania and southern Poland, and this converging advance put successful pressure on the German forces. During the bloody Battle for Budapest the Germans made major efforts to retain control of the city, launching counter-attacks in an attempt to relieve the besieged garrison, but they were defeated and the garrison surendered. The desperate, 110-day struggle for Budapest, during which more than 120,000 people died (including most of Budapest's surviving Jews, murdered by the Arrow Cross) and most of the buildings were damaged or destroyed, still marks the national consciousness of Hungary today.

ROMANIAN VIEWPOINTS

'The Development of the National Anti-Fascist Army Insurrection 23–31 August 1944', 1971.

Edited by Stefan Pascu, *Atlas Istoric* ('Historical Atlas') was published in Bucharest in 1971, at a time when Romania was pursuing an independent course under Nicolae Ceauşescu, who had become First Secretary of the Communist Party in 1965, then added the Presidency of the State Council in 1967. Ceauşescu was angry at the Soviet reorganisation of Comecon's production and trade priorities in 1964, which had seemed to relegate Romania to the position of an exporter of materials and an importer of finished goods as if it was in the Third World. *Atlas Istoric* almost represented a declaration of national independence. Upon publication *The Journal of Modern History* was complementary about the maps and their content, remarking on the atlas having been 'scarcely affected by Marxian doctrine'.

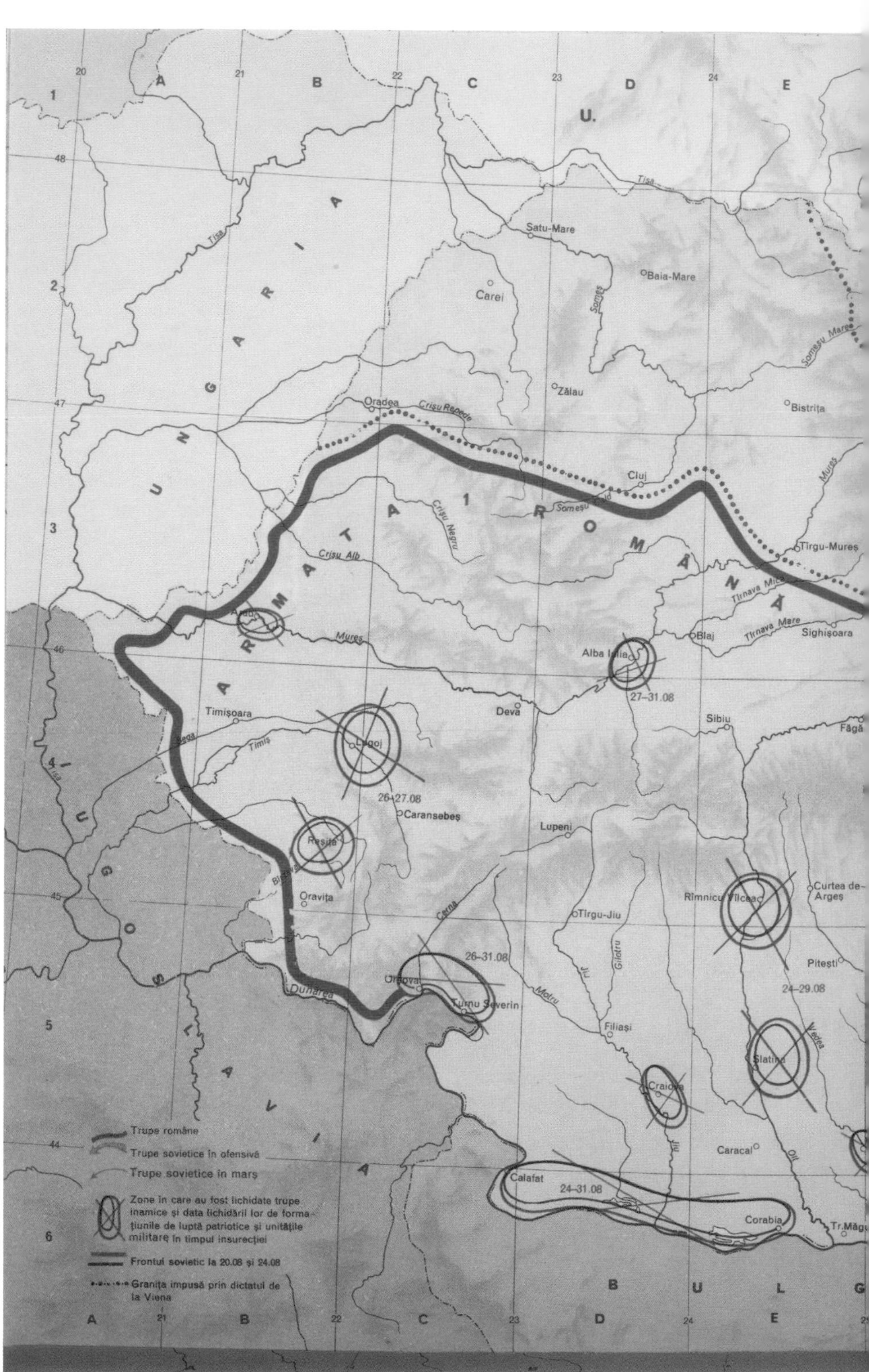

Germany's failure to knock the Soviet Union out of the war led to growing unease among its allies, including Romania. The concerns had begun with the Soviet counter-offensive in the winter of 1941–42, and resumed as the 1942 campaign ran out of steam, then became far more serious with the Soviet counter-offensive. This was seen with growing disagreement between the prime minister and dictator General (later Marshal) Ion Antonescu and Constantin Brătianu and Iuliu Maniu, the leaders of opposition parties with whom Antonescu maintained links. They urged him to withdraw Romania from the Soviet Union, but Romania was scarcely in a position to do that given the prospect of German intervention if it did. In early 1943, however, after terrible losses for the Romanian forces protecting the German flank at Stalingrad, Romanian interest in a change grew, with an approach to Italy for a joint proposal to the Allies for neutrality. Mussolini rejected this, and instead the Romanian government approached the Allies via the Papacy and Portugal, both of which were neutral. However, armistice negotiations were handicapped by Antonescu's suspicion of the Soviet Union and his unwillingness to abandon Hitler, combined with the Allied stress on unconditional surrender.

As an ally of Germany in the war, Romania was greatly affected by the Soviet advance in August 1944. In practice, the key move within Romania was that of the young King Michael who, on 23 August, had the dictator Antonescu arrested. This act was followed by the surrender of Romanian forces to the Soviet Union, with its army on 24 August turning against Germany and Hungary. Romania was to regain Transylvania

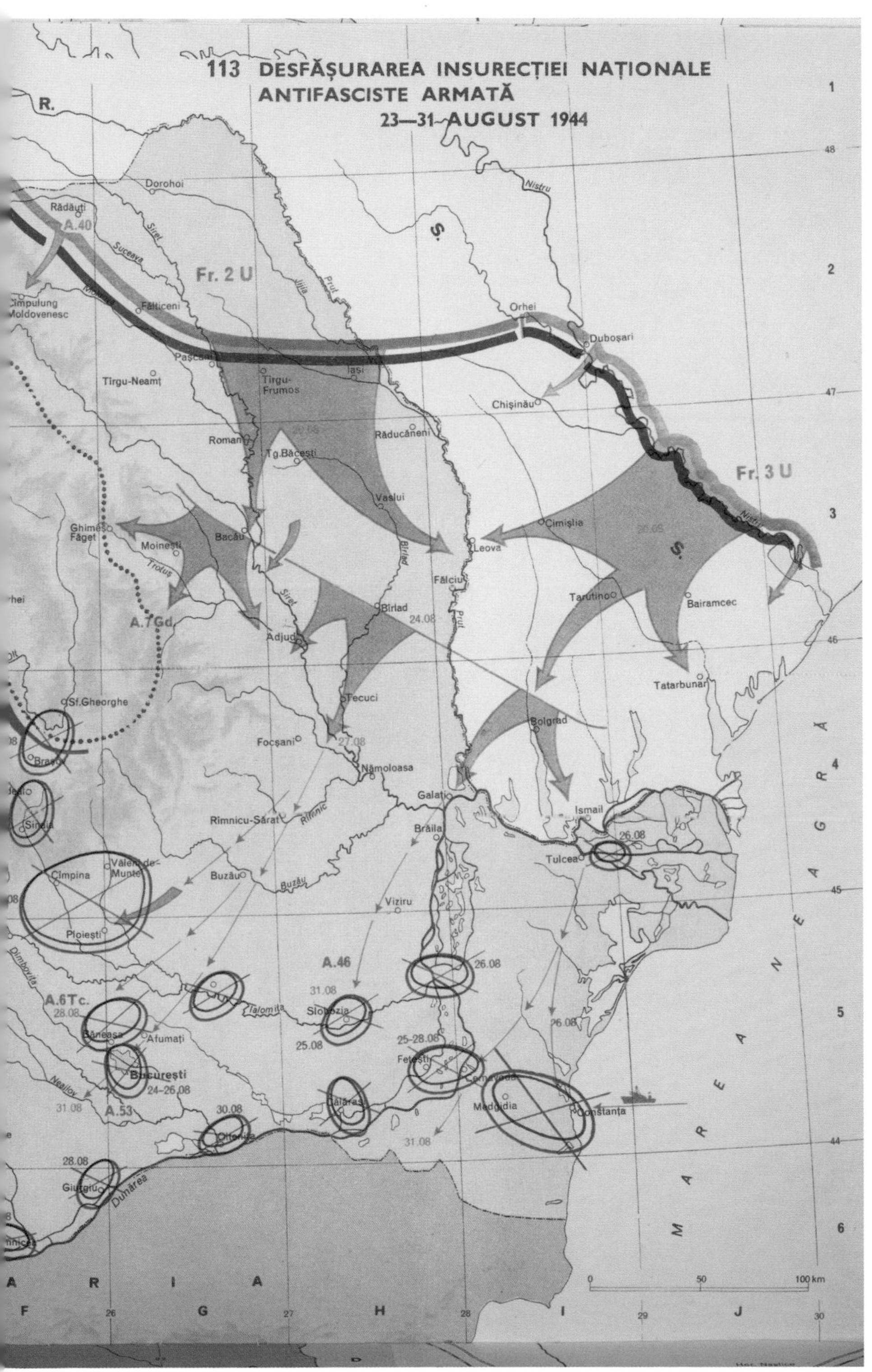

(annexed in 1940) from the latter. Antonescu was handed over to the Soviet occupation forces, which sent him back in 1946 (when he was tried and shot). In reality, the Romanian élite was more divided than was subsequently to be revealed by the communists. Antonescu had been moving away from Hitler but that narrative did not suit later commentators.

The communists entrenched an account of the war that justified their takeover in 1946. The key war memorials constructed after the fall of the Antonescu government in 1944 were monuments to the Red Army. Obelisks topped with a communist star were constructed in all Romania's major cities and also in many villages. In Bucharest a large statue of a Soviet Red Army figure was erected on a tall column in 1947 in the centre of Victory Square. In 1948 the communists destroyed a 1930 equestrian statue of King Carol I in Palace Square. However, in the 1950s the emphasis shifted to the construction of memorials for Romanians, albeit only for those soldiers who had fought alongside the Red Army in 1944–45. In practice, some of the soldiers buried had been killed fighting against the Soviets in 1941–44, but this point was ignored in the inscriptions.

In the 1960s, there was a new interest in honouring civilians, including an emphasis on civilian resistance to the Romanian war effort. This emphasis was an attempt to construct a notion of there having been widespread anti-Axis Romanian nationalism, which implied communist rule had come about as a sort of benign inheritance. Other nationalist themes were also pursued, notably with monuments in Transylvania to those killed for having opposed Hungarian wartime occupation, a theme designed to underline the anti-Romanian nature of Hungarian links. The nationalism of the Ceauşescu regime from the 1960s onwards, particularly its distancing from the Soviet Union, led to the Soviet Red Army statue being removed from Victory Square to an obscure setting elsewhere. There are now plans to erect in the square a statue of Maxim Pandelescu, a wartime general in the Romanian army and an anti-communist guerrilla leader.

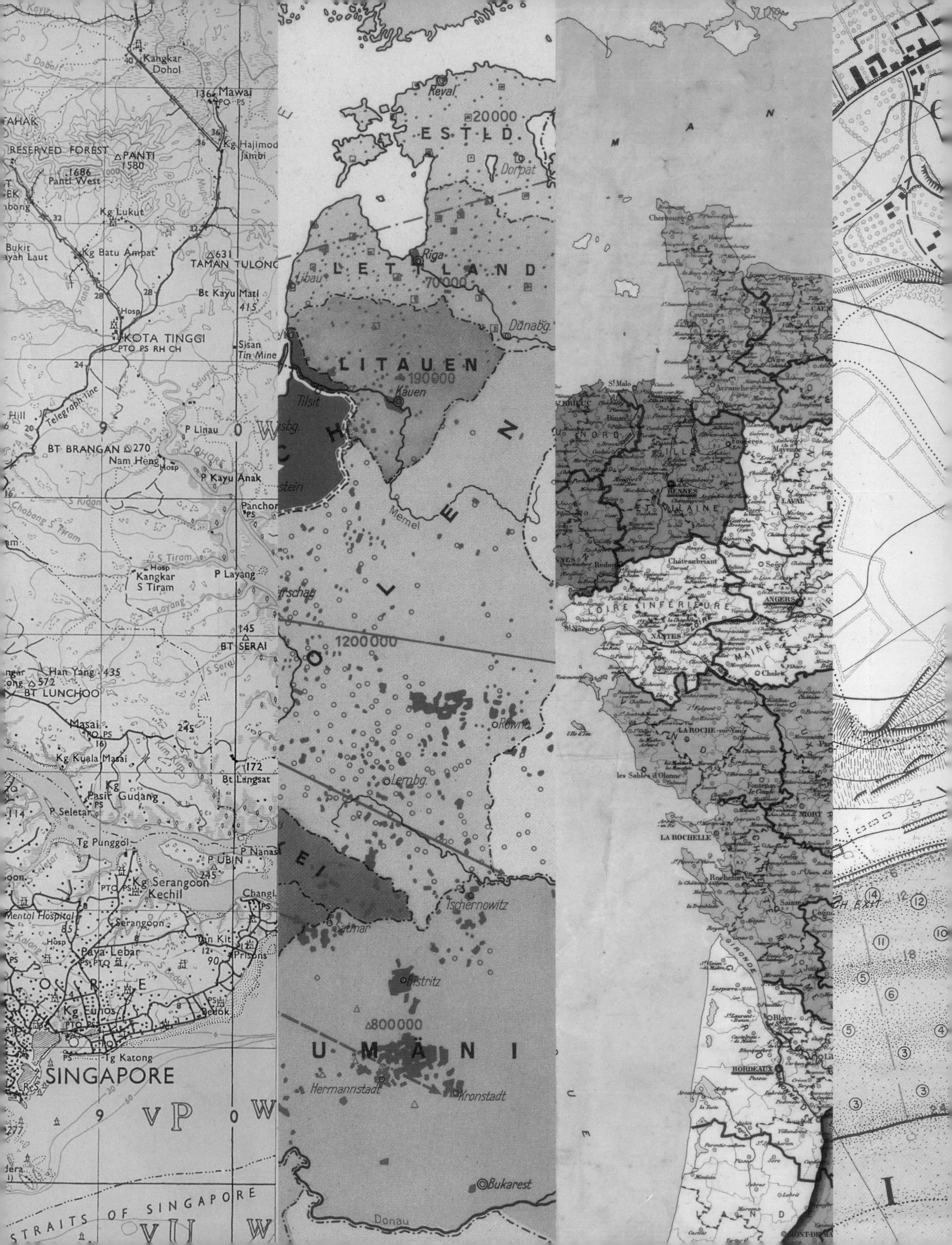

Kangkar Dohol
Mawai
RESERVED FOREST
PANTI
1580
Panti West
Kg Lukut
Kg Batu Ampat
TAMAN TULONG
Bt Kayu Mati
KOTA TINGGI
Sjsan Tin Mine
Telegraph line
BT BRANGAN
P Linau
Nam Heng
P Kayu Anak
Kangkar S Tiram
P Layang
BT SERAI
Han Yang
BT LUNCHOO
Masai
Kg Kuala Masai
Bt Langsat
Kg Pasir Gudang
P Seletar
Tg Punggol
P UBIN
Kg Serangoon Kechil
Changi
Mental Hospital
Serangoon
Paya Lebar
Tan Kit
Prisons
Kg Eunos
Bedok
Tg Katong
SINGAPORE
STRAITS OF SINGAPORE
Reval
20000
ESTLD.
Dorpat
Riga
LETTLAND
Libau
70000
Dünabg.
LITAUEN
190000
Kauen
Tilsit
Memel
Warschau
1200000
Rowno
Lembg.
Tschernowitz
Sathmar
Bistritz
800000
Hermannstadt
Kronstadt
Bukarest
Donau
Cherbourg
Coutances
St. Malo
Dinan
RENNES
Fougères
LAVAL
Mayenne
Redon
Chateaubriant
ANGERS
LOIRE INFÉRIEURE
St. Nazaire
NANTES
MAINE ET LOIRE
Cholet
LA ROCHE-sur-Yon
les Sables d'Olonne
Fontenay
NIORT
LA ROCHELLE
Rochefort
GIRONDE
BORDEAUX

Reference

EVILLE-S
BEACH EXIT
LINCOLN'S INN FIELDS
KINGSWAY
ST. CLEMENT DANES
ST. MARY LE STRAND
Thames Police Station
PRECINCT OF THE SAVOY
Waterloo Bridge
Cleopatra's Needle
Landing Stages
Charing Cross Pier
S. E. & C. RY.
Hungerford Bridge
Lion Brewery
Landing Stage
Union, Parly. & Met. Boro. By.
Roman Boat found A. D. 1910
Archbishop's Park
Lambeth Palace
MANCHUKUO
HSINKING
VLADIVOSTOK
KOREA
SEOUL
AOMORI
OMINATO
SENDAI
OSAKA
TOKYO
YOKOHAMA
YOKOSUKA
KAGOSHIMA
MITSUGO JIMA
MUKO JIMA
OGASAWARA
BONIN IS.
KAZAN RETTO
URUCAS
MARCUS
MARIANAS
PAGAN
ISLANDS
SAIPAN
ROTA
GUAM
YAP
NGULU
NAMONUITO IS.
HALL IS.
TRUK IS.
LOSAP
NOMOI IS.
CAROLINE ISLANDS
NUKUORO
150°
KAVIENG
GREENWICH
NEW GUINEA
MADANG
LAE
SALAMAUA
RABAUL
GASMATA
BUKA
KIETA
BUNA
BUIN
KOKODA
PORT MORESBY
CAPE YORK
MILNE BAY
LOUISADE ARCHIPELAGO
CAIRNS
TOWNSVILLE
BARRIER REEF
BRISBANE
PORT AUGUSTA
CANBERRA
ADELAIDE
SHIPS
COAL
BREST
LORIENT
ST.

FURTHER READING

Corps of Engineers, U.S. Army, *The Army Map Service: Its Mission, History, and Organization*, Washington, DC: Army Map Service, 1960.

Bishop, Chris, *The Military Atlas of World War II*, London: Amber Books, 2005.

Black, Jeremy, *Rethinking World War Two: The Conflict and its Legacy*, London: Bloomsbury, 2015.

Bond, Barbara, *Great Escapes: The Story of MI9's Second World War Escape and Evasion Maps*, London: HarperCollins, 2015.

Clough, A.B. Brigadier, *Maps and Survey*, London: The War Office, 1952.

Collier, Basil, *The Defence of the United Kingdom*, London: HMSO, 1957.

Douglas, Roy, *The World War 1939–45: The Cartoonists' Vision*, London: Routledge, 1990.

Herb, Guntram, *Under the Map of Germany: Nationalism and Propaganda, 1918–1945*, London: Routledge, 1997.

Keegan, John, *The Times Atlas of the Second World War*, London: Times Books, 1989.

Kirchubel, Robert, *Atlas of the Eastern Front, 1941–45*, London: Osprey Publishing, 2016.

Kries, John F. (ed.), *Piercing the Fog: Intelligence and Army Air Forces Operations in World War II*, Washington, DC: Air Force History and Museums Program, 1996.

Macksey, Kenneth, *Invasion: The Alternate History of the German Invasion of England, July 1940*, London: Arms and Armour Press, 1980.

Man, John, *The Penguin Atlas of D-Day and the Normandy Campaign*, London: Viking Books, 1994.

Messenger, Charles, *The D-Day Atlas: Anatomy of the Normandy Campaign*, London: Thames & Hudson, 2004.

Military High Command, Department for War Maps and Communications, *German Invasion Plans for the British Isles 1940*, Oxford: The Bodleian Library, 2007.

Monmonier, Mark, (ed.), *Cartography in the Twentieth Century* (Volume Six in *The History of Cartography*), London and Chicago: University of Chicago Press, 2015.

Pimlott, John, *The Historical Atlas of World War II*, New York: Henry Holt & Company, 1995.

Potter, John, *Pim and Churchill's Map Room: Based on the papers of Captain Richard Pim RNVR Supervisor of Churchill's Map Room 1939–1945*, Belfast: Northern Ireland War Memorial, 2014.

Schulten, Susan, *The Geographical Imagination in Americas, 1880–1950*, London and Chicago: University of Chicago Press, 2001.

Smurthwaite, David, *The Pacific War Atlas 1941–1945*, London: HMSO, 1995.

Snyder, John P. *Flattening the Earth: Two Thousand Years of Map Projections*, London and Chicago: University of Chicago Press, 1993.

Stanley II, Roy M. Col., *World War II Photo Intelligence: The first complete history of the aerial photoreconnaissance and photo interpretation operations of the Allied and Axis nations*, New York: Scribner, 1981.

Stewart, John Q. and Newton L. Pierce, *Marine and Air Navigation*, Boston: Ginn and Company, 1944.

Swanston, Alexander and Malcolm Swanston, *The Historical Atlas of World War II*, Secaucus, New Jersey: Chartwell Books, 2007.

Ward, Arthur, *Churchill's Secret Defence Army: Resisting the Nazi Invader*, Barnsley, South Yorkshire: Pen & Sword Military, 2013.

Zaloga, Steven, *Atlas of the European Campaign: 1944–45*, Oxford: Osprey Publishing, 2018.

LIST OF MAPS

40–41 'Carte administrative de la France, regions (Novembre 1941): regions administratives délimitées par les décrets des 19 et 26 Avril, 30 Juin, 18 Juillet, 4 Août, 9 et 24 Septembre 1941, 25 Mars 1942, places sous l'autorité des Préfets régionaux institués par la lois du 19 Avril 1941', 1942 (Paris: Institut Géographique National, 1942). Maps X.3588.

43 'Lage am 2.7.1941 abds', sheet 14 from a set of maps showing military positions on the Russian Front on a day by day basis (Berlin: Heer. Abteilung für Kriegskarten- und Vermessungswesen, 1941). Maps 35847.(43.). 118 x 59 cm.

44 'Lage am 19.7 1941 abds mit finnischer Front', sheet 31 from a set of maps showing military positions on the Russian Front on a day by day basis (Berlin: Heer. Abteilung für Kriegskarten- und Vermessungswesen, 1941). Maps 35847.(43.). 118 x 59 cm.

47 'Lage am 28.11.1941 abds', sheet 122 from a set of maps showing military positions on the Russian Front on a day by day basis (Berlin: Heer. Abteilung für Kriegskarten- und Vermessungswesen, 1941). Maps 35847.(43.). 118 x 59 cm.

48 Royal Engineers, Field Survey Company, 512th, 'Distribution of Axis Forces in the Balkans', December 1942 (Cairo: Survey Directorate, Middle East, 1942). Maps MOD MDR Misc 2172.

50–51 'Italy (South) Special Strategic Map, Scale 1:1,500,000, 1943' (Washington: Army Map Service, 1943). Library of Congress Geography and Map Division, Washington, DC. 46 x 57 cm.

52–53 Engineer Section 12th Army Group, 'Situation 2400 hrs. 6 June 1944' ([England?]: Twelfth Army Group, 1944). Library of Congress Geography and Map Division, Washington, DC. 52 x 54 cm.

54–55 Engineer Section 12th Army Group, 'Situation 1200 hrs. 2 October 1944' ([England?]: Twelfth Army Group, 1944). Library of Congress Geography and Map Division, Washington, DC. 52 x 54 cm.

56 United States Strategic Bombing Survey, 'Hiroshima before and after bombing. Area around ground zero. 1,000 foot circles' in *The Effects of Atomic bombs on Hiroshima and Nagasaki* (Washington DC.: U.S. Government Printing Office, 1946). A.S.760/4.(1.).

57 United States Strategic Bombing Survey, 'Atomic Bomb Damage, Nagasaki, Japan' in *The Effects of Atomic bombs on Hiroshima and Nagasaki* (Washington DC.: U.S. Government Printing Office, 1946). A.S.760/4.(1.).

66–67 'Stadtplan von Plymouth mit Militärgeographische Eintragungen, England 1:10,000', [overprinted on Ordnance Survey map of 1937], 1941, part of the *Militärgeographische Objetkarte von England* (Berlin: Generalstab des Heeres, Abteilung für Kriegskarten un Vermessungsswesen, 1941–42). Maps 47.g.14.

69 'Bildskizze Liverpool–Birkenhead', November 1940, in *Zielanhäufungen in britischen Hafen- und Industrie-Städten* (Berlin: Der Oberbefehishaber der Luftwaffe, Führungsstab Ic, 1940). Maps 46.d.1.

70–71 'Karte der Insel Kreta', 1941 ([S.I.]: Generalstab der Luftwaffe, 1941). Maps X.942.

72 'Aegean Sea Communications', 1941 (Great Britain: Indian General Staff Geographical Section, 1941). Maps MOD Greece Misc 4.

74–75 'Stalingrad-Süd', 1942 (Berlin: Heer. Generalstab, 1942). Library of Congress Geography and Map Division, Washington, DC. 49 x 78 cm.

76–77 'Buna, New Guinea', four-mile 'Strategical Series' New Guinea, 1942 (Melbourne: Australian Survey Corps, 1942–45). Maps 89885.(27.). 44 x 66 cm.

79 British copy of Field Marshal Rommel's annotated map of El Alamein, 23 October 1942. National Army Museum, NAM.2000-04-50-1.

80–81 Royal Engineers, Palestinian Field Survey Company, 524th, 'El Alamein C.B. Chart, Night 23/24 Oct', 1942 (Cairo: Survey Directorate, Middle East, 1943). Maps MOD MDR Misc 2245.

82, 83 'Italy 1:250,000 Army/Air Sheet 22 Siena, Water Supply Overprint Oct. 1943', River Discharge Overprint Oct. 1943 (London: War Office, General Staff, Geographical Section, 1943). Maps MOD GSGS 4230.

84 'Area "K": administrative map Operation Overlord South Wester Zone, 15 April 1944' (England: Headquarters XIX District, 1944). Library of Congress Geography and Map Division, Washington, DC. 59 x 49 cm.

86–87 U.S. Army Engineer Topographical Company, 663rd, 'Omaha Beach-East (Colleville-sur-Mer)', 1944 ([S.I]: Commander Task Force, 1944). Maps MOD US Misc 266.

88–89, 89 U.S. Army Engineer Topographical Company, 663rd, 'Omaha Beach-West (Vierville-sur-Mer)', 1944 ([S.I]: Commander Task Force, 1944). Maps MOD US Misc 266.

90–91 'Arromanches-Les-Bains, Defences Information as at May 1944' (London: War Office, 1944). Maps 14317.(241.). 40 x 60 cm.

93 U.S. Navy Flight, 8th, 'Panoramic Beach Section, AF 1463 to AF 4474', 1943 (Caserta: Allied Forces Headquarters, Italy, 1943). Maps MOD AF 4463.

94–95 'Übersichtskarte zur Geländebeurteilung von Mitteldalmatien (Raum Zara - Knin - Šibenik). Sonderausgabe Panzerkarte', 1944 ([S.I.]: Pz. AOK 2 Ia/Mess, 1944). Maps X.5433.

96–97, 97 U.S. Army Engineer Topographical Company, 663rd, 'Leyte Island', 1944. Maps MOD US Misc 311.

98–99 U.S. Army Map Service, *Rhine River Photomap*, 1944. Maps MOD US Misc 11.

100–101 'Frankreich 1:25000, Royan', captured and annotated by French Forces of the Interior (FFI), after October 1944. Collection: Alan Latter.

102 'Japanese Forces Burma, as known at 30 Mar 1945' ([S.I]: Supreme Headquarters Allied Expeditionary Force, 1945). Maps MOD SHAEF-F 262.

114–115 'Western Front Situation 05:00 hrs 28th May, 1940' (London: War Office, 1940). Maps MOD OR 5158.

116–117 'London Mayfair Square. Neues Luftfahrtministerium Techn. Amt', 1940 (Berlin: Luftwaffe, 1940). Maps CC.5.a.555. 30 x 31 cm.

118–119 'Weekly Diagram of U-Boat Warfare, Sept 30th – October 27th 1940.' [Printed by the Anti-Submarine Warfare Division.] The National Archives, MFQ 1/583 (26). 33 x 88 cm.

120–121 'Schaffhausen escape map', c.1940 (London: War Office). Maps CC.5.a.424.

122–123 'Bekanntmachung betr. die Bildung eines geschlossenen jüdischen Wohnbezirks in der Stadt Lublin' (poster; Lublin, 24 March 1941). AKG-Images.

124–125 'Befestigungsskizze Tobruk 1:50000, Stand: August 1941' ([S.I.]: OKH-Gen Std H-Abt. Fremde Here West III, 1941). Maps X.6667.

127 Mitsuo Fuchida, 'Hawai Kaisen tai suijō kansen senkazu, Shōwa jūrokunen jūnigatsu yōka' ('Estimated damage report against surface ships on the air attack of Pearl Harbor, December 8th, 1941'). Library of Congress Geography and Map Division, Washington, DC. 80 x 61 cm.

128–129 Lieut-Cmdr G.M.S. Stitt, 'The Battle of Matapan, 1941' (Cairo: Survey Directorate, Middle East, 1942). Maps MOD MDR Misc 876.

130, 131 'Plan of Town, Showing Objectives' and 'Dieppe "Jubilee" 19th August 1942" in *Combined report on the Dieppe Raid*, 1942 (London: Combined Operations Headquarters, 1942). ORB.30/5389.

132, 133 'Dieppe Area Defences' and 'Defences of Dieppe West (according to statements by prisoners of war)' in *Combined report on the Dieppe Raid*, 1942 (London: Combined Operations Headquarters, 1942). ORB.30/5389.

134–135 Royal Engineers, Palestinian Field Survey Company, 524th, 'Gibraltar', 1942 (Cairo: Survey Directorate, Middle East, 1942). Maps MOD MDR Misc 2087.

136–137 'Operation Husky' [map in the White House Map Room, August 1943]. Franklin D. Roosevelt Presidential Library and Museum.

138 Capt. Alfred J. Reid, 'Manuscript planning map for the Allied landing near Licata, Sicily, on July 10, 1943'. Library of Congress Geography and Map Division, Washington, D.C. 79 x 155 cm.

139 Second Army, 21st Army Group BR, 'Going map: Erkelenz', June 1944 (London: War Office, 1944). Maps MOD SA 9 GSGS 4414 Rhine.

140–141 Second Army, 21st Army Group BR, 'Going map: Mezidon', March 1944 (London: War Office, 1944). Maps MOD SA Misc 374.

142–143 'Stop Press Edition of 20 May 1944: Defences Isigny' (London: War Office, 1944). Maps 14317.(259).

144–145 U.S. Army Engineer Topographical Company, 995th, 'Marianas Islands Saipan–Tinian', June 1944 (Commanding General, Army Air Forces, 1944). Maps MOD US Misc 295.

146 'Holland 1:10,000 Defence Overprint Koudekerke-Vlissingen (Flushing)', 28 October 1944 ([S.I.]: 1st Canadian Army, 21st Army Group, 1944). Maps MOD CA 15 Def.

148, 149 Prisoners Press, 'Braunschweig Camp, and Braunschweig and environs', 1944. (Braunschweig, 1944). Maps C.25.a.23.

152–153 'Attacks of 100 tons & over by Eighth Air Force & RAF Bomber Command during January 1945' ([S.I.]: Supreme Headquarters Allied Expeditionary Force, 1945). Maps MOD SHAEF Air Staff no 100.

154 'Das War der Plan', 1945 ([S.I.]: Supreme Headquarters Allied Expeditionary Force, 1945). Maps CC.5.b.66.(1.). 22 x 14 cm.

155 'Westkämpfer! Das habt Ihr geschafft!', 1944. Cornell University – PJ Mode Collection of Persuasive Cartography. 30 x 21 cm.

156–157 64th Engineering Topographical Battalion, 'Iwo Jima: Special Air and Gunnery Target Map', November 1944 (Fleet Marine Force, Pacific and Intelligence Section, Amphibious Forces Pacific, 1944). Courtesy Tim Bryars, www.bryarsandbryars.co.uk.

158 Detail from 'The Pacific—Vital Area for United States and Japan', see 174–175.

165 Charles H. Owens, 'Save this map! It shows the theatre of war', 10 September 1939 (Los Angeles: *Los Angeles Times*, 1939). David Rumsey Map Collection Cartography Associates. 54 x 38 cm.

166 Robert M. Chapin Jr, 'Poland at War', 1939 (New York: Time Inc., 1939). P.P.6339.d.

169 Millet, 'An Airman's View of War Fronts', 9 January 1940 (Melbourne: *The Argus*, 1940). PJ Mode Collection of Persuasive Cartography. 42 x 29 cm.

170–171 Richard Edes Harrison, 'The World Divided' [*Fortune* Magazine August 1941], in *The Fortune Atlas for World Strategy*, 1944 (New York: Knopf, 1944). Maps 217.c.4. 36 x 46 cm.

172–173 'Dai Tōa senkyoku chizu', 1942 (Tokyo: Nobarasha, 1942). National Library of Australia. 24 x 34 cm.

174–175 Howard Burke, 'The Pacific—Vital Area for United States and Japan', November 1942 (Los Angeles: *Los Angeles Examiner*, 1942). David Rumsey Map Collection Cartography Associates. 51 x 40 cm.

176–177 Richard Edes Harrison, 'The Not-So-Soft Underside', January 1943 (New York: Army Information Branch, Army Service Forces 1942–1946). Maps 197.h.1. 88 x 119 cm.

178 F.E. Manning, 'Target Berlin', Newsmap October 25, 1943 (New York: Army Information Branch, Army Service Forces 1942–1946). Maps 197.h.1. 88 x 119 cm.

179 F.E. Manning, 'Target Tokyo', Newsmap October 18, 1943 (New York: Army Information Branch, Army Service Forces 1942–1946). Maps 197.h.1. 88 x 119 cm.

180–181 Newsmap, 8 November 1943 (New York: Army Information Branch, Army Service Forces 1942–1946). Maps 197.h.1. 88 x 119 cm.

182–183 Richard Edes Harrison, 'Four Approaches to Japan', Newsmap 8 May 1944 (New York: Army Information Branch, Army Service Forces 1942–1946). Maps 197.h.1. 88 x 119 cm.

184–185 Newsmap, 8 May 1944 (New York: Army Information Branch, Army Service Forces 1942–1946). Maps 197.h.1. 88 x 119 cm.

186–187 Esso War Map II featuring 'Fortress Europe – The World Island', 1944 (New York: General Drafting Co. Inc., 1944). Maps CC.5.a.591. 58 x 84 cm.

188–189 Esso War Map II featuring 'Fortress Europe – The World Island', 1944 (New York: General Drafting Co. Inc., 1944). Maps CC.5.a.591. 58 x 84 cm.

191 George H. Davis, 'D-Day', 1944 [published in *The Illustrated London News* 17 June 1944]. Maps CC.5.a.24. 61 x 49 cm.

194–195 Edwin L. Sundberg, 'Quadruple Threat to Japan', 3 September 1944 (New York: *New York Sunday News*, 1944). David Rumsey Map Collection Cartography Associates. 38 x 55 cm.

196–197 Robert M. Chapin, 'I Have Returned', 30 October 1944 (New York: Time Inc., 1944). David Rumsey Map Collection Cartography Associates. 72 x 105 cm.

204–205 'The World Moving Toward a New World Order' in *Schoolmates-Students on the Front Lines*, July 1940 (Tokyo, 1940). Cornell University – PJ Mode Collection of Persuasive Cartography. 17 x 23 cm.

206 'Britain's Vast Air Offensive—Over 700 R.A.F. Attacks on Germany', 12 October 1940 (London: *The Illustrated London News*, 1940). P.P.7611.

208–209 Ernst Adler, 'Englands Blockade 1939/40 wird blockeiert' and 'Die aussichtslose Fernblockade gegen Europa' in *Der Krieg 1939/40 in Karten* edited by Giselher Wirsing (Munich: Knorr and Hirth [n.d.]). Wq90/0169.

210–211 'Spontaneous Help in Britain's Air War', 1941 (London: Chromoworks Ltd., 1941). Museum of New Zealand Te Papa Tongarewa, Wellington/Bridgeman Images. 75.7 x 50.2 cm.

212–213 Louis Emile Manche, 'De Dollarpoliep', 1942 (Holland: Goedgekeurd door de Afd. Propaganda van het Departement von Volkslichting en Kunsten, 1942). Cornell University – PJ Mode Collection of Persuasive Cartography. 42 x 57 cm.

214 'Confiance... ses amputations se poursiuvent méthodiquement', c.1942 ([Paris?]: S.P.K., c.1942). Maps CC.5.a.546. 119 x 84 cm.

215 Patrick Cokayne Keely, 'Indie moet vrij! Werkt en vecht ervoor!', 1944 (London: James Haworth & Brother Ltd., 1944). Maps CC.6.a.77. 73 x 49 cm.

216–217 *Mit Uns im Osten: Eine Bildfolge vom Einsatz der Ulmer Infanterie-Division*, 1942 (Stuttgart, Chr. Belser Verlagsbuckhandlung, 1942). Cornell University – PJ Mode Collection of Persuasive Cartography. 50 x 112 cm.

218–219 'Das War Hamburg', 1943 ([London?], 1943). Cornell University – PJ Mode Collection of Persuasive Cartography. 22 x 50 cm.

220–221 Frederick Donald Blake, 'Britain – Spearhead of Attack', 1943. (London: Ministry of Information, 1943). B.S. 51/29. 50 x 76 cm.

222–223 Theodore Geisel, 'This is Ann... she drinks blood!', Newsmap 8 November 1944 (New York: Army Information Branch, Army Service Forces 1942–1946). Maps 197.h.1. 88 x 119 cm.

224–225 Frederick Donald Blake, 'The Battle of the Atlantic', 1944 (London: Ministry of Information, 1944). B.S. 51/29. 76 x 50 cm.

226 'Speaking of time-tables', 1944 ([Berlin?]: 1944). Cornell University – PJ Mode Collection of Persuasive Cartography. 21 x 15 cm.

227 'It's a Long Way to Rome', April 1944 ([France?], 1944). Cornell University – PJ Mode Collection of Persuasive Cartography. 117 x 80 cm.

228 'Shadow over England', September 1944 ([Berlin?]: 1944). Cornell University – PJ Mode Collection of Persuasive Cartography. 28 x 20 cm.

229 6th Army, Psychological Warfare Branch, 'The South Pacific is the South Pacific. Japan is Japan', 1944. Cornell University – PJ Mode Collection of Persuasive Cartography. 15 x 20 cm.

234–235 MacDonald Gill, 'The "Time & Tide" Map of the Atlantic Charter', 1942 (London: G. Philip & Son, 1942). Maps 950.(211.). 76 x 109.2 cm.

238–239 'Bataille de Normandie, Juin–Aout 1944', 1947 (Clermont Ferrand: Michelin, 1947). Maps X.6608. 72 x 86 cm.

240–241 *Atlas Bolgariskikh govorov v SSSR*, 1958 (Moscow: Izdatel'stvo Akademii nauk SSSR, 1958). Maps 28.e.36. 39 x 44.5 cm.

242–243 *Történelmi Atlasz*, 1961 (Budapest: Kartográfiai Vállalat, 1961). Maps 197.b.68.

244–245 'Desfăşurarea Insurectie Nationale Antifasciste Armàtă 23–31 August 1944' in Stefan Pascu, *Atlas Istoric*, 1971 (Bucureşti: Editura didactică şi pedagogică, 1971). L.45/119.

INDEX

Figures in **bold** refer to pages on which illustrations appear; those in *italic* refer to all other captions

PICTURE CREDITS

All images from the collections of the British Library except the following:

Page 11 © Historic England; 36, 108 Franklin D. Roosevelt Presidential Library and Museum; 37, 58, 111 Imperial War Museum, London; 64 National Archives and Records Administration (NARA); 104 Library of Congress Geography and Map Division Washington, DC; 126 TRH Pictures; 160, 162 David Rumsey Map Collection Cartography Associates; 198, 200 Cornell University – PJ Mode Collection of Persuasive Cartography.

ACKNOWLEDGEMENTS

Thanks to Chiaki Akimoto, Kristofer Allerfeldt, Mirela Altić, Anne Louise Antonoff, Knut Arstad, Stephen Badsey, Nick Baron, Pradeep Barua, Harry Bennett, Peter Brobst, Peter Caddick-Adams, Stan Carpenter, Olavi Fält, Heiko Werner Henning, Nick Lipscombe, Ciro Paoletti, Alexander Querengässer, Kaushik Roy, Frédéric Saffroy, Gary Sheffield, Donald Spiers, Ulf Sundberg, Gregory J. Urwin and Edward Westermann for advice on particular points. I have benefitted from the support of John Lee at BL Publications, from the map-finding skills of Sally Nicholls and, in particular, from the editorial work of Christopher Westhorp. All books are a joint effort, but this is especially so of map books, and my thanks are very much deserved by Chris, Sally and Georgie.